Smart Grid and Sustainable Energy

SMART GRID AND SUSTAINABLE ENERGY

Edited by:

Zoran Gacovski

ARCLER PRESS

www.arclerpress.com

Smart Grid and Sustainable Energy

Zoran Gacovski

Arcler Press
224 Shoreacres Road
Burlington, ON L7L 2H2
Canada
www.arclerpress.com
Email: orders@arclereducation.com

ISBN: 978-1-77469-527-2 (Hardcover)

DECLARATION

Some content or chapters in this book are open access copyright free published research work, which is published under Creative Commons License and are indicated with the citation. We are thankful to the publishers and authors of the content and chapters as without them this book wouldn't have been possible.

ABOUT THE EDITOR

Dr. Zoran Gacovski's current position is a full professor at the Faculty of Technical Sciences, "Mother Tereza" University, Skopje, Macedonia. His teaching subjects include Software engineering and Intelligent systems, and his areas of research are: information systems, intelligent control, machine learning, graphical models (Petri, Neural and Bayesian networks), and human-computer interaction. Prof. Gacovski has earned his PhD degree at Faculty of Electrical engineering, UKIM, Skopje. In his career he was awarded by Fulbright postdoctoral fellowship (2002) for research stay at Rutgers University, USA. He has also earned best-paper award at the Baltic Olympiad for Automation control (2002), US NSF grant for conducting a specific research in the field of human-computer interaction at Rutgers University, USA (2003), and DAAD grant for research stay at University of Bremen, Germany (2008 and 2012). The projects he took an active participation in, are: "A multimodal human-computer interaction and modelling of the user behaviour" (for Rutgers University, 2002-2003) - sponsored by US Army and Ford; "Development and implementation of algorithms for guidance, navigation and control of mobile objects" (for Military Academy – Skopje, 1999-2002); "Analytical and non-analytical intelligent systems for deciding and control of uncertain complex processes" (for Macedonian Ministry of Science, 1995-1998). He is the author of 3 books (including international edition "Mobile Robots"), 20 journal papers, over 40 Conference papers, and he is also a reviewer/ editor for IEEE journals and Conferences.

TABLE OF CONTENTS

Section 4: Intelligent implementation of Smart Grids

LIST OF CONTRIBUTORS

Kamal Kant Sharma
Department of EE, Chandigarh University, Mohali, India

Himanshu Monga
Department of ECE, JNGEC, Mandi, India

Peizhong Yi
Electrical and Computer Engineering Department, Illinois Institute of Technology, Chicago, IL 60616-3793, USA

Abiodun Iwayemi
Electrical and Computer Engineering Department, Illinois Institute of Technology, Chicago, IL 60616-3793, USA

Chi Zhou
Electrical and Computer Engineering Department, Illinois Institute of Technology, Chicago, IL 60616-3793, USA

Morris Brenna
Politecnico di Milano – Department of Energy, Milan, Italy

Ettore De Berardinis
CESI S.p.A., Milan, Italy

Federica Foiadelli
Politecnico di Milano – Department of Energy, Milan, Italy

Gianluca Sapienza
Politecnico di Milano – Department of Energy in Collaboration with ENEL Distribuzione S.p.A., Milan, Italy

Dario Zaninelli
Politecnico di Milano – Department of Energy, Milan, Italy

Salem Al-Agtash
Department of Computer Engineering, German Jordanian University, Amman, Jordan.
Department of Computer Science and Engineering, Santa Clara University, Santa Clara, CA, USA

Hossam Abdel Hafez
Department of Computer Engineering, German Jordanian University, Amman, Jordan.

Shiming Chen
East China Jiaotong University, Nanchang 330013, China

Kaiqiang Li
East China Jiaotong University, Nanchang 330013, China

Hermes José Loschi
Department of Communications, Faculty of Electrical and Computer Engineering, University of Campinas (UNICAMP), Campinas/SP, Brazil

Julio Leon
Department of Communications, Faculty of Electrical and Computer Engineering, University of Campinas (UNICAMP), Campinas/SP, Brazil

Yuzo Iano
Department of Communications, Faculty of Electrical and Computer Engineering, University of Campinas (UNICAMP), Campinas/SP, Brazil

Ernesto Ruppert Filho
Department of Communications, Faculty of Electrical and Computer Engineering, University of Campinas (UNICAMP), Campinas/SP, Brazil

Fabrizzio Daibert Conte
Department of Communications, Faculty of Electrical and Computer Engineering, University of Campinas (UNICAMP), Campinas/SP, Brazil

Telmo Cardoso Lustosa
Department of Communications, Faculty of Electrical and Computer Engineering, University of Campinas (UNICAMP), Campinas/SP, Brazil

Priscila Oliveira Freitas
Department of Communications, Faculty of Electrical and Computer Engineering, University of Campinas (UNICAMP), Campinas/SP, Brazil

M. D. Berni
Interdisciplinary Center on Energy Planning (NIPE), State University of Campinas (UNICAMP), Campinas, Brazil

P. C. Manduca
Interdisciplinary Center on Energy Planning (NIPE), State University of Campinas (UNICAMP), Campinas, Brazil

S. V. Bajay
Interdisciplinary Center on Energy Planning (NIPE), State University of Campinas (UNICAMP), Campinas, Brazil

J. T. V. Pereira
Interdisciplinary Center on Energy Planning (NIPE), State University of Campinas (UNICAMP), Campinas, Brazil

J. T. Fantinelli
Interdisciplinary Center on Energy Planning (NIPE), State University of Campinas (UNICAMP), Campinas, Brazil

Mohammed Mansur Ibrahim
Department of Mathematics & Computer Science, Federal University of Kashere, Kashere, Gombe State, Nigeria

Anas Ahmad Danbala
Trans. Access Planning Engineer, Mobile Telecom. Network (MTN) Abuja Switch Office, Abuja FCT, Nigeria

Mustapha Ismail
Department of Mathematics & Comp. Sci., Gombe State University, Gombe, Gombe State, Nigeria

Hwa Meei Liou
Graduate Institute of Technology Management, National Taiwan University of Science and Technology, Taipei, Taiwan

Vilas S. Bugade
Dr. Babasaheb Ambedkar Technological University, Lonere, India

Chunqiang Hu
Key Laboratory of Dependable Service Computing in Cyber Physical Society, Chongqing University, Ministry of Education, Chongqing, China
School of Software Engineering, Chongqing University, Chongqing, China

Department of Electrical Engineering & Computer Science, The Catholic University of America, Washington, DC, USA

Hang Liu
Department of Electrical Engineering & Computer Science, The Catholic University of America, Washington, DC, USA

Liran Ma
Department of Computer Science, Texas Christian University, Fort Worth, TX, USA

Yan Huo
School of Electronics and Information Engineering, Beijing Jiaotong University, Beijing, China

Arwa Alrawais
Department of Computer Science, The George Washington University, Washington DC, USA

Xiuhua Li
Department Electrical and Computer Engineering, University of British Columbia, Vancouver, BC, Canada

Hong Li
Beijing Key Laboratory of IOT Information Security Technology, Institute of Information Engineering, CAS, Beijing, China

Qingyu Xiong
Key Laboratory of Dependable Service Computing in Cyber Physical Society, Chongqing University, Ministry of Education, Chongqing, China
School of Software Engineering, Chongqing University, Chongqing, China

Hassan Mahmood
School of Computer Science and Communication Engineering, Jiangsu University, Zhenjiang, China
Department of Computer Science, SZABIST, Islamabad, Pakistan

Danish Mahmood
School of Computer Science and Communication Engineering, Jiangsu University, Zhenjiang, China
Department of Computer Science, SZABIST, Islamabad, Pakistan

Qaisar Shaheen
Department of Computer Science, Superior College, Lahore, Pakistan

Rizwan Akhtar
Department of IT and Computer Science, Pak-Austria Fachhochschule Institute of Applied Sciences and Technology, Haripur, Pakistan

Wang Changda
School of Computer Science and Communication Engineering, Jiangsu University, Zhenjiang, China

Tawfiq M. Aljohani
Ming Hsieh Department of Electrical Engineering, University of Southern California, Los Angeles, USA

Mohammed J. Beshir
Ming Hsieh Department of Electrical Engineering, University of Southern California, Los Angeles, USA

Navin Shenoy
School of Electrical and Computer Engineering, Oklahoma State University, Stillwater, USA

R. Ramakumar
School of Electrical and Computer Engineering, Oklahoma State University, Stillwater, USA

Muhammad Qamar Raza
Department of Electrical Engineering COMSATS, Institute of Information Technology, Abbottabad, Pakistan

Muhammad Ali
Department of Electrical Engineering COMSATS, Institute of Information Technology, Abbottabad, Pakistan

Nauman Tareen
Department of Electrical Engineering COMSATS, Institute of Information Technology, Abbottabad, Pakistan

Waheed ur Rehman
Department of Electrical Engineering COMSATS, Institute of Information Technology, Abbottabad, Pakistan

Asadullah Khan
Department of Electrical Engineering COMSATS, Institute of Information Technology, Abbottabad, Pakistan

Azam Ul Asar
Department of Electrical Engineering COMSATS, Institute of Information Technology, Abbottabad, Pakistan

Muhammad Ali
Electrical Engineering Department, COMSATS Institute of IT, Abbottabad, Pakistan

Zulfiqar Ali Zaidi
Department of Mathematics, COMSATS Institute of IT, Abbottabad, Pakistan

Qamar Zia
Electrical Engineering Department, NWFP University of Engineering & Technology, Peshawar, Pakistan

Kamal Haider
Electrical Engineering Department, Gandhara University, Peshawar, Pakistan

Amjad Ullah
Electrical Engineering Department, NWFP University of Engineering & Technology, Peshawar, Pakistan

Muhammad Asif
Electrical Engineering Department, CECOS University of IT, Peshawar, Pakistan

Md Multan Biswas
Department of Electrical and Electronic Engineering, Bangladesh University of Engineering and Technology, Dhaka, Bangladesh

Md Shafiul Azim
Department of Electrical and Electronic Engineering, Khulna University of Engineering and Technology, Khulna, Bangladesh

Tonmoy Kumar Saha
Department of Electrical and Electronic Engineering, Khulna University of Engineering and Technology, Khulna, Bangladesh

Umama Zobayer
Department of Electrical and Electronic Engineering, Stamford University Bangladesh, Dhaka, Bangladesh

Monalisa Chowdhury Urmi
Department of Electrical and Electronic Engineering, Stamford University Bangladesh, Dhaka, Bangladesh

Rosario Miceli
Dipartimento di Ingegneria Elettrica, Elettronica delle Telecomunicazioni, di Tecnologie Chimiche, Automatica e Modelli Matematici, Università di Palermo, Palermo, Italy

Salvatore Favuzza
Dipartimento di Ingegneria Elettrica, Elettronica delle Telecomunicazioni, di Tecnologie Chimiche, Automatica e Modelli Matematici, Università di Palermo, Palermo, Italy

Fabio Genduso
Dipartimento di Ingegneria Elettrica, Elettronica delle Telecomunicazioni, di Tecnologie Chimiche, Automatica e Modelli Matematici, Università di Palermo, Palermo, Italy

LIST OF ABBREVIATIONS

IEEE: Institute of Electrical and Electronics Engineering.
EPRI: Electric Power Research Institute.
SCADA: Supervisory Control and Data Acquisition.
SAIDI: System Average Interruption Duration Index.
SAIFI: System Average Interruption Frequency Index.
CAIDI: Customer Average Interruption Duration Index.
EUE: Expected Un-served Energy.
ASAI: Average Service Availability Index.
RBTS: Reliability Test System.
AMI: Advanced Metering infrastructure.
DG: Distributed Generation.
AR: Automatic Recloser.
AS: Automatic switch.
OD: Outage Duration.
OF: Outage Frequency.
U: Unavailability
A: Availability.
P: Probability of the component to be available.
Q: Probability of the component to be unavailable.
λ: Failure rate of an electrical component.
MTTR: Mean Time To Repair.
MTTS: Mean Time To Switch
d_a: Initial demand;
p_a: Initial Price;
Δp_t: Price change in period t;
Δd_s: Demand change in period s;
e: Elasticity;
$\Delta d(i)$: Change in demand in i-th hour after FEDRP;

$d_x(i)$:	Demand in i-th hour before FEDRP (KWh);
$d_y(i)$:	Demand in i-th hour after FEDRP (KWh);
Q(i):	Incentive in i-th hour ($/Kwh);
Z(i):	Penalty in i-th hour ($/Kwh);
L(i):	Contract level;
p_t:	Price before FEDRP in i-th hour ($/Kwh);
$p_{t'}$:	Price after FEDRP in i-th hour ($/Kwh);
e_{tt}:	Self Elasticity in i-th hour;
$e_{tt'}$:	Demand Cross Elasticity between i-th and j-th hour;
d_f:	Demand for must run load;
d_v:	Demand for variable load;
D_t:	Total demand of residential setup (MWH);
R:	Total Revenue to Utility ($);
G_k:	Supply generation of unit k;
C:	Generation cost;
y:	Profit to Utility ($).

PREFACE

The smart electricity network includes an information infrastructure and infrastructure for the entire power supply chain, including the generation, transmission, delivery and consumption of electric power, i.e. system that delivers electricity from the place of production to the place of consumption. It enables two-way communication between suppliers and consumers, which makes the real-time optimization of the electricity network feasible. End users are allowed to adjust consumption based on their own needs.

Electrical power is delivered through a distribution system, and information is exchanged over the Internet or a purpose-built network. Typical consumer groups are: residential, commercial, industrial. Applications in the smart power grid for end users are: automation of the house or building, automation in industry (control system for industrial processes, machines, warehouses, etc.), power supply of electric vehicles, micro generation (production system using solar panels, windmills, hydroelectric power plants).

Trade of power in the smart electricity network takes place on the market (national and international), where prices are set and a balance is created between production and consumption. Service providers in the smart electricity network provide services that should enable the operation of electricity producers, distributors and consumers. They are most often organized as utility companies that provide the following services: customer relationship management, installation and maintenance, smart home and building management, billing.

Smart grid operations in most countries are performed by state-controlled companies; parts of them are transferred to service providers or companies operating in the market. However, it is expected that planning and reliable operation will remain regulated even after the full implementation of the smart electricity network. Typical tasks in smart grids are: monitoring and control, troubleshooting, reporting, advanced analytics, operations planning, network development planning.

Smart meters are used as a basic part of the user's smart electricity network, which enable the consumer to have an accurate insight into consumption at all times. The smart meters store the measured data internally, and the competent service providers access it through the infrastructure (for meter's management). The infrastructure for smart meter management should enable the integration of meters from different manufacturers and connection to the NAN (Neighborhood Area Network).

This edition covers different topics from smart grids and sustainable energy, including: automation and control of smart grids, sustainability and energy efficiency, security and stability of smart grids, and intelligent implementation of smart grids.

Section 1 focuses on automation and control of smart grids, describing smart grids: future of electrical transmission and distribution, building automation networks for smart grids, voltage control in smart grids: an approach based on sensitivity theory, agents for smart power grids, and distributed optimal control of transient stability for a power information physical system.

Section 2 focuses on sustainability and energy efficiency, describing energy efficiency in smart grid: a prospective study on energy management systems, energy efficiency and renewable energy technologies using smart grids: study case on NIPE building at UNICAMP campus, towards attaining reliable and efficient green cloud computing using micro-smart grids to power internet data center, the development of electricity grid, smart grid and renewable energy in Taiwan, and optimal power flow approach for cognitive and reliable operation of distributed generation as smart grid.

Section 3 focuses on security and stability of smart grids, describing a secure and scalable data communication scheme in smart grids, S-DPS: an SDN-based DDOS protection system for smart grids, distribution system reliability analysis for smart grid applications, and approach to assess the resiliency of electric power grids.

Section 4 focuses on intelligent implementations of smart grids, describing intelligent load shedding using TCP/IP for smart grids, intelligent load management scheme for a residential community in smart grids network using fair emergency demand response programs, towards implementation of smart grid: an updated review on electrical energy storage systems, and a perspective on the future of distribution: smart grids, state of the art, benefits and research plans.

SECTION 1:
AUTOMATION AND CONTROL OF SMART GRIDS

Chapter 1

Smart Grid: Future of Electrical Transmission and Distribution

Kamal Kant Sharma[1], Himanshu Monga[2]

[1]Department of EE, Chandigarh University, Mohali, India

[2]Department of ECE, JNGEC, Mandi, India

ABSTRACT

For century, the need of change in the system of power grid is being felt and being considered as the most important change that we need in the modern era electrical system but before moving towards a new change the things in past should be kept in mind so that there are better chances of great and beneficial change. The purpose of this research is to investigate the changes need to be incorporated in a conventional system to make a system self-sufficient and automated in order to make Electrical Power system more reliable. This paper assesses the current one way power system that is needed to be changed and has tried to provide an overview of the changes that we need in the system. The paper has focused more on the smart grid system and has explained the importance of smart grid system in terms

Citation: Sharma, K. and Monga, H. (2020), "Smart Grid: Future of Electrical Transmission and Distribution". *International Journal of Communications, Network and System Sciences*, **13**, 45-54. doi: 10.4236/ijcns.2020.134004.

of efficiency, automation and decision making capability in case of faults occurred on primitive grids with the help of comparative studies. The paper also highlighted the results in form of comparison with conventional grids and threw some light on the vision, control and the application of the smart grid system that will provide a two way system to the electrical network of the country and will make the distribution and consumption of energy more efficient also which is going to increase the reliability and accuracy in the system.

Keywords: Smart Grid, Automatic Grid System, Vision of Smart Grid, Communication Technologies

INTRODUCTION

With the increase in the demand of less human power and decreased labor the technology is advancing towards the automation system to make things much easier, simpler and accurate [1]. Automation is the use of control systems and information technologies to reduce the need for human work in the production of goods and services. With the passing days the automation is taking over the industries and the desideratum of human sensory is decreasing. Since the world is getting smarter with advancement in the technology and automation is the foremost and most required technology as well as automation plays an increasingly important role in the world economy and in daily experience [1].

Since the technology is upgrading with the time the ease with which we are able to access our appliance and products whether it be home appliances or big industrial machineries is increasing. Automated system provides increased flexibility and security as compared to that of manual system [2]. In the same way the foremost and the basic need of human life is power that is being supplied with the help of grids. Before moving towards any other advancement in the technology with the electrical world we must need to think about the advancement in the electrical power distribution and the thing that we need to think the first and foremost is about termed as Smart Grid [3].

The idiom Smart Grid is being used to define a digitized electrical network that uses information as well as communication technology to perceive and react against the immediate changes in the use by the consumers. Smart grid is a fast and advanced wiring system that is changing the electrical world rapidly [3]. The idiom "smart" is being used for the grid, it's because the system is designed in a way that it uses communication system and is quite a smart system and is not being used in the current world. The crucial importance of the system is that it has capacity to store electrical power so that the consumer's demands can be fulfilled at the time of need.

The need to adapt smart grid as the system for distribution is that it has less dependency on operator that will reduce the need of human labor and most prominently it can respond immediately to the conditions that are being changed rapidly [4] [5] [6].

In today's era the power system which is purposefully designed in such a way that they support large scale generating stations and also they feed the need of the far distance consumers through transmission lines and that power is then being distributed to the consumer by a one way system but in the upcoming era of advanced world we should think about a two way system that is possible when the concept of smart grid will be introduced in the world. It has been observed from various case studies elaborated from various researches that Smart Grid is a need of system but certain roadblocks like dynamic security analysis, data encryption and technology interface are among the few which creates a wide gap between existing and automated grid and make research meaningful.

Stature of Smart Grid

Smart grid is something that is not only a technological advancement but is itself comprise of hundreds of benefits and the benefit of implementing smart grid as a system of transmission and distribution is very beyond the current power system network. Implementing smart grid is not as easy as it looks like because this is a technology which alone comprise of all the advances that have took place in century and is efficient more than those all advancements and modifications. To implement this system it needs a support from all governing companies as well as all advocacy groups. Also the technology which alone is capable of comprising all the advancements cannot be implemented in a very less fraction of time it will acquire quite extra time to implement although once it is being implemented is going to benefit all the stake holders as well as the consumers.

Some of the benefits that smart grid system will be comprised of are [7] - [13]:

- Give consumers a choice to get supply of power only for their use.
- The loss in power supplied that is not being consumed and is not needed to be supplied can be restricted.

- Provides better information to the consumers.
- Improvement in the system reliability.
- Reduction of environmental issues that are being observed due to current system of electrical networks.

Advanced Terminologies

Modern/current scenario is about need of Deregulated Environment and everything is getting updated or we can say that upgraded like from phones to smart phones, television to LED to smart LED (android television). Each and everything is getting faster, improvement in quality and many more. However, these cover basic amenities like seeing and living for a better sustainable life. Therefore, as a trend provided; electricity must be followed in a different way more smart oriented and method of electricity transmission and distribution should be flexible in parameters while opting for deliverables [14] [15] [16] [17]. The existing girds already have smart functionality but just to balance the supply or demand of consumers and it is one way process that firstly transmission and then distribution to the consumer so we need a advancement in technology to make these grid a smart grid. Qualitative Analysis is formulated in this paper citing key features of Smart Grid and steps required to fulfill a distant dream in terms of technology interface and making a system self sufficient.

The following table is showing the most appropriate fields which have a scope of advancement and the smart grid can be implemented somehow by using these criteria. Following table are observations citing in different literature citing need of smart Grid as evolving technology along with controlled parameters with dynamic security assessment. The need of smart grid is depicted by various scientists for decentralized power system as compared with today scenario making a system self sufficient to take decisions instead of manual intervention and increases its functionality in reducing mismatch between generation and distribution in larger context [18] - [23]. The main difference between conventional (traditional) and decision making grid (Smart Grid) about its operation and flexible characteristics in Hybrid power system with more penetration of renewable energy sources with a load factor not equal to 100%. Table 1 represents the overview of features and technical specifications in terms of deliverables.

Table 1: Technical gaps that to implement for system Self Decisive

S. NO.	TECHNICAL GAPS	EXPLANATION
1	Computerizing the Electrical Grid	Smart grid is a two way communication so we need to add up two ways channel to the devices connected to the grid.
2	Electric Meters	Advance meters are to be used like AMU (Advance Monitoring Unit) which can store data that how much power is consumed and how much is generated and supplied to other users.
3	Automation Technology	In normal grid if any fault arises it is very difficult to assess the point but in smart grid, fault assessment is easy.
4	Automatically repairing of fault without supply interruption	In normal grid if any fault arises, person is send to repair the fault but in case of smart grid if any fault arises in the line that will be automatically repaired
5	Load Information	ATC (Available Transfer Capability) can be assessed
6	Energy Flow	Transmission and distribution of next generation will be able to handle bidirectional energy flow, which allows distributed generation
7	More use of renewable resources	Distributed Generation can be emphasized
8	Advanced Services	Smart Grid is an decisive Grid with autonomous use of resources

TRADITIONAL GRID VS. SMART GRID

Different types of grids systems are available in different countries and continents depending upon centralized and decentralized power system. In Asia and most of the countries except Europe; Centralized System is being advocated and totally depended upon population and density of consumers [24] [25]. Traditional Grid incorporates different Grid levels and making Grid Architecture, but on the other hand; Smart Grid is a combination of embedded system and Optimization algorithms to make Grid self sufficient and controlling load flow in various bus system carrying active and reactive power compensations. Smart Grid has many other features like Interactive System, Two way approach reducing mismatch with Energy demand and consumption, leading to transparency which also make a structured framework which is essential for scheduling in Energy market and fulfill short term and long term goals. Smart Grid is a future of many nations which are densely populated like India where distribution losses are not optimal in comparison with transmission and generation losses due to mismanagement and less voltage profile. The main issues with distribution losses are as high as more than 30% due to Electrical Theft and non integration of renewable Energy sources with conventional Grid as lack of presence of deregulated Environment. Therefore, various advantages are being highlighted in this paper considering Smart Grid application and reduce the existing gap with generation and consumption. It has also been simulated in small distribution networks comprises of mesh structure that less voltage profile can be subjected to domestic users, but for industrial applications; high voltage profile is preferred with three system approach, hence smart grid is also beneficial in providing access to different phase systems altogether and provide a constructive framework with prefect optimized values of voltage and current values keeping losses minimum and optimized for a structured environment. Necessity of Smart Grid is felt in developed countries and developing countries also leading towards future of transmission and distribution in near future. Some of key features are being depicted in Table 2.

CHALLENGES TO BE FACED

Smart Grid is an idea with transforming a convention system with digital transformation with real time indices so that sufficient and transparent system framework can be designed altogether so that essential features can be structured out to make a system automated with decision making capability. It's said that incoming of ideas are fast but implementation of that takes time and even though we have to face various challenge to implement that idea. So smart grid is also a much more advance idea in transmission and distribution but it have to face various challenges for its implementation.

Table 2: Comparison between the Traditional Grid and the proposed Smart Grid

S. NO	SMART GRID	TRADITIONAL GRID
1	Smart Grid is Intelligent Grid comprises of user interactive system.	Electric power distribution is made possible by the power distribution grid, a system that allows electricity to be transferred from point of generation to our homes.
2	It is two way communication that can network one or more parts of smart grid via secure, high speed, high bandwidth Connections.	It is one way communication that can network only one part of the grid i.e. utility and its customers.

3	More efficient transmission of electricity With the help of sensing devices.	Less efficient transmission of electricity, no sensors are there in the grid.
4	Smart Grid Involves Automation.	Manual Intervention Required.
5	It will reduce operations and management costs for utilities ,and ultimately lower Power cost for domestic and industrial consumer leading to transparency in a system.	In this we have large number of operations so management cost is also high and that ultimately increase the power costs for consumers and increases a gap with Energy Generation and consumption leading to non efficient system with improper sustainable indices.
6	A smarter grid are prepared to address emergencies such as storm, earthquakes and its two way interactive capacity allow Rerouting when equipments fail.	Environment issues can lead to series of failures that can affect banking, traffic and all other electric equipments.
7	Energy Scheduling.	Losses High due to Sudden Contingencies.
8	Smart grid will take greater advantage of customer-owned power generators to produce Power when it is not available from utilities.	Power can only be generated from utilities not by customers and if utilities have some fault then customer will face problems.
9	It will help customers to save money by managing the electricity use and customers Can purchase electricity when they need.	Customers have constant supply to use they can't purchase only when they need so we have to pay more.

- Initial cost.
- 2 way communication.
- Lack of knowledge of costumer.
- Advanced Metering Infrastructure.
- Power issues.
- Existing Infrastructure.
- Sensing Mechanism.
- Grid Integration with RES.
- Optimization Algorithms.

Initial Cost

Smart gird is the idea that will take transmission and distribution technology to next advance level but to implement that there is a huge amount to be paid for the investment which may take much more time to come in action. As lot of things are from conductors to AMU, connecting the provider and consumer and many more. Despite of initial investment, existing framework must be improved in consideration with Smart Grid, therefore additional cost is also subjected to transformation of existing system to proposed system of Smart Grid. It has been analyzed in year 2019 that more than 70% of existing cost must be required to develop a testing framework of Smart Grid.

2 Way Communication

As already discussed smart grid is a two way communication so provider and consumer both need to be connected to each other but the population of INDIA is very large so connecting each and everyone to a service provider is much a challenging task. So it might take much a long time to make this communication part strong. Smart Grid makes a two way communication in terms of generation and consumption with communication protocols like PS2 or WSN leading to complex structure with inbuilt analysis for Energy forecasting considering short term and long terms goals subjected to number of consumers, number of devices and number of distribution and power generating companies to incorporate. It has been predicted in studies using HOMER software that with integration of Smart Grid, distribution losses will reduce to 20% of existing one.

Lack of Knowledge of Costumer

Since smart grid is a new idea. So many a people do not have proper idea of it. They don't actually know what may be the advantage of this smart grid. In India, most of the population lives in rural and suburban areas, therefore type of load in terms of voltage profile and current carrying capacity, amount of connected load with variable load factor is unknown. It has been observed that 70% of the population in India doesn't have a requisite knowledge of its load characteristics.

Advanced Metering Infrastructure

Smart meters our AMU (advance monitoring unit) in which both the information i.e. consumption of power and transmission of power both will be noted and final bill be generate. But this concept is very much costly and will take many years to come in action. This kind of system is automatic and requires no human intervention in determining sustainability indices in consideration of existing load.

Power Issues

As to implement smart grid new things are to be added which may also lead to power cuts may a time which may cause problem to consumers. Power dealing with existing authorities need to be replaced among inter and intra state purchase and selling requisite. It has been analyzed with Government of India that more than 60% manpower would be reduced in comparison with existing one.

Existing Infrastructure

India is densely populated country with generation capacity is only limited maximum to convention generation like hydro power plants, thermal power plants and nuclear power plants. Therefore, with existing power framework, India is limited to centralized power system and selling and purchase of electricity at an individual capacity is not allowed. Smart Grid will incorporate more generation capacity at generation end so that burden on power management companies would be reduced.

Sensing Mechanism

Conventional Power systems are subjected to theft and single variable approach is used with constant voltage profile. Smart Grid will incorporate WSNs at remote level and can be globally recognized with current carrying capacity of every customer so that analysis can be qualitatively with an objective of data collection and availability of power. Smart Grid will have to incorporate data encryption mechanism for securing of data collectively with multi variable approach.

Grid Integration with RES

RES (Renewable Energy Sources) like Wind, Tidal, Solar, Biomass and other potential sources have a very limited share in generating electricity, therefore capital investment is more and less operations and dwelling into electricity market. Across globe, various countries like Canada and USA, Electricity deals with share market and various TSO and GSO are available and deregulation of Electricity occurs. This is a very promising challenge in existing framework.

Optimization Algorithms

Various Algorithms (Systematic or Aligned) can be applied in a system to improve the sustainability indices but without digital transformation; this kind of change is not possible. However, Smart grid poses a variable optimization approach for individual parameter.

It has been studied by various studies and churn out various challenges posed by Smart Grid as compared with conventional system but these challenges must be overcome efficiently so that relative change can be maintained and efficient system can be developed.

CONCLUSIONS

Smart grid is the most comprehended research in the electrical system from a long period of time and it is developing day by day. From all the data the author has observed that the smart grid system will provide a different scenario for the distribution of power as well as it is neither centered to consumers nor to the suppliers but is equally beneficial to both of them. Electrical energy distribution through smart grid will reduce consumption by 10% - 20% with emission of CO_2 by 30% (source: electrical power research institute).

By the implementation of smart grid in the power system the consumers will realize a greater reliability of the system as well as will have a better control over power consumed and supplied.

The operators will have an improved monitoring system as well as better control capabilities that will help them to supply in the overgrowing power demands to fulfill the needs.

REFERENCES

1. Zahra, M. (2013) Smart Grid Technology, Vision, Management and Control. WSEAS TRANSACTIONS on SYSTEMS, 12, 11-21.
2. Isere, M., Saunter, T. and Hung, J.Y. (2010) Future Energy Systems: Integrating Renewable Energy Sources into the Smart Power Grid through Industrial Electronics. IEEE Industrial Electronics Magazine, 4, 18-37.
3. Sadeghi, M. and Gholami, M. (2011) Advanced Control Methodology for Intelligent Universal Transformers Based on Fuzzy Logic Controllers. Recent Researches in Multimedia Systems, Signal Processing, Robotics, Control and Manufacturing Technology, Canary Islands, Spain, 24-26 March 2011, 58-62.
4. Al-Khannak, R. and Bitzer, B. (2008) Grid Computing Technology Enhances Electrical Power Systems Implementations. 3rd IASME/WSEAS International Conference on Energy & Environment, University of Cambridge, UK, 23-25 February 2008, 324-329.
5. Al-Khannak, R. and Yet, L. (2008) Integrating Grid Computing Technology for Developing Power Systems Reliability and Efficiency. 12th WSEAS International Conference on SYSTEMS, Heraklion, Greece, 22-24 July 2008, 491-497.
6. Zahra, M., Atria, Y. and Ahmed AbulMagd. (2011) A Developed SCADA for Remote PV Systems. Engineering Research Journal, 32, 1-10.
7. Sharma, K.K. and Sharma, S. (2020) Congestion in Deregulated Environment Using FACTS Devices. International Journal of Scientific & Technology Research, 9, 4199-4204.
8. Laverty, D.M., Morrow, D.J., Best, R. and Crossley, P.A. (2010) Telecommunications for Smart Grid: Backhaul Solutions for the Distribution Network. IEEE Power and Energy Society General Meeting, Rhode Island, USA, 25-29 July 2010, 1-6. https://doi.org/10.1109/PES.2010.5589563
9. Luan, W.P., Sharp, D. and Lancashire, S. (2010) Smart Grid Communication Network Capacity Planning for Power Utilities. IEEE PES Transmission and Distribution Conference and Exposition, New Orleans, LA, 19-22 April 2010, 1-4.
10. Perishing, Y., Iwayemi, A. and Zhou, C. (2011) Developing Zigsbee Deployment Guideline under Wi-Fi Interference for Smart Grid

Applications. IEEE Transactions on Smart Grid, 2, 110-120. https://doi.org/10.1109/TSG.2010.2091655

11. Gezer, C. and Buratti, C. (2011) A ZigBee Smart Energy Implementation for Energy Efficient Buildings. 2011 IEEE 73rd Vehicular Technology Conference (VTC Spring), Yokohama, Japan, 15-18 May 2011, 1-5. https://doi.org/10.1109/VETECS.2011.5956726
12. Lewis, R.P., Igict, P. andZhou, Z.F. (2009) Assessment of Communication Methods for Smart Electricity Metering in the U.K. 2009 IEEE PES/IAS Conference on Sustainable Alternative Energy (SAE), Valencia, Spain, 28-30 September 2009, 1-4. https://doi.org/10.1109/SAE.2009.5534884
13. Sharma, K.K., Vishvakarma, S. and Manzoor, G. (2018) PSO-GSA Based MPPT Algorithm for Photovoltaic Systems. International Journal of Recent Technology and Engineering, 7, 259-263.
14. Sharma, K.K. and Singh, B. (2018) Optimization and Sensitivity Analysis of Hybrid Power System in HOMER for Study Area. International Journal of Pure and Applied Mathematics, 118, 2381-2396.
15. Paruchuri, V., Durresi, A. and Ramesh, M. (2008) Securing Power Line Communications. 2008 IEEE International Symposium on Power Line Communications and Its Applications, Jeju City, South Korea, 2-4 April 2008, 64-69. https://doi.org/10.1109/ISPLC.2008.4510400
16. Yang, Q., Barria, J.A. and Green, T.C. (2011) Communication Infrastructures for Distributed Control of Power Distribution Networks. IEEE Transactions on Industrial Informatics, 7, 316-327. https://doi.org/10.1109/TII.2011.2123903
17. Sauter, T. and Lobashov, M. (2011) End-to-End Communication Architecture for Smart Grids. IEEE Transactions on Industrial Electronics, 58, 1218-1228. https://doi.org/10.1109/TIE.2010.2070771
18. Sharma, K.K. and Singh, B. (2016) Review of Grid Integration with Conventional and Distributed Generation Sources. International Journal of Control Theory and Applications, 9, 6497-6512.
19. Bo, R. and Li, F.X. (2009) Probabilistic LMP Forecasting Considering Load Uncertainty. IEEE Transactions on Power Systems, 24, 1279-1289. https://doi.org/10.1109/TPWRS.2009.2023268
20. Ferreira, H., Lampe, L., Newbury, J. and Swart, T., Eds. (2010) Power Line Communications. Wiley, New York.

21. Sharma, K.K. and Singh, B. (2013) Smart Grid—An Indian Perspective. Trends in Electrical Engineering, 3, 62-67.
22. Sharma, K.K. and Singh, B. (2012) Smart Grid—Approach to Improve Existing Electrical Power System. International Journal of Computer Science and Communication Engineering, 1, 3-7.
23. Sharma, K.K. and Gupta, K. (2015) Literature Study on Hybrid Wind/ Mini hydro Power Plants. International Journal of Advanced Research in Computer Engineering and Technology, 4, 1988-1994.
24. Biagi, M. and Lampe, L. (2010) Location Assisted Routing Techniques for Power Line Communication in Smart Grids. 2010 First IEEE International Conference on Smart Grid Communications, Gaithersburg, MD, 4-6 October 2010, 274-278. https://doi.org/10.1109/ SMARTGRID.2010.5622056
25. Sanchez, J. Ruiz, P.M. and Marin-Perez, R. (2009) Beacon-Less Geographic Routing Made Particle: Challenges, Design Guidelines and Protocols. IEEE Communications Magazine, 47, 85-91. https:// doi.org/10.1109/MCOM.2009.5181897

Chapter 2

Building Automation Networks for Smart Grids

Peizhong Yi, Abiodun Iwayemi, and Chi Zhou

Electrical and Computer Engineering Department, Illinois Institute of Technology, Chicago, IL 60616-3793, USA

ABSTRACT

Smart grid, as an intelligent power generation, distribution, and control system, needs various communication systems to meet its requirements. The ability to communicate seamlessly across multiple networks and domains is an open issue which is yet to be adequately addressed in smart grid architectures. In this paper, we present a framework for end-to-end interoperability in home and building area networks within smart grids. 6LoWPAN and the compact application protocol are utilized to facilitate the use of IPv6 and Zigbee application profiles such as Zigbee smart energy for network and application layer interoperability, respectively. A differential service medium access control scheme enables end-to-end connectivity between 802.15.4 and IP networks while providing quality of service

Citation: Peizhong Yi, Abiodun Iwayemi, Chi Zhou, "Building Automation Networks for Smart Grids", International Journal of Digital Multimedia Broadcasting, vol. 2011, Article ID 926363, 12 pages, 2011. https://doi.org/10.1155/2011/926363.

guarantees for Zigbee traffic over Wi-Fi. We also address several issues including interference mitigation, load scheduling, and security and propose solutions to them.

INTRODUCTION

The smart grid is an intelligent power generation, distribution, and control system. It enhances today's power grid with intelligence, bidirectional communication capabilities and energy flows [1]. These enhancements address the efficiency, stability, and flexibility issues that plague the grid at present. In order to achieve its promised potential, the smart grid must facilitate services including the wide-scale integration of renewable energy sources, provision of real-time pricing information to consumers, demand response programs involving residential and commercial customers, and rapid outage detection. All these tasks demand the collection and analysis of real-time data. This data is then used to control electrical loads and perform demand response.

In order to obtain the full benefit of smart grids, their communication infrastructure must support device control and data exchanges between various domains which comprise the smart grid. The smart grid must be allied with smart consumption in order to achieve optimum power system efficiency. This necessitates the integration of smart buildings, appliances, and consumers in order to reduce energy consumption while satisfying occupant comfort. Building automation systems (BASs) already provide this intelligence, enabling computerized measurement, control and management of heating, ventilation, air-conditioning (HVAC), lighting, and security systems to enhance energy efficiency, reduce costs, and improve user comfort. Buildings consume 29% of all electricity generated in the United States [2]; therefore, the ability of BASs to communicate and coordinate with the power grid will have a tremendous effect on grid performance. Home area networks (HANs) provide similar capabilities for residential buildings. They facilitate the interconnection of smart appliances with smart meters to automatically regulate residential electricity usage and respond to pricing signals from the utility [3].

Zigbee is a low cost, low power, low data rate and short-range communication technology based on the IEEE 802.15.4 standard. United States National Institute for Standards and Technology (NIST) has defined Zigbee and the Zigbee smart energy profile (SEP) as the one of the communication standards for use in the customer premise network domain of the smart grid [4]. However, due to Zigbee's limited transmission range, it must be a combined with longer-range communication technologies such as IEEE802.11 in order to provide end-to-end connectivity across the smart grid.

In this paper, we discuss the different issues relevant to communication infrastructures for building automation system in smart grid. We begin with an introduction of whole system architecture of a smart grid system based on a perfect power system [5] including premises networks, field area networks, and a power system controller. We designed and implemented a Zigbee-based building energy management testbed system. Our system integrates a Zigbee-enhanced building automation system with the smart grid to harness energy management schemes such as demand response, real-time power pricing, peak load management, and distributed generation. We also propose a quality of service (QoS) aware 802.15.4/802.11 interoperability framework for home area network and building area network (BAN) which prioritizes wireless sensor network (WSN) traffic over Wi-Fi networks. In our scheme, WSN packets are classified according to their QoS requirements. They are then aggregated and tunneled over the Wi-Fi to the BASs server. We also proposed a frequency agility-based interference mitigation scheme to avoid interference from neighboring Wi-Fi networks. Distributed load scheduling based on optimal stopping rules [6] was proposed in the paper which can reduce the peak load and adjust utility operation time based on electricity pricing and waiting time. We also discuss open issues including security and data compression.

The rest of paper is organized as follows. Section 2 describes a smart grid system architecture. In Section 3, the Zigbee-based building energy management system was introduced. Our proposed QoS-aware 802.15.4/802.11 interoperability framework is presented in Section 4. Frequency-agility-based interference mitigation algorithm is proposed in Section 5. Section 6 presents our proposed optimal stopping rule-based distributed load scheduling scheme. Several open issues including smart grid security and data compression discussed in Section 7. Finally, the paper is concluded in Section 8.

SYSTEM ARCHITECTURE

The smart grid is the convergence of information technology, communications, and power system engineering to provide a more robust and efficient electrical power system [7]. Smart grids consist of sensing, communication, control, and actuation systems which enable pervasive monitoring and control of the power grid [8]. These features enable utilities to accurately predict, monitor and control the electricity flows throughout the grid. They also transform the power grid into a bidirectional power system in which customers can supply as well as receive power from the grid, converting the grid into a distributed power generation system [9].

The smart grid utilizes the hierarchical structure detailed in [8] and displayed in Figure 1. The foundation of this structure is the power system infrastructure consisting of power conversion, transportation, consumption, and actuation devices. They include power plants, transmission lines, transformers, smart meters, capacitor banks, reclosers, and various devices. Smart meters enable bidirectional power flows between utilities and consumers, enabling consumers to produce and supply energy to the grid, thereby becoming "prosumers". This development promises significant improvements in power system reliability, as alternative power sources can supply the grid during utility power outages. It also increases system efficiency, as line losses due to long-distance transmission are eliminated. These smart grid capabilities will foster greater incorporation of renewable energy sources such as wind and solar power into the grid, thereby reducing the dependence on fossil-fuel power generation and reducing greenhouse gas emissions.

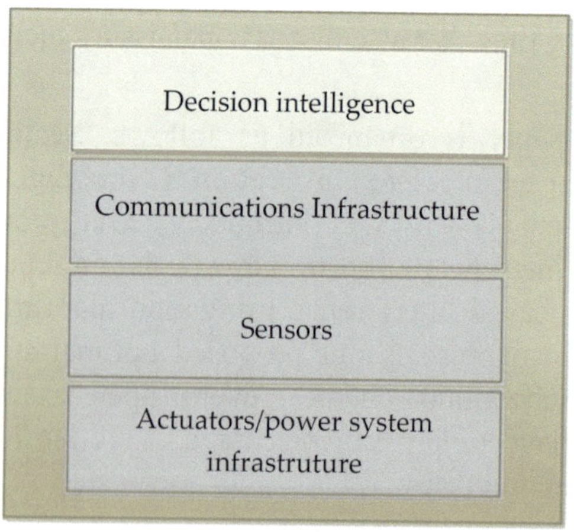

Figure 1: Smart grid structure.

The second layer of the smart grid architecture is the sensors. Power system reliability is significantly improved via embedded sensors distributed throughout nodes within the power system. These sensors enable real-time fault detection and isolation via bidirectional digital communication links. They also provide granular system health data that can be used for rapid system analysis, fault preemption, and trending. Smart meters also provide users and utilities with real-time power consumption data and enable the remote monitoring and control of building loads and home appliances. Consumers can also receive real-time pricing information to facilitate informed decision making.

The communications infrastructure is the glue that binds all these various layers together and consists of wide, local building and home area networks. They consist of broadband technologies such as 802.16 WiMAX, 802.11 Wi-Fi, optical fiber, 802.15.4/Zigbee, and power line carrier schemes. Zigbee has found great application in smart metering, home, and building automation control due to its low-cost, flexibility, wide-spread support, and intervendor interoperability.

At the top of the system is the decision intelligence block which encompasses substation automation, fault-management, load distribution, and other control strategies deployed to guarantee power system stability and balance power demand and supply.

The smart grid concept has been extended to smaller smart grid networks known as smart microgrids. A smart microgrid is a localized smart grid covering specific geographical regions, such as suburban neighborhoods or university campuses, and incorporating local or onsite power generation.

Building automation systems provide centralized and automated management of major or critical loads within building. Building automation aims to reduce energy costs, improve energy efficiency and facilitate off-site building management [10–12]. The primary requirements for building automation applications are low cost, ease of installation, and flexibility/reconfigurability.

ZIGBEE-BASED HOME AUTOMATION

Zigbee/IEEE 802.15.4

Zigbee is a low-rate, low-power, wireless personal area networking scheme [10] based on the IEEE 802.15.4 standard. It is designed for short-distance communication and supports a maximum data rate of 250 kbps without encryption.

Zigbee devices are ideal for smart grid and building automation applications, because they are wireless, low cost, and robust. Wireless nodes also provide flexibility, easy re-deployment, and reconfiguration. The integration of Zigbee radios with light switches, occupancy sensors, temperature sensors, and smoke detectors enables measurement and control of all the building loads. The low power consumption of Zigbee is achieved by very low system duty cycles, with typical Zigbee nodes having duty cycles of less than 5%. The result is significant energy savings and greater comfort for building occupants [13, 14]. Details of Zigbee's radio frequency characteristics, frequency bands, and modulation schemes are provided in Table 1.

Table 1: Zigbee radio frequency characteristics

Frequency	Region	Modulation scheme	Bit rate (kbps)	Channels	Channel spacing
868 MHz	Europe	BPSK	20	1	N/A
915 MHz	America and Asia	BPSK	40	10	2 MHz
2.4 GHz	Global	O-QPSK	250	16	5 MHz

Home Automation System

We developed a Zigbee-based home automation system [15] in order to demonstrate the utility of Zigbee-based home automation networks. Two-way communication was used to transmit readings from Zigbee end nodes to a data collection and control center (DCCC) and to pass control messages from the DCCC to the end nodes. Each end node is able to relay the collected data to the DCCC via distributed Zigbee routing nodes. The test bed architecture is shown in Figure 2. The Zigbee coordinator aggregates received data for display and processing and transmits control signals to the end nodes according to the selected power management strategy.

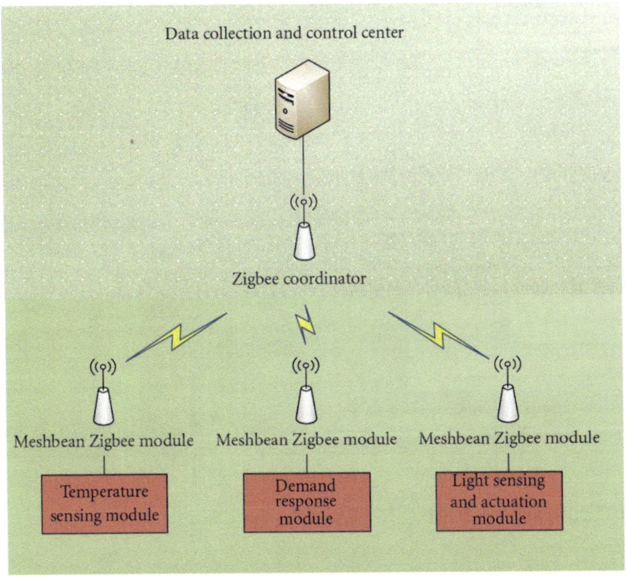

Figure 2: Zigbee HAN demonstration system architecture

The Data Collection and Control Center (DCCC)

The DCCC serves as the system controller, receiving input from the various sensors along with real-time power pricing. It also manages the loads for energy efficiency, demand response, and cost savings. A screenshot of the DCCC's user interface is shown in Figure 3. The DCCC is developed in MATLAB and utilizes a GUI front end to communicate directly with the Zigbee network coordinator and remote actuator modules. The DCCC provides the following functions:(i)the display of received sensor data (temperature, light levels, room occupancy, etc.),(ii)remote control of Zigbee modules,(iii)user configuration of timing, pricing, and sensor data threshold values,(iv)control of externally connected loads on the basis of user-determined price thresholds, time of day, and sensor readings,(v) lighting control based on room occupancy and other variables.

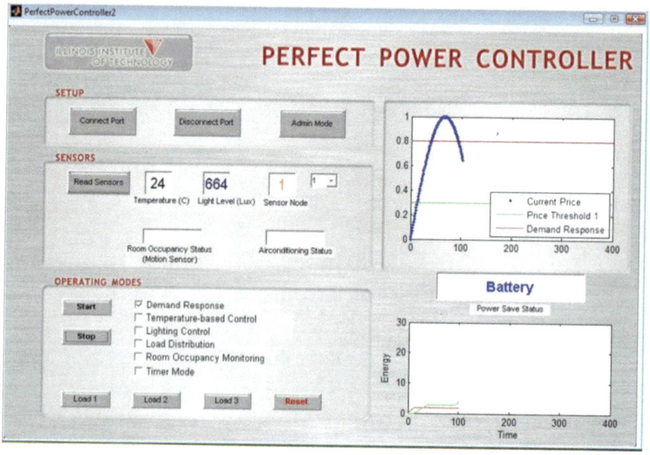

Figure 3: Perfect power controller GUI.

Hardware System

Our hardware system consists of several meshbean Zigbee motes which we programmed to support the following functions:(i)demand response,(ii) lighting control,(iii)ambient temperature sensing and control.

As shown in Figure 4, these modules combine an ATMEL 1281 V low-power microcontroller with 8 K of RAM and 128 kB of flash memory, an ATMEL RF230 Zigbee radio, onboard light and temperature sensors in a single battery-powered module with a USB interface. More details of our scheme can be found in [15].

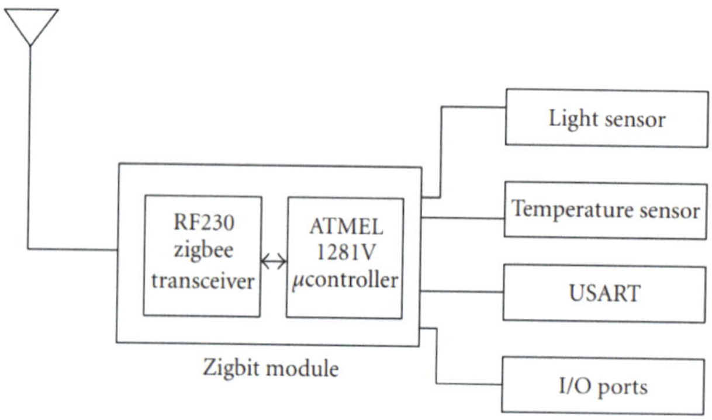

Figure 4: Meshnetics meshbean module block diagram.

INTEROPERABILITY OF ZIGBEE AND WI-FI

Building and home area networks are only one of a variety of networks that make up the smart grid. Due to the multiplicity of networks and protocols within the smart grid, interoperability is a key issue. The availability of an interoperability framework is essential to end-to-end communication across and within smart grid domains, so a significant amount of work is being invested in interoperability frameworks for the smart grid.

The usage of IP within wireless sensor networks facilitates easy interconnectivity with existing networks, enables the re-use of existing TCP/IP protocols, tools, and programming paradigms, and permits the usage of IP friendly protocols such as BACnet and Modbus over WSN nodes. These goals sparked research into the use of IPv6 over WSNs, as the ability to connect even tiny wireless sensor nodes to the internet would facilitate ubiquitous computing in the home and throughout the smart grid.

Interconnection between WSNs and TCP/IP networks has primarily been by means of gateways [16], as it had been assumed that TCP/IP was too memory and bandwidth intensive for usage in resource constrained wireless sensor networks [17]. However, the development of uIP, the first lightweight IP stack for WSNs [18] demonstrated the viability of IP for wireless sensor networks and led to a flurry of work into on the use of IP for WSNs. The 6LoWPAN IETF standard defines a framework for deployment of IPv6 over IEEE 802.15.4 networks [19] by means of header compression and routing and forwarding at layers 3 and 2, respectively. This work is extended in [20] to address issues such as link duty cycling, network bootstrapping and node discovery to create a complete IPv6 architecture for WSNs.

The primary interconnection schemes proposed for connecting Zigbee WSNs to the Internet are proxy-based gateways [16, 21, 22] and sensor stack schemes [20, 23]. The issue of the inability of Zigbee to natively support IP is addressed by the compact architecture protocol (CAP) [24] in which the authors create a framework to enable the usage of Zigbee application layer protocols over any IP-capable network. We extend their work by creating a framework for interworking between Zigbee and Wi-Fi networks in HANs and BANs while providing QoS guarantees. The proposed interoperability network architecture is shown in Figure 5. Taking into consideration BASs application requirements, reliability and short delay are two most important factor related to the performance. In [25], the authors present an architecture for a medical information system which integrates WLAN and WSNs. In

[26, 27], several QoS-enabling mechanisms present in the IEEE 802.11e provide us some ideas to design the frame work of the integration system. A two-tiered WSN and WLAN scheme with QoS guarantees is provided in [28], but the authors do not address IP-based interoperability.

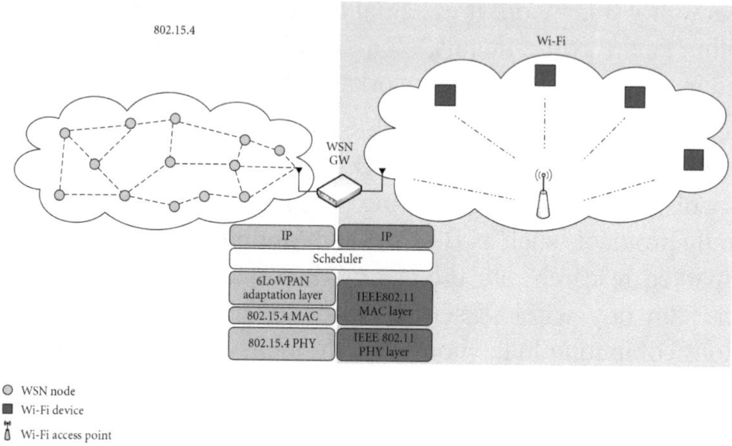

Figure 5: Interoperability network architecture.

Interworking

Interoperability is "The capability of two or more networks, systems, devices, applications, or components to exchange and readily use information—securely, effectively, and with little or no inconvenience to the user" [29]. The grid-wise architecture council (GWAC) [30] has defined an 8-layer interoperability framework encompassing all the facets of interoperability. Our primary focus is the 4 lowest layers of this framework (Figure 6), and we utilize it to develop an interoperability framework for HANs and BANs.

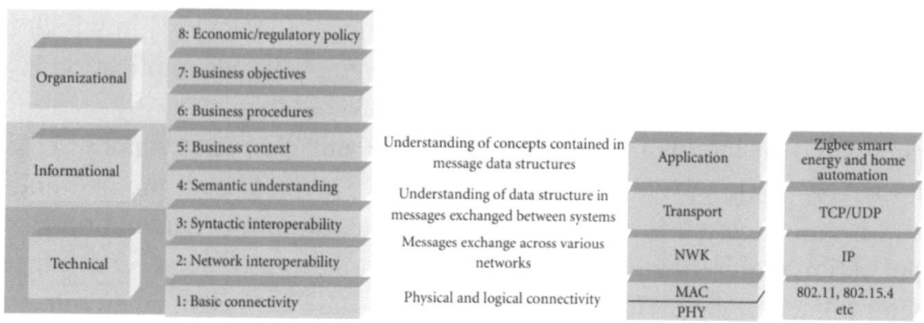

Figure 6: Interoperability framework.

The internet engineering task force (IETF) 6LoWPAN working group defined the IPv6 over low-power wireless personal area networks (6LoWPAN) protocol to facilitate the use of IPv6 over low power and low data rate WSNs [31]. It was initially designed for usage over the 802.15.4 physical (PHY) and medium access layer (MAC) layers but can be extended for use over other PHY and MAC architectures (Figure 6).

In order to use IPv6 over 802.15.4 networks, an adaptation layer between the 802.15.4 data link and network layers [17] was developed to provide the following functions:(i)stateless compression of IPv6 headers by means of HC1 compression [19] to reduce their size from 40 bytes to approximately 4 bytes, thereby reducing transmission overhead,(ii)fragmentation and reassembly scheme to support the transmission of IPv6 packets over 802.15.4 frames. This is required as the minimum MTU of IPv6 packets is 1280 bytes, while the maximum size of a 802.15.4 frame is 127 bytes.

The benefits of 6LoWPAN over competing WSN implementations are its ease of connectivity with IP networks and its large addressing space (2^{128} compared with 2^{16} for Zigbee). In addition, the concept of device roles found in Zigbee is not applicable, with each device serving as a router for its neighbor's traffic. Unlike Zigbee, 6LoWPAN permits duty cycling of routers, thereby extending device lifetime. The primary drawback of 6LoWPAN is its incompatibility with Zigbee, Zigbee's significant industry support, and very strong device interoperability guarantees across multiple vendors. A combination of the flexibility of IP networking and 6LoWPAN's power saving schemes with Zigbee's application profiles would marry the best features of both implementations to provide an industry standard, interoperable framework for HANs and BANs [32].

The compact application protocol (CAP) details a mapping of the Zigbee application layer to UDP/IP primitives [32, 33], permitting the usage of Zigbee application profiles over any IP capable network [24]. This removes the Zigbee application layer (ZAL) dependency on the Zigbee network layer and the 802.15.4 PHY and MAC layers. As shown in Figure 7, it preserves the excellent application layer interoperability features of public Zigbee application profiles while enabling end to end interoperability across the HAN/BAN using Wi-Fi, 802.15.4, and Ethernet. Rather than transmitting APS frames to the Zigbee network (NWK) layer for transmission to other nodes across the network using Zigbee addresses, the APS frames are now carried over UDP frames, necessitating modification to the addressing scheme to support communication with IP hosts using IP addresses and port numbers.

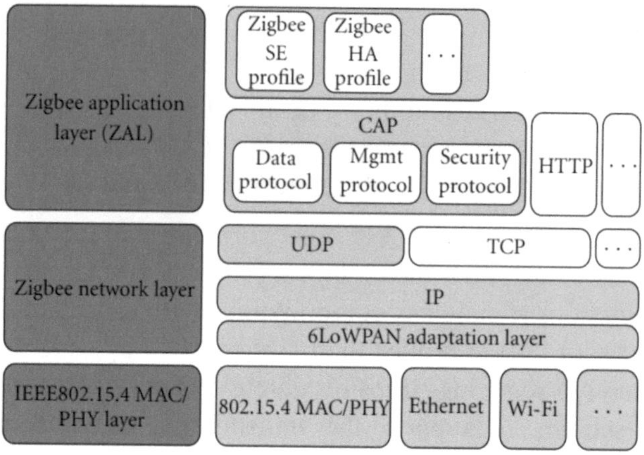

Figure 7: Zigbee, 6LoWPAN, and CAP stacks protocol stack [34, 35].

CAP is composed of four modules which correspond to the Zigbee application support sublayer (APS), Zigbee device objects (ZDO), Zigbee cluster library (ZCL), and APS security modules. The lowest layer of CAP is the Core module, which corresponds to the Zigbee APS layer. It frames data packets for transmission across the network, but now APS layer frames will be sent in UDP datagrams rather than in Zigbee NWK layer frames. In order to achieve this, Zigbee application profiles are rewritten to replace each Zigbee short (16-bit) and long (64-bit) address entry with a CAP address record. This consists of an IP address and UDP port pair, or a fully qualified domain name and UDP port number.

The data protocol is used to exchange data items and commands between communicating peer nodes. It encapsulates the ZCL and allows it to be used without modification, providing full ZCL support. The management protocol encapsulates Zigbee device profile (ZDP) command messages which are handled by the ZDO module, and provides service and device discovery and binding functionality. The final module is the security module which provides the same services as the APS security layer and is used to encrypt APS frames for secure transmission.

Gateway Router

Zigbee networks are primarily used for periodic data collection of low-bandwidth sensor and alarm data, while Wi-Fi networks support a variety of services with varying quality of service requirements. Based on this, a differential service medium access control scheme [25] is required to guarantee timely and reliable delivery of Zigbee traffic over building Wi-Fi networks. Thus, we design an enhanced distributed channel access-(EDCA-) based QoS model to achieve this.

Our framework facilitates the interconnection of the WSN to the BAN server via the in-building Wi-Fi system. This interconnection is achieved using a dual-stack gateway router (GR) node which performs QoS classification and packet aggregation on Zigbee application layer packets before tunneling them to the BAN server over Wi-Fi. As seen in Figure 6, we utilize the 802.11 and 802.15.4 MAC and physical layer protocols in conjunction with 6LoWPAN, the compact application protocol, and Zigbee application layer application profiles to provide end-to-end interoperability within HANs and BANs. Physical layer interoperability is provided by means of the GR's dual stack and 802.11 and 802.15.4 interfaces. Network layer interoperability is provided using IPv6 and the use of 6LoWPAN to enable the WSN to communicate using IP. Syntactic interoperability is achieved by the use of the CAP, which allows us to utilize publically defined Zigbee application profiles such as the smart energy or home automation profiles to provide application layer interoperability across multivendor devices. This frees us to use Zigbee application profiles across the HAN, on PCs, routers, and over any IP-capable device nodes all over the home or commercial building, rather than only over Zigbee 802.15.4 networks. In addition, the ability of our system to schedule 802.15.4 and 802.11 MAC frames enables us to provide quality of service prioritization to emergency Zigbee traffic.

QoS of Service Framework

The GR facilitates interconnection of the 802.15.4 and Wi-Fi networks to enable end-to-end communication. The GR contains an MAC scheduler in which can communicate with 802.15.4 MAC and 802.11 MAC layer. On the basis of these assumptions, we divide the queuing model into three parts. In the first part, traffic from WSN end nodes to the coordinator is considered; in the second part, packets from the Wi-Fi access point (AP) to the GR are discussed; finally, we will focus on the queuing model of packets sent from GR. In this scheduling scheme, the use of guaranteed time slots (GTSs) can combine the task of scheduling uplink and downlink flows of a naturally distributed carrier sense medium access with collision avoidance (CSMA/CA) environment into a central scheduler residing in the GR.

As shown in Figure 8, each WSN node has two traffic queues, one for emergency or alarm traffic and the second for normal traffic [25]. Class 0 (alarm packets) are higher priority emergency/control data, while Class 1 (normal packets) contains routine data. Nodes will typically transmit two message types. The first are GTSs requests to reserve slots in CFP, and the second types are data packets containing sensor data.

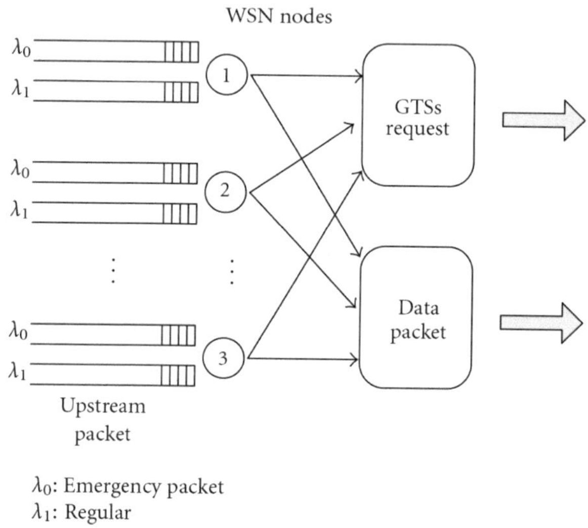

Figure 8: Queuing model from WSN end nodes to WSN coordinator.

Data frames are assigned to their respective queue and contend for transmission over the channel. A node contends per information frame and can only send one packet each time. If the node has an emergency message in its queue, it will request a shorter back off exponent value to enable prompt transmission of emergency traffic. Nodes which do not have emergency traffic utilize the regular value of the back off exponent, resulting in longer wait times.

Traffic differentiation at the GR is performed on the basis of destination ports. As seen in Figure 9, we use different ports for normal and emergency traffic and map them to EDCA video (AC_VI) and voice (AC_VO) access categories, respectively, before transmitting over the Wi-Fi network. A dedicated BASs server is the final recipient of the entire off network WSN traffic, and this server filters traffic based on the ports the data is received on.

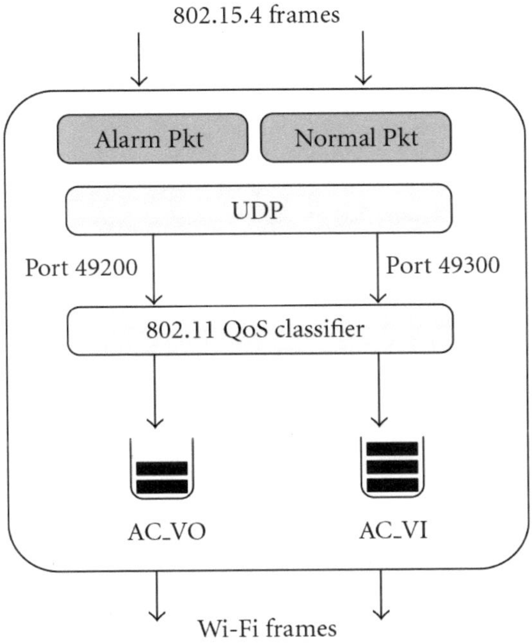

Figure 9: Queuing model from Gateway Router to Wi-Fi AP.

Due to the significant size difference between 802.15.4 and Wi-Fi frames, traffic aggregation is required for delay-tolerant traffic, while time sensitive WSN traffic is transmitted immediately. The encapsulation of individual 802.15.4 packets is very inefficient as the Wi-Fi header frame overhead is often larger than the useful information, necessitating packet aggregation to improve efficiency.

A hybrid scheduling model is used in the GR as shown in Figure 10. All packets received at the GR can be transmitted in either contention access period (CP) or contention-free period (CFP) modes. During the CP mode, nodes use a slotted CSMA/CA scheme to compete for the channel with other nodes. In CFP mode, up to seven GTS can be reserved and allocated by the coordinator. Devices which require the allocation of a new GTS transmit a GTS request command to the WSN coordinator and coordinator will assign GTS to each device. Our hybrid scheduling model adopts both EDCA in CP and point coordination function (PCF) controlled channel access (PCCA) in the CFP to achieve fairness and provide service guarantees. The GR assigns packets to different MAC schemes based on message type. Emergency/control messages which need to be sent out immediately will use EDCA contention access with smaller back off exponent. On the other hand,

PCCA is used for routine messages, as these can wait for aggregation to be performed and are subsequently transmitted in reserved times slots.

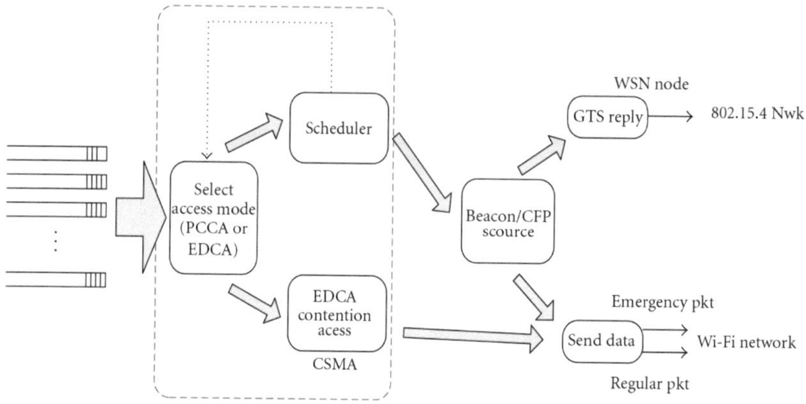

Figure 10: Queuing model in the WSN Coordinator.

Routine messages sent from the 802.15.4 PAN to the Wi-Fi access point are initially sent to the scheduler, where they are enqueued and time-stamped while a countdown timer initialized. The queue size is set to maximum size of a Wi-Fi payload, and if the queue is filled with routine messages before timer expiry, the scheduler reserves a GTS, aggregates all the enqueued traffic, and transmits them over the Wi-Fi radio. If the queue is not filled by timer expiry, then the GR reserves the number of GTSs required to transmit the queue and sends the accumulated data. The primary benefits of message aggregation with PCCA are collision and delay reduction for routine traffic.

INTERFERENCE AVOIDANCE SCHEME

Zigbee networks operate in the license-free industrial, scientific and medical (ISM) frequency band, making them subject to interference from various devices that also share this license-free frequency band. These devices range from IEEE 802.11 wireless local area networks or Wi-Fi networks and Bluetooth devices to baby monitors and microwave ovens. Studies have shown that Wi-Fi is the most significant interference source for Zigbee within the 2.4 GHz ISM band [36, 37]. Zigbee and Wi-Fi networks are used extensively for BAN in smart grid applications, leading to coexistence problems as seen in Figure 11.

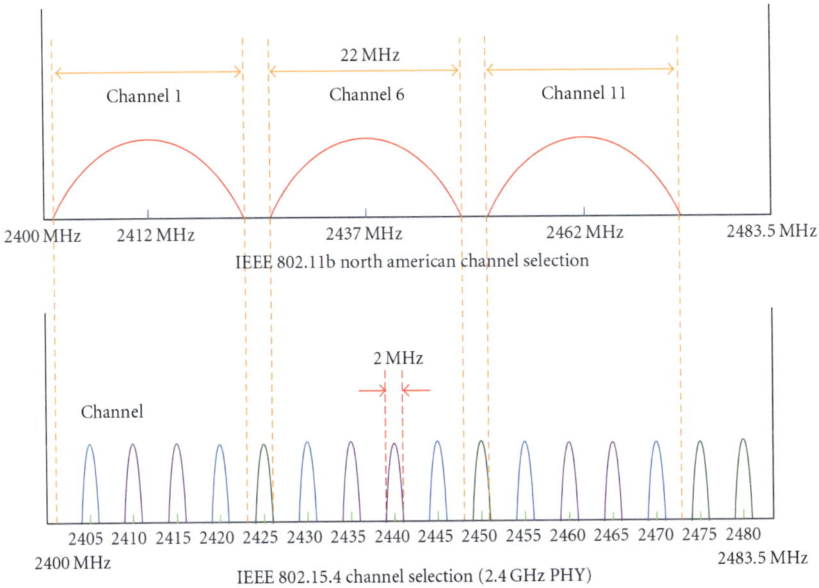

Figure 11: Zigbee and Wi-Fi channels in the 2.4 GHz band.

Therefore, we have performed a large amount of experiments to identify the "safe distance" and "safe offset frequency" to guide the Zigbee deployment [38]. The performance of Zigbee in the presence of IEEE 802.11 is defined and analyzed in terms of bit error rate (BER) and packet error rate (PER) by comprehensive approach including theoretical analysis, software simulation, and empirical measurement. Based on the concepts of "safe distance" and "safe offset frequency," we propose a frequency-agility-based interference mitigation algorithm [39]. PER, link quality Indication (LQI), and energy detection mechanisms are used to detect the presence of significant levels of interference within the current channel. Once interference is detected, the coordinator instructs all the routers to perform an energy detection scan on channels and then send a report to the coordinator. The coordinator selects the channel with the lowest noise levels and then requests all nodes in the PAN to migrate to this channel. In order to improve the detection time and power efficiency, all Zigbee channels are divided into three classes based on the offset frequency. The energy detection scan will be performed from high-priority class to low-priority class channels to quickly identify the channel with acceptable interference level. The testbed implementation shows that the proposed frequency-agility-based algorithm is simple but efficient, fast, and practical.

OPPORTUNISTIC LOAD SCHEDULING

Demand response is the technology that manages customers' electricity usage to reduce electricity expenditure. Since customers are provided with the real-time power price by smart metering devices, load scheduling must incorporate real price in order to perform load control. The real-time price is an indicator of the system load. In general, the price is high when the load demand is high and vice versa. Some level of peak demand reduction may be automatically achieved by rational customers who aim to minimize the electricity cost. Naturally, the customers will choose to operate their flexible loads when the real-time price reaches the minimum. In this way, those flexible loads are shifted to the low demand time period, and consequently, the peak demand is reduced.

Nowadays, most existing load scheduling schemes are based on the assumption that future electricity prices are known or predictable. We propose to apply the optimal stopping rule [40] to perform distributed load scheduling. Our scheme to determines when to operate the flexible loads under the assumption that price signals are unknown and considered as random processes optimal stopping rule is proved to perform excellently in communication networks [41]. Thus, we extend the application of optimal stopping rule to power grids [42]. The time requirement of the load is taken into the consideration. If a user does not have time requirement, it will always choose to operate at the time when the electricity price is the lowest to minimize the electricity cost. However, many appliances, such as washing machine, are sensitive to the waiting time. Therefore, the spent time (which includes waiting time and service time) must be taken into consideration. The cost is modeled as the wait cost plus the electricity cost, and the objective is to minimize the total cost by choosing the best operating time. We show that the optimal scheduling scheme is a pure threshold policy; that is, each user needs to turn on the load when the electricity price is lower than a certain value; otherwise, the load remains idle. Simulation results show that the proposed low-complexity distributed scheduling scheme can dramatically reduce the cost. In other words, the loads are effectively shifted to low-demand time period. More details can be found in [6].

OPEN ISSUES AND FUTURE WORK

Smart Grid Security

The smart grid requires detailed energy usage information in order to facilitate services such as real-time pricing and billing, customer energy management, and system load prediction. Unfortunately, as is the case with many other complex systems, the smart grid falls foul of the law of unintended consequences. The availability of such detailed usage data from every household every 5–15 minutes has created a massive security problem [43]. Smart meter data analysis provides the ability to determine which appliances are in use at any given time period. This has led to the fear that users can be spied upon by their meters, negatively impacting smart meter deployment [44]. The networking of smart meters with the electricity grid also raises the specter of smart meter fraud and increases the vulnerability of these devices to malicious attacks such as denial of service (DoS) attacks.

Privacy Issues

Research into nonintrusive appliance load monitoring technology (NALM) [45–47] has enabled the identification of appliances by means of their unique fingerprint or "appliance load profiles." By means of software analysis, it possible to determine which appliances are in use and at what frequency. It provides access to information including the types of appliances a resident possesses, when he/she has their shower each day (by monitoring extended usage of the heater), how many hours they spend using their PC, or whether they cook often or eat microwave meals. This has led to the very valid fear that customers can be profiled and monitored by means of their smart meter. In addition, improper access to such data can lead to violations of privacy or even make one open to burglary.

Smart Meter Fraud

The desire for lower electricity bills provides a compelling incentive for smart meter fraud. The ability to report inaccurate data to the utility means that customers can reduce their bills by falsely claiming to supply power the grid or consume less power than their actually do. The possibility of commercially available smart meter hacking kits is also a reality [48].

Malicious Attacks

The internetworking of smart meters makes them especially vulnerable to denial of service attacks in which several meters are hijacked in order to flood the network with data in order to shut down portions of the power grid or report false information which can result in grid failures.

Smart Grid Security Solutions

Smart grid security issues can only be solved by a combination of regulatory and technological solutions. A regulatory framework is required to specify who has access to smart meter data and under which conditions as well as enforcement of penalties for data misuse [48]. Two technological solutions have been proffered. The first is to aggregate residential data at the neighborhood transformer and then anonymize it by stripping it off its source address before transmitting it to the utility [43]. Kalogridis et al. [42] propose the use of a third party escrow service which receives the detailed meter data, anonymizes it by stripping off any information that could be used to identify a specific household, then sends the utility the aggregate data required for billing and monthly energy usage for each customer.

We propose a digital rights management system-(DRMS-) based scheme which extends that proposed in [49]. Users license permission to the utility to access their data at varying levels of granularity. By default, the utility would have access to monthly usage and billing data, but customers have to grant the utility permission to access their data at higher levels of granularity in exchange for rebates or other incentives. Such a system eliminates the need for an intermediary between the utility and the consumer but requires a means of guaranteeing that the utility cannot access restricted customer data.

Data Compression

Mitigating data surges and traffic congestion due to catastrophic events is an open research area. When emergencies such as power blackouts occur, hundreds to thousands of smart meter flood the data collection center with traffic. Reliability is an important issue, since the data needs to be transmitted effectively and efficiently, and network coding is a promising approach to improving the reliability of the wireless networks under such conditions. By means of network coding, we could potentially introduce intraflow network coding into the data transmission in Zigbee networks; that is, routers mix packets heading to the same destination. As a result of this mixing, each received packet contains some information about all packets in the original file, and thus, no coded packet is special.

Conventionally, without coding, a transmitter needs to know which exact packets the destination misses so that it can retransmit them. When the network is unreliable, communicating this feedback reliably consumes significant bandwidth. In the presence of coding, no specific packet is indispensable, and as a result, a transmitter does not need to learn which particular packet the destination misses; it only needs to get feedback from the destination once it has received enough packets to decode the whole file. The reader may have noticed that the above applies to erasure-correcting coding applied at the source too. Indeed, source coding is just a special case of intraflow network coding, where the source is the only node allowed to mix the packets in the flow.

CONCLUSION

In order for the smart grid to achieve its potential, we need the resolve the problem of interoperability between the different communications technologies deployed in the grid. In this paper, we proposed an HAN architecture for energy management within smart grid environments. Zigbee-based building energy management was demonstrated to enhance building automation systems and permit granular control of electrical and HVAC systems in a smart grid context. An open architecture of an interoperability frame work for HANs and BANs was presented in the paper. Physical layer interoperability is provided by means of a router platform with 802.11 and 802.15.4 interfaces. Network layer interoperability is provided using IPv6 and the usage of 6LoWPAN to enable the WSN to communicate using IP. Syntactic interoperability is achieved by the use of the CAP. In the QoS framework, emergency/control message need to compete with routine traffic from other nodes. The prioritized contention algorithm ensured the high priority access the channel for these messages. Use of compression and scheduling increases the efficiency of the data transferred from Zigbee to Wi-Fi frames. A frequency-agility-based interference mitigation algorithm was introduced in the paper to guarantee the performance of Zigbee and Wi-Fi coexistence. Optimal stopping rule base load scheduling scheme as a distributed load control was present in the paper. More open issues including security and data compression were discussed in the paper.

REFERENCES

1. P. Yi, A. Iwayemi, and C. Zhou, "Developing ZigBee deployment guideline under WiFi interference for smart grid applications," *IEEE Transactions on Smart Grid*, vol. 2, no. 1, pp. 98–108, 2011.
2. US Department of Energy, "Buildings Energy Data Book," March 2009, http://buildingsdatabook.eren.doe.gov/.
3. T. J. Lui, W. Stirling, and H. O. Marcy, "Get smart," *IEEE Power and Energy Magazine*, vol. 8, no. 3, pp. 66–78, 2010.
4. Office of the National Coordinator for Smart Grid Interoperability, NIST, "Framework and Roadmap for Smart Grid Interoperability Standards, Release 1.0," National Institute of Standards and Technology, 2010.
5. "The Galvin Path to Perfect Power—A Technical Assessment: Report on the Galvin Electricity Initiative, Phase II, Tasks 1 & 2," Galvin Electricity Initiative, 2007.
6. P. Yi, X. Dong, and C. Zhou, "Optimal energy management for smart grid systems—an optimal stopping rule approach," in *IFAC World Congress Invited Session on Smart Grids*, 2001, accepted.
7. H. Farhangi, "The path of the smart grid," *IEEE Power and Energy Magazine*, vol. 8, no. 1, pp. 18–28, 2010.
8. E. Santacana, G. Rackliffe, LE. Tang, and X. Feng, "Getting smart: with a clearer vision of the intelligent grid, control emerges from chaos," *IEEE Power and Energy Magazine*, vol. 8, no. 2, pp. 41–48, 2010.
9. S. Amin, "For the good of the grid," *IEEE Power and Energy Magazine*, vol. 6, pp. 48–59, 2008.
10. J. A. Gutiérrez, "On the use of IEEE Std. 802.15.4 to enable wireless sensor networks in building automation," *International Journal of Wireless Information Networks*, vol. 14, no. 4, pp. 295–301, 2007.
11. A. C. W. Wong and A. T. P. So, "Building automation in the 21st century," in *Proceedings of the 4th International Conference on Advances in Power System Control, Operation and Management*, vol. 2, pp. 819–824, February 1998.
12. D. Snoonian, "Smart buildings," *IEEE Spectrum*, vol. 40, no. 8, pp. 18–23, 2003.
13. W. Kastner, G. Neugschwandtner, S. Soucek, and H. M. Newman, "Communication systems for building automation and

control," *Proceedings of the IEEE*, vol. 93, no. 6, pp. 1178–1203, 2005.

14. K. Gill, S. H. Yang, F. Yao, and X. Lu, "A ZigBee-based home automation system," *IEEE Transactions on Consumer Electronics*, vol. 55, no. 2, pp. 422–430, 2009.
15. A. Iwayemi, P. Yi, P. Liu, and C. Zhou, "A perfect power demonstration system," in *Innovative Smart Grid Technologies Conference (ISGT '10)*, pp. 1–7, January 2010.
16. J. J. P. C. Rodrigues and P. A. C. S. Neves, "A survey on IP-based wireless sensor network solutions," *International Journal of Communication Systems*, vol. 23, no. 8, pp. 963–981, 2010.
17. J. W. Hui and D. E. Culler, "Extending IP to low-power, wireless personal area networks," *IEEE Internet Computing*, vol. 12, no. 4, pp. 37–45, 2008.
18. A. Dunkels, "TCP/IP for 8-bit architectures," in *Proceedings of the 1st International Conference on Mobile Systems, Applications, and Services (MOBISYS '03)*, 2003.
19. G. Mulligan, "The 6LoWPAN architecture," in *Proceedings of the 4th Workshop on Embedded Networked Sensors (EmNets '07)*, pp. 78–82, June 2007.
20. J. W. Hui and D. E. Culler, "IP is dead, long live IP for wireless sensor networks," in *Proceedings of the 6th ACM conference on Embedded Network Sensor Systems*, pp. 15–28, 2008.
21. M. Sveda and R. Trchalik, "ZigBee-to-internet interconnection architectures," in *Proceedings of the 2nd International Conference on Systems (ICONS '07)*, p. 30, April 2007.
22. R. Wang, R. Chang, and H. Chao, "Internetworking between ZigBee/802.15.4 and IPv6/802.3 network," in *SIGCOMM Workshop "IPv6 and the Future of the Internet"*, Kyoto, Japan, August 2007.
23. G. Mulligan et al., "Seamless sensor network IP connectivity," in *Proceedings of the 6th European Conference on Wireless Sensor Networks (EWSN '09)*, 2009.
24. G. Tolle, "A UDP/IP adaptation of the ZigBee application protocol," http://tools.ietf.org/html/draft-tolle-cap-00.
25. X. Yuan, S. Bagga, J. Shen, M. Balakrishnan, and D. Benhaddou, "DS-MAC: differential service medium access control design for wireless medical information systems," in *Proceedings of the 30th Annual International Conference of the IEEE Engineering in Medicine and*

Biology Society (EMBS '08), pp. 1801–1804, August 2008.

26. A. Banchs, A. Azcorra, C. García, and R. Cuevas, "Applications and challenges of the 802.11e EDCA mechanism: an experimental study," *IEEE Network*, vol. 19, no. 4, pp. 52–58, 2005.
27. Y. P. Fallah and H. Alnuweiri, "Hybrid polling and contention access scheduling in IEEE 802.11e WLANs," *Journal of Parallel and Distributed Computing*, vol. 67, no. 2, pp. 242–256, 2007.
28. J. Leal, A. Cunha, M. Alves, and A. Koubâa, "On a IEEE 802.15.4/ ZigBee to IEEE 802.11 gateway for the ART-WiSe architecture," in *Proceedings of the 12th IEEE International Conference on Emerging Technologies and Factory Automation (ETFA '07)*, pp. 1388–1391, September 2007.
29. U. S. Department of Energy, Office of Electricity Delivery and Energy Reliability, "Recovery Act Financial Assistance, Funding Opportunity Announcement. Smart Grid Investment Grant Program Funding Opportunity Number: DE-FOA-0000058," June 2009.
30. GridWise Architecture Council, "GridWise Interoperability Context-Setting Framework (v1.1)," 2008.
31. IETF, "RFC 4944: Transmission of IPv6 Packets over IEEE 802.15.4 Networks," http://www.rfc-editor.org/rfc/rfc4944.txt.
32. D. E. Culler and G. Tolle, "Compact Application Protocol (CAP): Uniting the Best of IP and ZigBee," http://www.rtcmagazine.com/articles/print_article/101065.
33. Z. Shelby and C. Bormann, *6LoWPAN: The Wireless Embedded Internet*, Wiley, 2010.
34. Parallax Inc., "Parallax PIR Sensor (#555-28027)," March 2010, http://www.parallax.com/dl/docs/prod/audiovis/pirsensor-v1.2.pdf.
35. "Panasonic AQH Solid State Relay," http://pewa.panasonic.com/assets/pcsd/catalog/aq-h-catalog.pdf.
36. Zensys, "White Paper: WLAN interference to IEEE 802.15.4," 2007.
37. S. Y. Shin, H. S. Park, S. Choi, and W. H. Kwon, "Packet error rate analysis of ZigBee under WLAN and Bluetooth interferences," *IEEE Transactions on Wireless Communications*, vol. 6, no. 8, pp. 2825–2830, 2007.
38. P. Yi, A. Iwayemi, and C. Zhou, "Frequency agility in a ZigBee network for smart grid application," in *Innovative Smart Grid Technologies Conference (ISGT '10)*, pp. 1–6, January 2010.

39. G. Thonet and P. Allard-Jacquin, "ZigBee-WiFi Coexistence White Paper and Test Report," Schneider Electric White Paper, 2008.
40. T. S. Ferguson, "Optimal Stopping and Applications," http://www.math.ucla.edu/~tom/Stopping/Contents.html.
41. D. Zheng, W. Ge, and J. Zhang, "Distributed opportunistic scheduling for ad hoc networks with random access: an optimal stopping approach," *IEEE Transactions on Information Theory*, vol. 55, no. 1, pp. 205–222, 2009.
42. G. Kalogridis, C. Efthymiou, S. Denic, T. Lewis, and R. Cepeda, "Privacy for smart meters: towards undetectable appliance load signatures," in *Proceedings of the 1st IEEE International Smart Grid Communications (SmartGridComm '10)*, pp. 232–237, 2010.
43. E. L. Quinn, "Privacy and the New Energy Infrastructure," *SSRN eLibrary*, February 2009, http://papers.ssrn.com/sol3/papers.cfm?abstract_id=1370731.
44. "IEEE Spectrum: Privacy on the Smart Grid," http://spectrum.ieee.org/energy/the-smarter-grid/privacy-on-the-smart-grid.
45. G. W. Hart, "Nonintrusive appliance load monitoring," *Proceedings of the IEEE*, vol. 80, no. 12, pp. 1870–1891, 1992.
46. J. G. Roos, I. E. Lane, E. C. Botha, and G. P. Hancke, "Using neural networks for non-intrusive monitoring of industrial electrical loads," in *Proceedings of the 10th IEEE Instrumentation and Measurement Technology Conference*, vol. 3, pp. 1115–1118, 1994.
47. H.-H. Chang, C.-L. Lin, and J.-K. Lee, "Load identification in nonintrusive load monitoring using steady-state and turn-on transient energy algorithms," in *Proceedings of the 14th International Conference on Computer Supported Cooperative Work in Design (CSCWD '10)*, pp. 27–32, 2010.
48. P. McDaniel and S. McLaughlin, "Security and privacy challenges in the smart grid," *IEEE Security and Privacy*, vol. 7, no. 3, pp. 75–77, 2009.
49. Z. Fan, G. Kalogridis, C. Efthymiou, M. Sooriyabandara, M. Serizawa, and J. McGeehan, "The new frontier of communications research: smart grid and smart metering," in *Proceedings of the 1st International Conference on Energy-Efficient Computing and Networking (e-Energy '10)*, pp. 115–118, 2010.

Chapter 3

Voltage Control in Smart Grids: An Approach Based on Sensitivity Theory

Morris Brenna[1], Ettore De Berardinis[2], Federica Foiadelli[1], Gianluca Sapienza[3], Dario Zaninelli1

[1]Politecnico di Milano – Department of Energy, Milan, Italy
[2]CESI S.p.A., Milan, Italy
[3]Politecnico di Milano – Department of Energy in Collaboration with ENEL Distribuzione S.p.A., Milan, Italy

ABSTRACT

Due to the development of Distributed Generation (DG), which is installed in Medium-Voltage Distribution Networks (MVDNs) such as generators based on renewable energy (e.g., wind energy or solar energy), voltage control is currently a very important issue. The voltage is now regulated at the MV busbars acting on the On-Load Tap Changer of the HV/MV transformer. This method does not guarantee the correct voltage value in the network nodes when the distributed generators deliver their power. In this paper an approach based on Sensitivity Theory is shown, in order to control the node

Citation: M. Brenna, E. Berardinis, F. Foiadelli, G. Sapienza and D. Zaninelli, "Voltage Control in Smart Grids: An Approach Based on Sensitivity Theory," *Journal of Electromagnetic Analysis and Applications*, Vol. 2 No. 8, 2010, pp. 467-474. doi: 10.4236/jemaa.2010.28062.

voltages regulating the reactive power exchanged between the network and the dispersed generators. The automatic distributed voltage regulation is a particular topic of the Smart Grids.

Keywords: Voltage Regulation, Reactive Power Injection, Distributed Generation, Smart Grids, Sensitivity Theory, Renewable Energy

INTRODUCTION

Due to the development of Distributed Generation (DG), which is installed in Medium-Voltage Distribution Networks (MVDNs) such as generators based on renewable energy (e.g., wind energy or solar energy), voltage control is currently a very important issue.

The voltage of MVDNs is now regulated acting only on the On-Load Tap Changer (OLTC) of the HV/MV transformer [1]. The OLTC control is typically based on the compound technique, and this method does not guarantee the correct voltage value in the network nodes when the generators deliver their power [2,3].

When a generator injects power in the network, the voltage tends to rise. In HV networks this phenomenon happens mainly when reactive power is injected, because the resistance is negligible if compared with the inductive reactance [4]. Instead in MVDNs the resistance is not negligible and the result is that an injection of active power also increases the voltage.

In other words the so-called Pq - QV decoupling [5], which is a typical of HV networks, is inexistent in MVDNs. The P variations are "coupled" with the voltage variations.

If no precautions are taken, in particular network conditions the overcome of the maximum admissible voltage can happen in any nodes.

When a generator injects power, the voltage rises in all network nodes, but some nodes are mainly influenced than others by the power injection. This influence can be obtained using a Sensitivity method.

In this paper an approach based on Sensitivity Theory is shown, in order to control the network voltage using the reactive power exchanged between network and the distributed generators. This approach allows to control the voltage in the long term period. Besides, fastdynamic voltage disturbances are not taken into account [6].

After the theoretical analysis, a numerical example is shown, in order to validate the proposed theory.

The proposed method differs from the others used in HV networks analysis, based on the Jacobian Matrix [1,2-4] and its application is easy.

The topological proprieties that results from the theoretical analysis imply that the proposed sensitivity method can be easily implemented in automatic voltage control devices, in order to obtain the distributed voltage regulation.

The automatic voltage regulation in a distributed manner is a typical topic of the Smart Grids context.

The paper is structured in the following way. In Section 1, the proposed voltage control method is shown, and an overview on the voltage profiles with DG, are given. In Section 3, the proposed Sensitivity approach is studied, referring to a MV test network, composed by four nodes. Finally, in Section 4, a numerical application is presented, in order to validate the proposed theory.

THE PROPOSED CRITERIA TO CONTROL THE NETWORK VOLTAGE WITH DISTRIBUTED GENERATION

Many methods can be used to control the voltage in network nodes (network voltages). The proposed method varies the reactive power exchanged between the generators and the network while maintaining the OL-TC in a fixed position for a particular load condition.

Let us suppose that the Automatic Voltage Regulator (AVR) that controls the OLTC maintains the MV bus-bar voltage at the rated value (1 p.u.), assuming that the transformer taps are adequate.

For passive grids, when no generators are connected to the MVDN, the voltage profile (VP; i.e., the voltage values along a line) decreases monotonically (see profile a in Figure 1) due to the load absorptions. When the generators are connected and inject power into the MVDN, the nodal voltages increase and the VP is no longer monotonic, as shown in profile b in Figure 1 (profile b). This phenomenon also occurs if generators work at unitary power factor (i.e., only active power is injected due to the non-negligible network resistance) [7].

It is important to note that, in steady-state, the condition maintained at the MV busbar by the AVR decouples the MV feeders, and the result is that each feeder works without the influence of the other lines. In other words, the loads and generators connected to other feeders do not influence the VP of the considered line.

Typically, the generators installed in Smart Grids are based on renewable energy; therefore, their power-time profiles are unknown. Due to the high generated power and a possibly low load condition, the voltage in some nodes can thus exceed the maximum admissible value (V_{max}; i.e., the voltage threshold [8]) defined by the standards.

Of course the voltage threshold is strictly related with the settings of the voltage relays installed in the network, e.g. at the generator nodes [9].

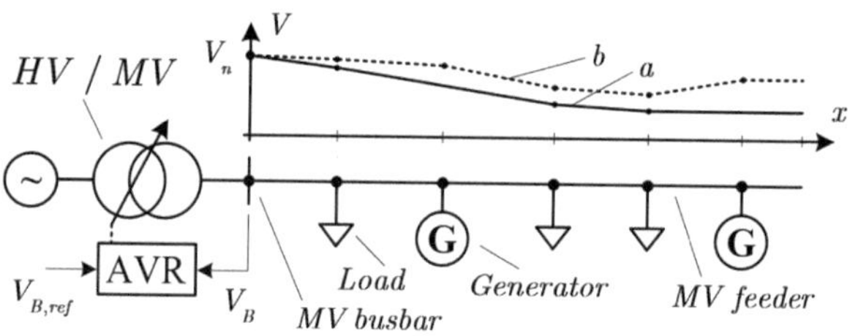

Figure 1: Voltage profiles in a MV feeder with and without Distributed Generators

If the generators are able to control the injected or absorbed reactive power, the network voltage profiles can be modified by acting on the reactive powers. It is clear that each controllable generator needs a Generator Remote Terminal Unit (GRTU) that is connected to a central control system to set the generator reactive power, (i.e., to control the exciter of the synchronous generators [1] or act on the inverter control if the generator is inverter-based) [10,11]. In this work, the central control is called the Generator Control Centre (GCC). In addition, we use a hierarchical control structure [12,13].

Let us suppose that the voltage is measured only in the generator nodes by the GRTUs. This assumption does not affect the generality of the proposed method because a Measuring Remote Terminal Unit connected to the GCC can be installed in each node that must be controlled.

When the voltage in the ith node exceeds V_{max}, the GRTU installed in the same node sends the signal "Voltage Threshold Overall" (VTO) to the GCC using a communication channel. The GCC then selects the generator in the jth node that has the maximum influence on the voltage of the ith node, the "Best Generator" (BG), and switches it to the reactive power absorption (RPA) mode. Therefore, the voltage in the ith node tends to decrease.

The problem is thus to determine the best generator and ensure that the GCC chooses it. In this work, a sensitivity-based method is proposed to select the BG.

Moreover, we suppose that the generators can only be switched in the RPA mode by the GCC by a constant power factor. Therefore, if P_j is the active power injectted by the generator connected to the jth node, then it absorbs the reactive power $Q_j = P_j \tan\varphi_j$ (where $\cos\varphi_j$ is the minimum power factor of the generator) when it is switched during RPA. In other words, we assume that no continuous reactive power modulation is possible.

An example of the procedure described above is shown in **Figure 2**. Let us suppose that load Ld suddenly decreases its power (for example, due to a trip) and V_2 exceed V_{max}.

The GRTUs of G2 send the signal VTO to the GCC that must choose the BG using the sensitivity method. Assuming that the BG is G1, it will be switched by the

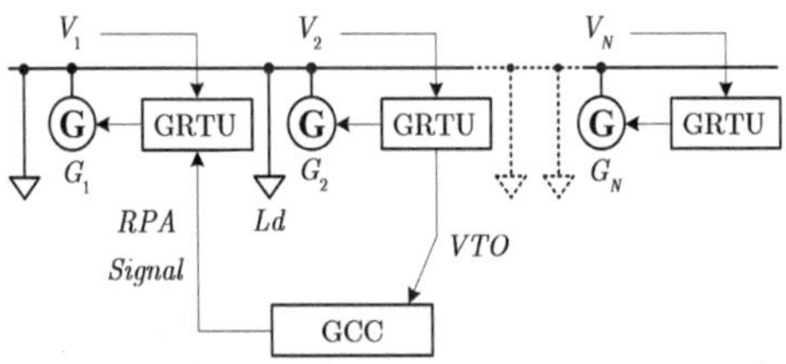

Figure 2: Voltage control using GRTU and GCC.

GCC in the RPA mode; therefore, the reactive power absorbed by G1 becomes $Q_1 = P_1 \tan\varphi_1$.

As explained in the following, the GCC must know the reactive power that each controllable generator can absorb in order to choose the BG. We suppose that this information is acquired by the GCC using a polling technique on each GRTU.

THE PROPOSED SENSITIVITY APPROACH

Classical Sensitivity Theory Overview

The classical sensitivity theory used in HV network analysis to perform primary and secondary voltage regulation [14] is based on the Jacobian Matrix and reveals the relationships between the nodal voltages (magnitude and phase) and the nodal power injections (active and reactive). The relationships mentioned above are represented by the following matrix expression [2]:

$$\begin{bmatrix}[\Delta E]\\ [\Delta \vartheta]\end{bmatrix} = \begin{bmatrix}\left[\frac{\partial P}{\partial V}\right] & \left[\frac{\partial P}{\partial \vartheta}\right]\\ \left[\frac{\partial Q}{\partial V}\right] & \left[\frac{\partial Q}{\partial \vartheta}\right]\end{bmatrix}^{-1} \begin{bmatrix}[1] & [0]\\ [0] & [1]\end{bmatrix} \begin{bmatrix}[\Delta P^*]\\ [\Delta Q^*]\end{bmatrix} \tag{1}$$

where $[\Delta E]$ and $[\Delta \vartheta]$ are, respectively, the nodal voltage magnitudes (rms) and phase variations corresponding to the nodal active or reactive power injections $[\Delta P^*]$ and $[\Delta Q^*]$ ($[1]$is the identity matrix).

Equation (1) can be rewritten in the following compact form:

$$\begin{bmatrix}[\Delta E]\\ [\Delta \vartheta]\end{bmatrix} = [s] \begin{bmatrix}[\Delta P^*]\\ [\Delta Q^*]\end{bmatrix} \tag{2}$$

where:

$$[s] \triangleq \begin{bmatrix}\left[\frac{\partial P}{\partial V}\right] & \left[\frac{\partial P}{\partial \vartheta}\right]\\ \left[\frac{\partial Q}{\partial V}\right] & \left[\frac{\partial Q}{\partial \vartheta}\right]\end{bmatrix}^{-1} \begin{bmatrix}[1] & [0]\\ [0] & [1]\end{bmatrix} \tag{3}$$

is the (injection) sensitivity matrix. The method descries above is generally valid, but its computational complexity is too high for practical voltage analysis in MVDNs. For radial networks, only the voltage magnitude is needed to control the nodal voltages. The proposed theory is easier than classical theory, and it is suitable for radial MVDNs.

The Proposed Theory

In this section, the proposed theory for choosing the BG is outlined. The method is first described in general and considers the possibility of reactive power regulation for all nodes.

After the general treatment, the analysis focuses on a realistic network in which the reactive power can only be controlled in some nodes (generator nodes).

Let us consider the network depicted in Figure 3, which is a four-node test MVDN.

The general loads $Ld_1 \ldots Ld_4$ are represented using constant PQ models. Positive P (or Q) corresponds to the absorbed power by the load. Negative P (or Q) corresponds to the injected power in the network (i.e., the general load is really a generator). The per-phase equivalent circuit is shown in Figure 4.

The lines L_{01}…L_{24} are modeled using the RL-direct sequence equivalent circuit [15], but the shunt admittances are neglected. The node 0 represents the MV busbar, which is regulated at a constant voltage value E_0 by the AVR of the OLTC. This reference voltage coincides with the rated value $E_0 = V_n/\sqrt{3}$.

Because the busbar is regulated at E_0, we can characterize the generic node i using the difference V_{0i} between the magnitude of the busbar voltage and the node voltage E_i. In other words, we can write:

$$V_{0i} = E_0 - E_i \quad (4)$$

In radial networks, (4) can be calculated as the sum of the voltage differences between adjacent nodes from the ith node toward the MV busbar. For example, if $i = 3$ (see Figure 4), (4) becomes:

$$V_{03} = E_0 - E_3 \quad (5)$$

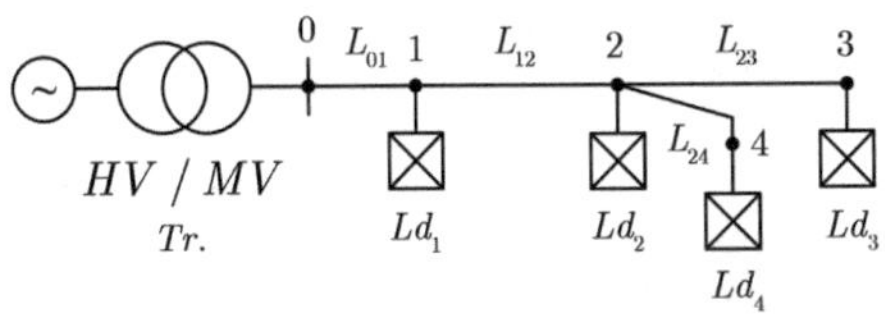

Figure 3: The considered four nodes test MVDN.

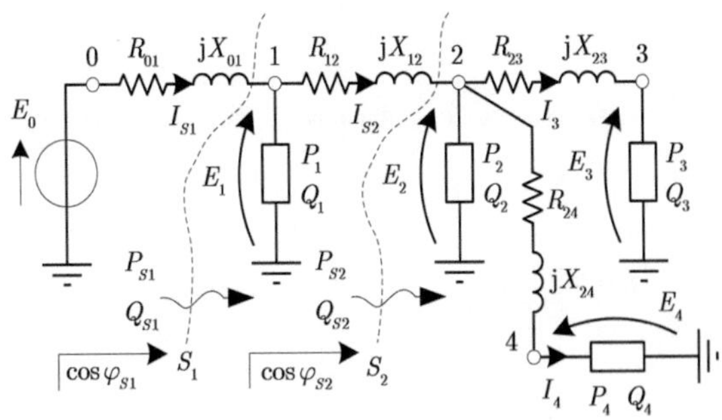

Figure 4: The per-phase equivalent circuit.

By adding and subtracting E_1 and E_2 in (5), we obtain:

$$\begin{aligned} V_{03} &= (E_0 - E_1) + (E_1 - E_2) + (E_2 - E_3) \\ &= V_{01} + V_{12} + V_{23} \end{aligned} \quad (6)$$

where V_{03} is the sum of the voltage differences V_{01}, V_{12} and V_{23}.

V_{23} can be calculated considering the network parameters and the line power flows as follows:

$$
\begin{aligned}
V_{23} &= E_2 - E_3 \\
&= R_{23} I_3 \cos\varphi_3 + X_{23} I_3 \sin\varphi_3 \\
&= \frac{R_{23} E_3 I_3 \cos\varphi_3 + X_{23} E_3 I_3 \sin\varphi_3}{E_3} \\
&= \frac{R_{23} P_3 + X_{23} Q_3}{E_3}
\end{aligned} \tag{7}
$$

where $\cos\varphi_3$, P_3 and Q_3 are the power factor and the active and reactive (per-phase) powers of the load Ld_3, respectively. I_3, R_{23} and X_{23} are the current, resistance and reactance of the line L_3.

Normally, the nodal voltages are close to the rated voltage E_n. Applying this assumption to (7) leads to:

$$
V_{23} \cong \frac{R_{23} P_3 + X_{23} Q_3}{E_n} \tag{8}
$$

Similarly, considering nodes 1 and 2, we can write:

$$
\begin{aligned}
V_{12} &= E_1 - E_2 \\
&= R_{12} I_2 \cos\varphi_{S2} + X_{12} I_2 \sin\varphi_{S2} \\
&= \frac{R_{12} E_2 I_2 \cos\varphi_{S2} + X_{12} E_2 I_2 \sin\varphi_{S2}}{E_2} \\
&\cong \frac{R_{12} P_{S2} + X_{12} Q_{S2}}{E_n}
\end{aligned} \tag{9}
$$

where P_{S2} and Q_{S2} are the active and reactive powers through the section S_2 and $\cos\varphi_{S2}$ is the power factor for the same section. For P_{S2} and Q_{S2}, we can write:

$$
P_{S2} = P_2 + P_3 + P_4 + P_{R23} + P_{R24} \tag{10}
$$

$$
Q_{S2} = Q_2 + Q_3 + Q_4 + Q_{X23} + Q_{X24} \tag{11}
$$

where P_{R23} and P_{R24} are the power losses in R_{23} and R_{24}, while Q_{X23} and Q_{X24} are the reactive powers absorbed by X_{23} and X_{24}. These active and reactive losses are negligible compared to the load powers. Applying this assumption to (9), (10) and (11) leads to:

$$
P_{S2} \cong P_2 + P_3 + P_4 \tag{12}
$$

$$
Q_{S2} \cong Q_2 + Q_3 + Q_4 \tag{13}
$$

and:

$$V_{12} \cong \frac{R_{12}(P_2+P_3+P_4)+X_{12}(Q_2+Q_3+Q_4)}{E_n} \quad (14)$$

Finally, the voltage difference V_{01} is:

$$V_{01} = E_0 - E_1 \cong \frac{R_{01}P_{S1}+X_{01}Q_{S1}}{E_n} \quad (15)$$

where:

$$P_{S1} \cong P_1+P_2+P_3+P_4 \quad (16)$$

$$Q_{S1} \cong Q_1+Q_2+Q_3+Q_4 \quad (17)$$

are the powers through section S_1.

Using (6) with (15), (9) and (8), we can say that V_{03} is a function of all loads and active and reactive powers, i.e., $P_1 \ldots P_4$ and $Q_1 \ldots Q_4$. The same observation is valid for E_3:

$$E_3 = E_0 - V_{03} = E_0 - (V_{01}+V_{12}+V_{23}) \quad (18)$$

because E_0 is constant. In other words, we can write:

$$E_3 = f(P_1,\ldots,P_4,Q_1,\ldots,Q_4) \quad (19)$$

Equation (19) shows that an active/reactive power variation (in the general j node) that is defined as:

$$\Delta P_j = P_j^f - P_j^0 \quad (20)$$

$$\Delta Q_j = Q_j^f - Q_j^0 \quad (21)$$

where P_j^f (Q_j^f) and P_j^0 (Q_j^0) are the final and initial power values, respectively, produces a voltage variation in node 3 that is defined as:

$$\Delta E_3 = E_3^f - E_3^0 \quad (22)$$

In this treatment, we only consider the reactive power variations (i.e., $\Delta P_j = 0$) because we assume that only the reactive power can be used to control the node voltages.

The variation ΔE_3 can be calculated by linearizing (19) and considering only the reactive power variations. In particular, we can write:

$$\Delta E_3 = \frac{\partial E_3}{\partial Q_1}\Delta Q_1 + \frac{\partial E_3}{\partial Q_2}\Delta Q_2 + \frac{\partial E_3}{\partial Q_3}\Delta Q_3 + \frac{\partial E_3}{\partial Q_4}\Delta Q_4 \quad (23)$$

The terms $\partial E_i / \partial Q_j$ in (23) indicate the "gain" from the voltage variation ΔE_i in node i when a reactive power variation ΔQ_j occurs in node j. In other words, they are sensitivity terms.

According to (18), we can obtain:

$$\frac{\partial E_3}{\partial Q_2} = -\frac{X_{01} + X_{12}}{E_n}$$
$$\frac{\partial E_3}{\partial Q_3} = -\frac{X_{01} + X_{12} + X_{23}}{E_n}$$
$$\frac{\partial E_3}{\partial Q_4} = -\frac{X_{01} + X_{12}}{E_n} \tag{24}$$

Substituting equation group (24) into (23) has important implications. If we have a reactive injection in any node, i.e., $\Delta Q_j < 0$ (in this case $j = 1 \ldots 4$), then $\Delta E_3 > 0$ in node 3 (i.e., the voltage increases). Then, if we were to reduce the voltage in any node, we must absorb reactive power from the network (i.e., $\Delta Q_j > 0$) by using, for example, the distributed generators.

If the above analysis that focuses on node 3 is extended to all network nodes, (23) has a general matrix relationship:

$$\begin{bmatrix} \Delta E_1 \\ \Delta E_2 \\ \Delta E_3 \\ \Delta E_4 \end{bmatrix} = \begin{bmatrix} \frac{\partial E_1}{\partial Q_1} & \frac{\partial E_1}{\partial Q_2} & \frac{\partial E_1}{\partial Q_3} & \frac{\partial E_1}{\partial Q_4} \\ \frac{\partial E_2}{\partial Q_1} & \frac{\partial E_2}{\partial Q_2} & \frac{\partial E_2}{\partial Q_3} & \frac{\partial E_2}{\partial Q_4} \\ \frac{\partial E_3}{\partial Q_1} & \frac{\partial E_3}{\partial Q_2} & \frac{\partial E_3}{\partial Q_3} & \frac{\partial E_3}{\partial Q_4} \\ \frac{\partial E_4}{\partial Q_1} & \frac{\partial E_4}{\partial Q_2} & \frac{\partial E_4}{\partial Q_3} & \frac{\partial E_4}{\partial Q_4} \end{bmatrix} \begin{bmatrix} \Delta Q_1 \\ \Delta Q_2 \\ \Delta Q_3 \\ \Delta Q_4 \end{bmatrix} \tag{25}$$

which in a compact form yields:

$$[\Delta E] = [s_Q][\Delta Q] \tag{26}$$

where $[s_Q]$ is the reactive sensitivity matrix, $[\Delta Q]$ is the reactive power-variations vector and $[\Delta E]$ is the nodal voltages vector.

Calculating the partial derivatives contained in $[s_Q]$, we have Equation (27).

$$[s_Q] = -\frac{1}{E_n} \begin{bmatrix} X_{01} & X_{01} & X_{01} & X_{01} \\ X_{01} & X_{01} + X_{12} & X_{01} + X_{12} & X_{01} + X_{12} \\ X_{01} & X_{01} + X_{12} & X_{01} + X_{12} + X_{23} & X_{01} + X_{12} \\ X_{01} & X_{01} + X_{12} & X_{01} + X_{12} & X_{01} + X_{12} + X_{24} \end{bmatrix} \tag{27}$$

After analyzing this form of (27), we can say that this matrix can be built using the following inspection rule:

"The element i, j is the arithmetic sum of the reactance of the branches in which both the powers absorbed by node i and node j flow multiplied by $-1/E_n$".

For example, in (27), the element 2, 4 is

$$-(X_{01} + X_{12})/E_n$$

because the powers delivered by node 2 and node 4 flow in branches 01 and 12.

The Choice of the Best Generator

The BG is the generator that has the greatest influence on node i, which is the node where the voltage exceeds the threshold.

Thus, after analyzing (25), we can say that the BG is the generator that maximizes the following product, which we call the "sensitivity product":

$$\frac{\partial E_i}{\partial Q_j}\Delta Q_j \qquad (28)$$

For example, if the node with a voltage that exceeds V_{max} is $i=2$ and the BG is connected to node $j=4$, the sensitivity product $(\partial E_2/\partial Q_4)\Delta Q_4$ is the highest compared to the other products contained in row 2 of the sensitivity matrix. In addition, in order to choose the BG, it is necessary to evaluate the single products (28) of the row that represents node i. Thus, the value ΔQ_j is needed and is acquired as the GCC polls the GRTUs, as stated previously.

The procedure described above suggests a way of defining the "sensitivity table" $[T_S]$ that contains the single sensitivity products. For the MVDN represented in Figure 4, $[T_S]$ takes the following form

$$\begin{bmatrix} \frac{\partial E_1}{\partial Q_1}\Delta Q_1 & \frac{\partial E_1}{\partial Q_2}\Delta Q_2 & \frac{\partial E_1}{\partial Q_3}\Delta Q_3 & \frac{\partial E_1}{\partial Q_4}\Delta Q_4 \\ \frac{\partial E_2}{\partial Q_1}\Delta Q_1 & \frac{\partial E_2}{\partial Q_2}\Delta Q_2 & \frac{\partial E_2}{\partial Q_3}\Delta Q_3 & \frac{\partial E_2}{\partial Q_4}\Delta Q_4 \\ \frac{\partial E_3}{\partial Q_1}\Delta Q_1 & \frac{\partial E_3}{\partial Q_2}\Delta Q_2 & \frac{\partial E_3}{\partial Q_3}\Delta Q_3 & \frac{\partial E_3}{\partial Q_4}\Delta Q_4 \\ \frac{\partial E_4}{\partial Q_1}\Delta Q_1 & \frac{\partial E_4}{\partial Q_2}\Delta Q_2 & \frac{\partial E_4}{\partial Q_3}\Delta Q_3 & \frac{\partial E_4}{\partial Q_4}\Delta Q_4 \end{bmatrix} \qquad (29)$$

Row i represents the node in which we want to control the voltage, and column j represents the nodes in which we can control the reactive power. The BG is the generator connected to node j that has the maximum absolute value of the sensitivity product in position i,j. By finding the maximum sensitivity product in row i, we automatically choose the BG because the location corresponds to column j of the maximum sensitivity product.

It is clear that, for a general network with N nodes, the sensitivity table takes the following form:

$$\begin{bmatrix} \frac{\partial E_1}{\partial Q_1}\Delta Q_1 & \frac{\partial E_1}{\partial Q_2}\Delta Q_2 & \cdots & \frac{\partial E_1}{\partial Q_N}\Delta Q_N \\ \frac{\partial E_2}{\partial Q_1}\Delta Q_1 & \cdots & \cdots & \cdots \\ \cdots & \cdots & \cdots & \cdots \\ \frac{\partial E_N}{\partial Q_1}\Delta Q_1 & \cdots & \cdots & \frac{\partial E_N}{\partial Q_N}\Delta Q_N \end{bmatrix} \quad (30)$$

It is important to note that, if it is not possible to regulate the reactive power (e.g., if in that node there is a load or a non-controllable generator) in a node j, then $\Delta Q_j = 0$ and, consequently, the sensitivity product in the position i,j of the sensitivity table is 0.

Comparing (29) with (25), we can say that each element i,j of $[T_S]$ represents the line-to-ground voltage variation in node i when a reactive power variation occurs in node j.

In the following section, a numerical example of the sensitivity method application is shown.

APPLICATION OF THE PROPOSED METHOD

The network considered in this numerical application is represented in Figure 5.

During normal network operation, we have four generators and eight loads. The generator and load characteristics are summarized in Table 1 (S is the apparent power) and Table 2, respectively (three-phase powers are represented in these tables).

We suppose that the generators normally operate with a unitary power factor (i.e., no reactive power is injected in the nodes).

The per-kilometer reactance of the cable lines is $x = 0.17\,\Omega/km$, which is a typical value for Italian MVDNs. The line lengths and parameters are summarized in Table 3.

Let us suppose that each generator is connected to its GRTU that measures the nodal voltage and communicates with the GCC. Moreover, let us suppose that G_5 cannot regulate the reactive power because it is not designed for this purpose. The MV busbar is regulated at the rated voltage (1 p.u.), which is 20 kV (line-to-line). In this example, the voltage threshold V_{max} is 1.05 p.u.

Using load-flow software, we calculated the voltage E in the generator nodes (nodes 4, 5, 6, and 7) for normal network operation. The results are shown in Figure 6 (Normal Operation).

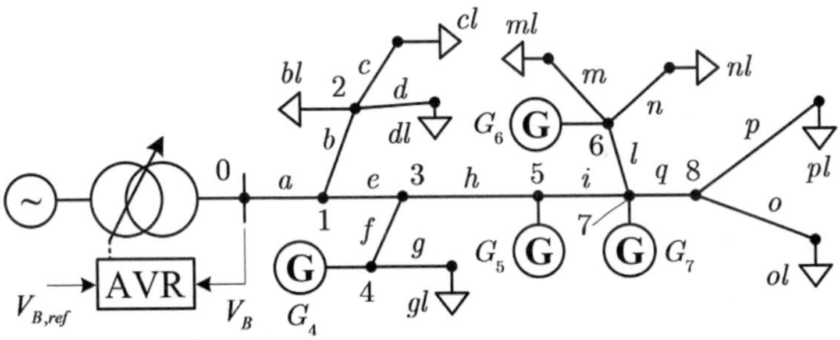

Figure 5: The network considered in the numerical application.

Table 1: Loads characteristics

Load	*S [MVA]*	*cos*	*P [MW]*	*Q [MVAR]*
bl, nl	2	0.95	1.9	0.62
cl	7	0.92	6.44	2.74
dl	3	0.92	2.76	1.18
gl	2.08	0.95	1.99	0.62
ml	2.58	0.93	2.4	0.94
ol	1.98	0.96	1.9	0.57
pl	1.5	0.92	1.38	0.59

Table 2: Generators characteristics

Generator	*P [MW]*
G_4	6
G_5	1.75
G_6	4.5
G_7	3.75

Table 3: Lines parameters

Line Name	*L [km]*	*X [Ω]*
a, l, q	2	0.34
b, f, g, i, m, n	1	0.17
c, d, o, p	0.5	0.085
e	15	2.55
h	5	0.85

If line b trips (e.g., due to a fault), loads bl, cl and dl are cut off from the supply, which causes the voltage to increase in the network. In particular, if the load-flow is re-computed to take into account the new network configuration, we obtain the results shown in Figure 6 (Tripped Line).

It is important to note that, if the voltage exceeds the maximum threshold V_{max} in node 5, the GRTU connected to G_5 sends the VTO signal to the GCC that must choose the BG using the sensitivity table.

We suppose that the three-phase reactive powers absorbable by each generator that were collected from the last poll are those summarized in Table 4, which also contains the corresponding power factors cosφ. To calculate the sensitivity table, we need the single-phase powers. Therefore, the reactive powers shown in Table 4 have to be divided by three. It is important to note that the reactive powers calculated this way correspond to ΔQ_j because Q_j^0 is zero (see (21)). The ΔQ_j values are shown in Table 5.

The voltage exceeds the threshold in node 5. Thus, we only consider the fifth row of the sensitivity table. According to the inspection rule mentioned above, this row is as follows:

$$\begin{bmatrix} 0 & 0 & 0 & t_{S5,4} & t_{S5,5} & t_{S5,6} & t_{S5,7} & 0 \end{bmatrix} \quad (31)$$

where the single sensitivity products $t_{Si,j}$ are:

$$t_{S5,4} = -\frac{1}{E_n}(X_a + X_e)\Delta Q_4 = -164.53\ V \quad (32)$$

$$t_{S5,5} = -\frac{1}{E_n}(X_a + X_e + X_h)\Delta Q_5 = 0\ V \quad (33)$$

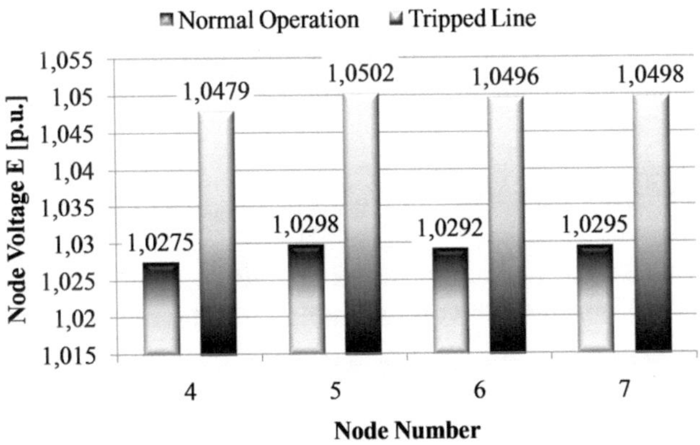

Figure 6: Load-Flow results with the Network Normal Operation.

Table 4: Reactive powers absorbable by the generators

Generator	*Q* [*MVAR*]	*cosφ*
G_4	1.97	0.95
G_5	0.00	1
G_6	0.91	0.98
G_7	1.23	0.95

Table 5: Reactive Power Variations in the Generator Nodes

Generator	ΔQ_j [*MVAR*]
G_4	0.66
G_5	0.00
G_6	0.30
G_7	0.41

$$t_{S5,6} = -\frac{1}{E_n}(X_a + X_e + X_h)\Delta Q_6$$
$$= -98.65\ V \quad (34)$$

$$t_{S5,7} = -\frac{1}{E_n}(X_a + X_e + X_h)\Delta Q_7$$
$$= -133.07\ V \quad (35)$$

The maximum sensitivity product (in absolute value) corresponds to generator 4, (i.e., $j = 4$). Thus, the BG is G_4.

Equation (32) provides important information. If G_4 performs the considered reactive power variation, the line-to-ground voltage variation in node 5 is:

$$\Delta E_5 = -164.53\ V \rightarrow -0.0142\ p.u. \quad (36)$$

Then, considering (22) (rewritten for node 5), and $E_5^{\ 0} = 1.0502\ p.u.$ from Figure 6, we can say that the voltage value after the reactive power variation is:

$$E_5^{\ f} = E_5^{\ 0} + \Delta E_5 = 1.036\ p.u. \quad (37)$$

which is less than the voltage threshold V_{max}.

Equation (37) shows the theoretical result obtained using the proposed method. We checked this value using load-flow software:

$$E_5^{\ f}{}_{load\ flow} = 1.034\ p.u. \quad (38)$$

The percentage error between (38) and (37) is:

$$\varepsilon_{\%} = \frac{E_5^{\ f}{}_{load\ flow} - E_5^{\ f}}{E_5^{\ f}{}_{load\ flow}} \cdot 100 = -0.19\ \% \quad (39)$$

which is negligible and demonstrates the validity of the proposed approach.

CONCLUSIONS

The proposed sensitivity method allows the voltage within network acting on single generators to be regulated by choosing the most effective generator on the controlled node (i.e., the Best Generator). This is a very important feature in grids that have distributed generation (e.g., in a Smart Grid context).

The proposed method uses a topological approach. Moreover, the sensitivity table can be constructed automatically.

In addition to the BG choice, the proposed method also evaluates the voltage in all network nodes after a reactive power variation.

After choosing the BG, but before its commutation during RPA, it is possible to verify that the voltage variation in the other nodes is tolerable for the connected loads. Moreover, it is necessary to verify that the threshold settings of the voltage relay installed in the same nodes.

When a generator is switched during RPA, it works with a non-unitary power factor; the reactive power flow increases along the lines and increases the power loss [16].

This phenomenon is negligible in HV networks because the line resistance is typically smaller than the line reactance, but is important to consider in MV networks.

Therefore, if network analysis reveals that the RPAswitching produces high losses, voltage control using the reactive power variation must only be used for temporary voltage variation mitigation (i.e., during emergency conditions).

Morris Brenna, Ettore De Berardinis, Federica Foiadelli, Gianluca Sapienza, Dario Zaninelli The possible future develops of this work could be focused on the optimization of the forecasted power-time profiles of the loads and generators applying both the se nsitivity approach and distributed voltage measurement.

REFERENCES

1. R. Marconato, "Electric Power Systems," Vol. 2, CEI, Milano, 2008.
2. P. Kundur, "Power System Stability and Control," McGrawHill, New York, 1994.
3. Y. Rosales Hernandez and T. Hiyama, "Distance Measure Based Rules for Voltage Regulation with Loss Reduction", Journal of Electromagnetic Analysis and Applications (JEMAA), Vol. 1, No. 2, June 2009, pp. 85-91.
4. F. Saccomanno, "Electric Power Systems" Wiley-Interscience IEEE Press, Piscataway, 2003.
5. G. Andersson, "Modeling and Analysis of Electric Power Systems" Lecture 227-0526-00, ITET ETH Zürich, Zürich, 2008.
6. Y. Abdel-Rady, I. Mohamed and E. F. El-Saadany "A Control Scheme for PWM Voltage-Source DistributedGeneration Inverters for Fast Load-Voltage Regulation and Effective Mitigation of Unbalanced Voltage Disturbances" IEEE Transactions on Industrial Electronics, Vol. 55, No. 5, May 2008, pp. 2072-2084.
7. P. M. S. Carvalho, P. F. Correia and L. A. F. M. Ferreira, "Distributed Reactive Power Generation Control for Voltage Rise Mitigation in Distribution Networks," IEEE Transactions on Power Systems, Vol. 23, No. 2, 2008, pp. 766-772.
8. P. N. Vovos, A. E. Kiprakis, A. R. Wallace and G. P. Harrison, "Centralized and Distributed Voltage Control: Impact on Distributed Generation Penetration," IEEE Transactions on Power Systems, Vol. 22, No. 1, 2007, pp. 476-483.
9. P. M. Anderson, "Power System Protection," IEEE Press, Piscataway, 1999.
10. N. Mohan, T. M. Undeland and W. P. Robbins, "Power Electronics: Converters, Applications, and Design", Wiley, 1995.
11. M. H. Rashid, "Power Electronics Handbook", Academic Press-Elsevier, 2007.
12. L. L. Grigsby, "Electric Power Generation, Transmission and Distribution," CRC Press-Taylor & Francis Group, Boca Raton, 2006.
13. F. A. Viawan and D. Karlsson, "Coordinated Voltage and Reactive Power Control in the Presence of Distributed Generation," PES General Meeting - Conversion and Delivery of Electrical Energy in the 21st Century, IEEE, Pittsburgh, 2008, pp. 1-6.

14. S. Corsi, "Wide Area Voltage Regulation & Protection" 2009 IEEE Bucharest Power Tech Conference, Bucharest, June 28 -July 2, Bucharest, pp. 1-7.
15. A. Gandelli, S. Leva and A. P. Morando, "Topological Considerations on the Symmetrical Components Transformation", IEEE Transactions on Circuits and Systems I: Fundamental Theory and Applications, Vol. 47, No. 8, August 2000, pp. 1202-1211.
16. H. M. Ayres, L. C. P. da Silva, W. Freitas, M. C. de Almeida and V. F. da Costa, "Evaluation of the Impact of Distributed Generation on Power Losses by Using a Sensitivity-Based Method," IEEE Power & Energy Society General Meeting, Calgary, 2009, pp. 1-6.
17. A. Kishore and E. F. Hill, "Static optimization of Reactive Power Sources by use of Sensitivity Parameters", IEEE Transactions on Power Apparatus and Systems, Vol. PAS-90, No. 3, 1971, pp. 1166-1173.

Chapter 4

Agents for Smart Power Grids

Salem Al-Agtash[1,2], Hossam Abdel Hafez[1]

[1]Department of Computer Engineering, German Jordanian University, Amman, Jordan.

[2]Department of Computer Science and Engineering, Santa Clara University, Santa Clara, CA, USA

ABSTRACT

The future of electricity systems will compose of small-scale generation and distribution where end-users will be active participants with localized energy management systems that are able to interact on a free energy market. Software agents will most likely control power assets and interact together to decide the best and safest configuration of the power grid system. This paper presents a design of agents that can be deployed in real-time with capabilities that include optimization of resources, intensive computation, and appropriate decision-making. Jordan 51-bus system has been used for simulation with a total generation capacity of 4050 MW of which 230 MW

Citation: Al-Agtash, S. and Hafez, H. (2020), "Agents for Smart Power Grids". *Energy and Power Engineering*, **12**, 477-489. doi: 10.4236/epe.2020.128029.

represents renewable energy. The economic analyses demonstrated the use of smart grid technologies with 2016 generation—load profiles for nominal liquified gas (NLG) prices and ±20% sensitivity analysis. The results have shown variations in the range of 1% in the price of MWh with smart grid technologies. These variations are mainly driven by the fact that agents shift power generation to renewable power plants to produce maximum power at peak hours. As a result, there is a positive economic impact in both NLG ± 20% sensitivity analysis, due to the fact that agents coordinate to better displace expensive thermal generation with renewable generation. It is evident that renewable resources compensate for power at peak times and provide economic benefits and savings.

Keywords: Agents, Electric Power System, Smart Power Grid

INTRODUCTION

The development of smart grids has been progressing rapidly in recent years. Traditional electric power grids began to adopt smart grid technologies [1] [2]. Agents have demonstrated their applicability as innovative tools to manage and operate pervasive elements of smart grids. They act as smart self-organizing hardware and software structures in an integrated power system environment that is comprised of generators, distribution substations, transformers, and transmission lines [3]. Intuitively, a smart grid is considered to evolve with grid reconstructive technologies and tools to dynamically optimize grid operations and resources and to incorporate end-user demand response participations [4]. Multiagent systems (MAS) may be designed to inherit SCADA (supervisory control and data acquisition) architecture while accommodating autonomous and intelligence attributes. MAS technologies have been used in diverse power system applications like disturbance diagnosis, restoration, secondary voltage control, and visualization [5] [6] [7]. Their wider implementation has been demonstrated in solar power generation batteries, controllable loads, converters and inverters, demand-side management, data acquisition and grid planning [5] [8]. Agents are implemented as software units which accept signals and interactively prioritize loads, specify time and control status of loads [9].

The specific agent-based solutions for smart grids that are composed of various passive and active system modules have been presented [10]. Decentralized grid arrangements are demonstrated with generation patterns and demand configurations modeled as self-healing agents characterizing transmission, transformers, generation plants, and end-user metering devices. Smart capabilities for visualization, smart sensing automation, and self-healing digitized elements represent the main attributes [11] [12]. The complications of reliability and reassurance, which arise in smart grid, measure agent actions, security and dependability in a multifaceted and vibrant environment. It has been concluded that MAS inherits profits of tractability and extensibility and employs autonomous resolutions like smooth conversion of a grid-networked system to island mode, load shedding, fortifying loads termed as critical, and much more [6]. The aggregate composition of MAS in grid system implementation that considers renewable resources poses a new challenge.

In this paper, we investigate MAS as an appropriate technology for smart grids, while investigating distributed elements necessitating self-sufficiency in their operation and interaction. A smart grid model is presented as a mesh network of micro-grids, which are composed of diverse distributed energy resources with self-possessed smart ICT sub-systems [13]. The Java Agent Development (JADE) Framework is used for agent implementation representing generation, transmission, and distribution elements [14]. The Jordan's 52-bus electric power system is used for illustration. The potential solar and wind energy sources contribute to significant encounters to reduce imports of fuel and preserve environmental and climatic assets. These demonstrations provide in-depth understanding of MAS strategy and development perspectives in the smart grid context with distributed energy resources. Control processes for handling partial distribution from distributed energy resources are used to protect critical loads throughout crises while still assisting prioritized non-critical loads belonging to numerous consumers.

The remaining sections of this paper are organized as follows: Section II presents emerging technologies of smart micro-grid systems and integration in agent automation. The design and deployment architecture of agents are given in Section III. An illustrative example and implementation are presented in Section IV. Finally, the paper is concluding in Section V.

SMART GRID

Power grids are empowered by smart grid technologies and distributed generation, including photovoltaic and wind energy resources [7] [15]. It has been noted that smart grids automate generation, transmission, and distribution to guarantee stable stream of power with minimum blackouts

[9]. Smart grids provide efficient tools for power grid control in real-time while maintaining secure and cost-effective operation [16]. Figure 1 shows a model agent-based smart grid. Consumers actively participate in the generation market. They generate power from renewable resources and interconnect with the power grid either for feeding or consuming power. Ultimately, transmission, distribution, and end users are integrated so that power flow is controlled and that environmental concerns are preserved.

Intelligent Agents

Agents represent elements of power system and use standard plug-in interfaces to interact with the energy management system. The system consists of scheduling algorithms, power flow optimization, day-ahead load prediction and generation planning, monitoring and decision making, and supervisory control. Local and remote databases are maintained for power generation, transmission, and distribution. The design of intelligent distribution and operation of agents is tractable and would enable a high-level degree of automation in the power industry. Consumer agent collects information on carbon emissions and energy use from smart meters. It supports the use of energy storage devices as essential ways to decrease peak demand and facilitate integration of intermittent renewable resources. The monitoring and control system manage and operate smart grid system [17].

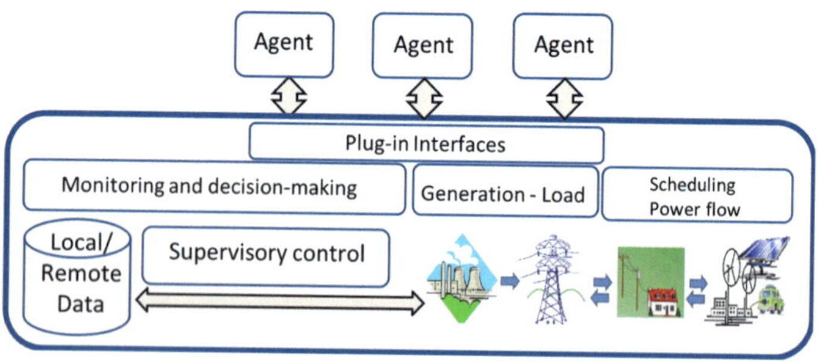

Figure 1: Smart grid proposition.

It integrates various tools that are used to identify instabilities and to integrate renewable energy resources.

Smart Technologies

Intelligent agents may ember smart technologies for automatic generation control, static Var compensators, high voltage direct current control, energy management systems, demand side response, and energy storage. Integration of information and communication technologies (ICT) is essential for real time operation, monitoring, and control of the electric power system and its integration with variety end-user devices and appliances. Users receive real-time price information on cost and carbon intensity. This information is used to forecast forthcoming power demand and supply. Smart meters represent the core components for automating energy use within homes and offices. They are used to manage two-way power flow, where end users reduce energy waste and gain good control over how they consume power [7]. Smart meters interact with residential demand management, self-healing and outage management, and electric vehicle charging. They communicate with load control devices and data repositories using secure communication on the Internet. They integrate into existing utility operations and asset management processes and may reduce grid vulnerabilities and enhance energy efficiency.

Power Flow

The power flow on transmission lines is managed by control centers and sub-stations. Intelligent agents are potential technologies to empower features similar to analytical proficiencies for analysis and monitoring at control centers. A variety of devices are used. The devices of the flexible alternating current transmission system (FACTS), for instance, are generally used to enhance controllability and increase power transfer capability. The high-voltage direct current (HVDC) is used to supply power over extensive distances at minimum losses. It also enables connection between asynchronous grids [17]. The supervisory control and data acquisition (SCADA) system provides analysis of situations and information in real time. Dynamic line ratings deliver a way of classifying existing transporting ability of a unit of network in real-time scenario. Demand response includes various actions that are taken in case of risks due to load shortage or excess. Such contingencies can make demand and supply unbalanced. One of the primary actions is to minimize the usage of electricity when prices are high by shifting load shifting. Agents operate smart power transmission and coordinate with control centers to enhance power utilization, security, and quality and reliability.

AGENT IMPLEMENTATION

Agents are implemented with functionalities that include auto recovery, immune to cyber-attacks, cooperation and coordination with generation mix, and monitoring and control in real-time.

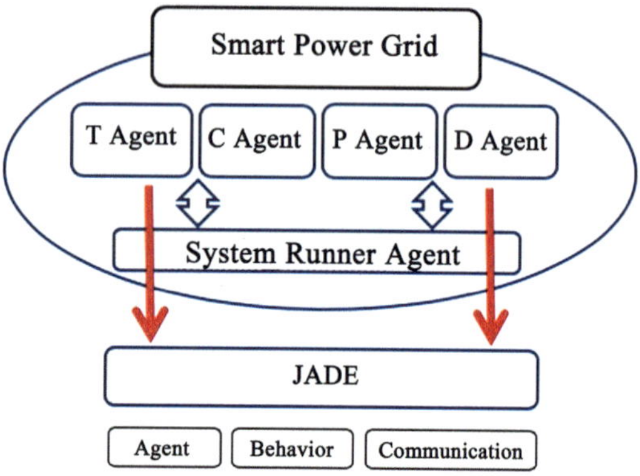

Figure 2: JADE agents component architecture.

Complex interaction patterns like cooperation, coordination, and negotiation are embedded [13]. We use JADE platform as a generic robust agent architecture to design and implement generation, transmission, and distribution agents of the smart grid [14] [18] [19]. It allows distributed intelligence and simplifies the construction of interoperable agents. The main features of JADE include standardization, portability, dynamism, and versatility. Ready-made functionalities and interfaces are provided for custom and application-dependent tasks. JADE is fully compatible with FIPA standards and enables agent platforms to be distributed across different machines, and supports portability [13].

Agent Design

0Each agent embeds operational intelligence, decision making, and interactive communication. Its state attributes are common to all agents, including identity, owner, agent type, license, and authorization. An agent reacts to a spectrum of events using embedded behavioral functions and coordinates with its peer agents to meet operational requirements, and to maintain security and reliability. Figure 2 presents JADE agent design with components that include transmission agent (T Agent), customer agent (C Agent), generation agent (P Agent), and distribution agent (D Agent). All agents start and create their operational logic defined in SystemRunnerAgent. The JADE library contains all subcomponents necessary for the simulation process, including agent, behavior and communication. The packages "Agent" and "behavior" are used for the purpose of enabling interaction between various JADE agents. The data package carries classes that help in serializing the payload messages from agents. Properties and entities are stored in the Classes from the "model" package. The Java swing libraries and classes contained in the "monitoring" package are used for the implementation of application's front-end. Snippet of an Agent code is given as follows:

```
public class NewAgent extends MonitorableAgent {
private AgentBehavior agentBehaviors;
private DemandRequest demandRequest;
public NewAgent () { }
protected void setup() { }
public void resetDemand() {
if ((!isPaused()) && (this.demandRequest != null)) {
addDemand(this.demandRequest);
}
}
}
```

The primitive or composite behaviors are extended by typical Behavior code, defined as follows:

```
public class NewBehavior extends CyclicBehavior {
private final Logger log = LoggerFactory.getLogger(getClass());
private final MessageTemplate messageTemplate;
private final MonitorableAgent agent;
public NewBehavior(MonitorableAgent agent) { }
public void action() { }
}
protected void setup () {
super.setup();
addBehavior(new ReceivePowerRequestsBehavior(this));
addBehavior (new HandleGeneratorListBehavior ());
addBehavior(new ReceiveGenerationResponsesBehavior());
```

```
sniffMe ();
}
```

Agent Behavior

Agents perform their tasks independently and operate autonomously [19]. Agent behavior as defined by JADE implements a series of concurrent threads. Whenever a new message arrives, the agent invokes the relevant behavior object [13], eventually derived from one of the following behavior classes:

- CyclicBehavior: handles messages.
- OneShotBehavior: performs only one task.
- WakerBehavior: performs a task after a predefined time.
- TickerBehavior: performs continuously and repetitively.
- ReceiverBehavior: responses to an action.
- Composite Behaviors: combines composite tasks.
- ParallelBehavior: performs tasks simultaneously.
- SequentialBehavior: performs tasks sequentially.

The specific behavior of an agent is mainly driven by the operational characteristics of the underlying device or entity. In studying the Jordan electric power grid, agents are designed to represent generating units, distribution substations, end user customers, and transmission system.

A generation agent i represents the ith generating unit that injects electric power g_i through its kth connected generation bus b_k to the transmission network. It operates under a number of constraints, formally given by $h(g_i)$. The power generation at bus k, during hour t, is $g_k(t)$. The cost of generation is given by $C_k[g_k(t)]$ including start-up and other costs. Power distribution and consumption agents operate at load substation busses. D and C agents predict power demand at bus k during hour t as $d_k(t)$. The power demand is assumed inelastic, i.e. fixed for any price signal. Renewable generation is becoming widely used through small-scale technologies that produce low-cost, reliable, and clean electricity close to customers. The system total demand, during hour t, is computed as the sum of $d_k(t)$ and is given by $D(t)$. Agents are assumed to implement dynamic scheduling of dispatchable generation, demand-side management techniques, consolidation of load balances for separate power zones represented at bus k, and flexible energy storage. The transmission agent coordinates the physical operation and controls the entire system in real-time in coordination with SCADA control center. It maintains [1]:

- generation-load balance at time t, $\sum k,tgk(t)=D(t)\sum k,tgk(t)=D(t)$;
- enough spinning reserve r at bus k, time t, $\sum i,tgi(t)-gk(t)\leq rk(t)\sum i,tgi(t)-gk(t)\leq rk(t)$;
- min-max available bus generation, at time t $gk,min\leq gk(t)\leq gk,maxgk,min\leq gk(t)\leq gk,max$;
- flow limits on transmission lines: $|zl(t)|=zl,max|zl(t)|=zl,max$.

Agent Classes

The generation, distribution, customer, and transmission agents are setup with adaptable properties [18]. As shown in Figure 3, each agent extends from the abstract class MonitorableAgent and integrates with variety of other classes including MonitoringConstants and SystemConstants interfaces. The systemConstants Interface include: sniffer_agent, system_runner_agent, distribution_agent, transmission_agent, generation_agent, and customer_ agent. T Agent represents behaviors that interact with generation and load balance and maintain real-time power flow.

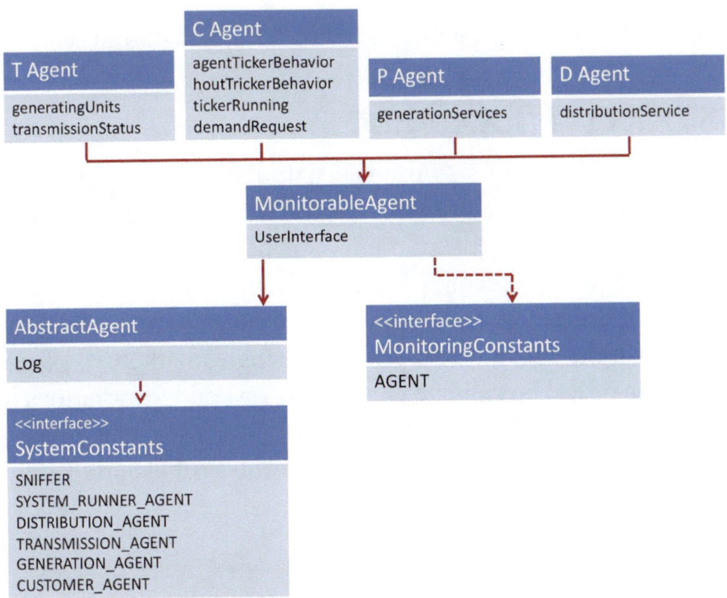

Figure 3: Agent class diagram.

C Agent has agentTickerBehaviour and hourTickerBehaviour fields that store custom agent behavioral characteristics. Fields tickerRunning and demandRequest are used to track current status and important power demand updates. P Agent has a generation services field that includes information for each generation unit on power production and spinning reserves for different time intervals. D Agent has a distribution services field that includes information on power demand and load profiles for different time intervals and for each substation/load bus. The systemRunnerAgent represents an important class for all agents. It facilitates a setup and registration to defining agent description and startup phase.

Agent Communication

Agents interact and communicate asynchronously with their peers. Each agent has a mailbox which notifies the agent when a new message arrives. Figure 4 shows the JADE messaging system. Inter-agent communication is implemented as an object derived from the agent transparency class. It selects the best route and finds a compromise on message semantics and format to make their communications easier. All messages adhere to ACL standards, namely:

Message source and communicative intention ("performativity").

Message content, language, conversation ID, and ontology.

ILLUSTRATIVE EXAMPLE

The Jordan 51-bus system was used for simulation. 230 MW of the 4050 MW total generation capacity represented renewable energy [18] [19] [20] [21]. The growing power demand, high costs of imported fuel, lack of water cooling, and frequent cascading blackouts represent inevitable challenges to the Jordanian electric power system. The economic analyses were demonstrated on the use of smart grid technologies with 2016 generation—load profiles. In order the validate how agents respond to variations in power costs, ±20% of liquefied natural gas (LNG) prices for sensitivity analysis were used. The analyses provide insights on the robustness of agents for smart power grids.

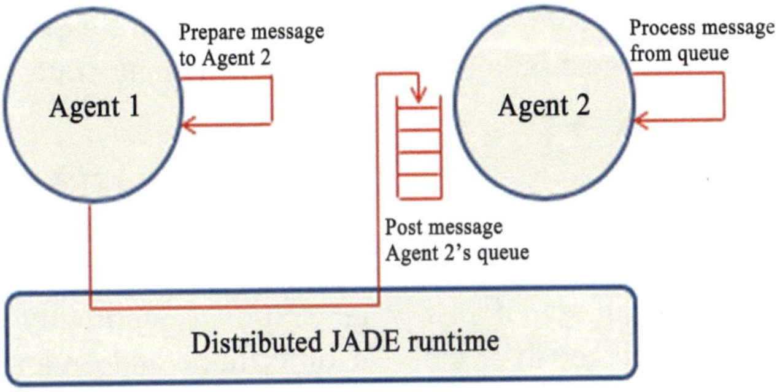

Figure 4: Message transmission on JADE platform.

Operating Costs

Operation planning requires day-to-day load profiles for unit commitment and power dispatch of available resources provided that security, stability, and reliability constraints are met. Generation changes on hourly basis and schedules of unit commitment and power dispatch are computed based on estimated load profiles and operational constraints. The presence of a significant amount of non-dispatchable and non-predictable renewable generation implies an update of the system operational requirements, mainly in regard to operating reserves. To assess the impact of MAS technology on the production costs of electricity, a day-ahead simulation has been used.

Renewable energy resources produce electricity to compensate for energy that could otherwise be produced by conventional power plants. Comparing energy production with and without renewable resources, it is possible to evaluate in quantitative terms the differences in electricity production costs, marginal costs, and displaced energy resources for different technologies. The increased penetration of renewable energies in the Jordanian generation mix contributes to economic costs and savings. These costs and savings are calculated on the basis of different scenarios of MAS simulations.

Operational costs of generation units are assumed quadratic and a centralized unit commitment and power dispatch mechanisms are implemented by the T Agent. The simulation results give the optimum schedule of resources with minimum cost of power generation. An equivalent network model is used with interconnected zones transferring power at maximum capacity. Interconnected neighboring electric systems are modelled with equivalent generating units. Negative production means that energy has been exported while positive production means that energy has been imported with neighboring grids. Thermal generating units are characterized by their company owner, zone, min-max capacity, fuel mix and prices, quadratic cost function, average forced outage rate, and start-up/shut-down constraints.

Simulation Results

MAS is implemented with agents representing power generation, distribution, and transmission. The set of power generation is optimally coordinated with renewable resources in a one-year time frame and on an hourly base discretization. This gives a total of 8760 hours that are considered for analysis. A day-ahead hourly energy system is simulated while considering system-wide marginal pricing and congestion management. The base case is investigated with different scenarios for year 2016. The average hourly prices of power are shown in Figure 5 for both agent-based and no-agent power systems for a nominal liquified natural gas (LNG) prices for a summer day. Hourly prices are higher for real operation during peak hours. Sensitivity analysis is performed for LNG with ±20% of fuel prices. Figure 6 and Figure 7 give comparison of average hourly prices of power for both agent-based and no-agent power systems for LNG ± 20%. The one-day analysis shows that energy prices are higher during peak hours with no MAS technologies integrated to accommodate real-time control of renewable energies (shown as dotted lines in the graph).

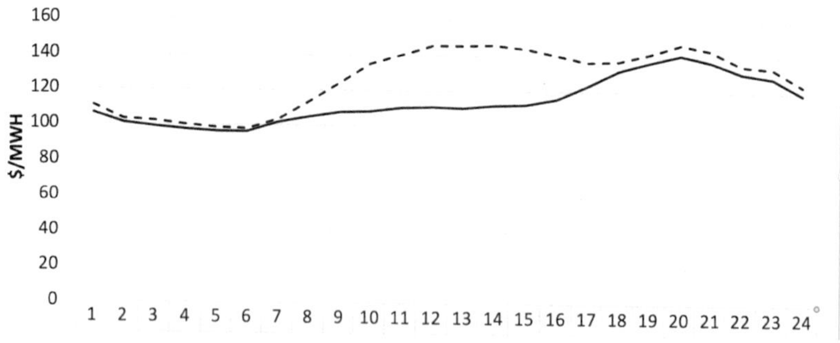

Figure 5: Average hourly prices for agent-based and no-agent LNG power system.

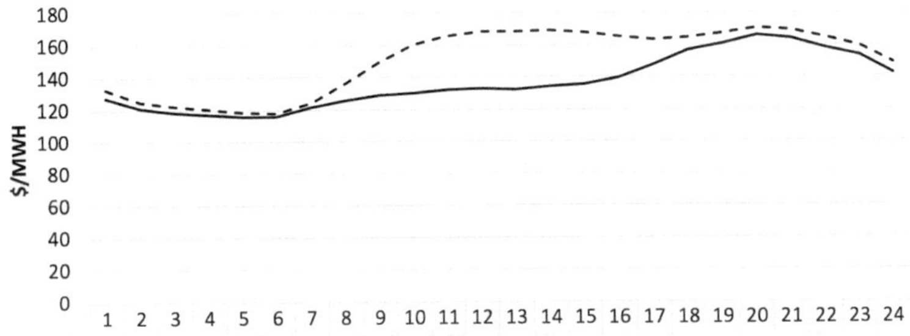

Figure 6. Average hourly prices for agent-based and no-agent LNG +20% power system.

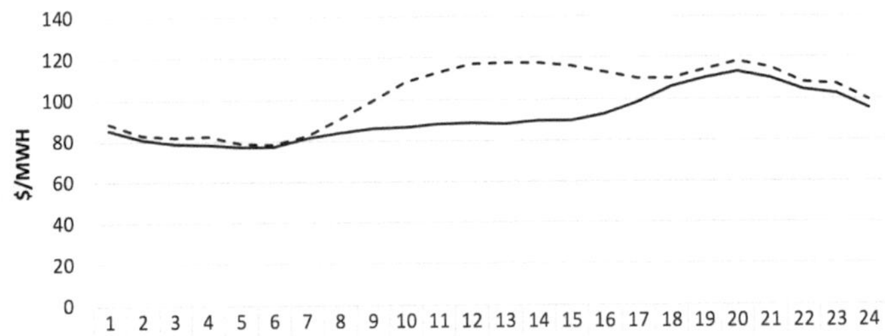

Figure 7: Average hourly prices for agent-based and no-agent LNG −20% power system.

The simulation has also been run for a full year 2016. Table 1 shows a summary of fuel costs and average energy prices for LNG and LNG ± 20% operation scenarios. The total LNG fuel cost is $2056.4 million with agent-based simulation compared to $2333.1 million with real operation. This gives an average Megawatt hour price of $120.8 with agent-based simulation compared to $122.5 with real operation.

The total LNG +20% fuel cost is $2456.6 million with agent-based simulation compared to $2748.3 million with real operation.

Table 1: Summary of fuel costs and average prices for different types of operation scenarios

Scenarios	Type of Operation	Fuel Cost (M$)	Average Price ($/MWh)
LNG Base case	MAS based operation	2056.4	120.8
	Real operation	2333.1	122.5
LNG +20%	MAS based operation	2456.6	143.6
	Real operation	2748.3	144.6
LNG −20%	MAS based operation	1645.1	96.7
	Real operation	1866.5	101.2

The average Megawatt hour price increased from $120.8 to $143.6 with agent-based simulation compared to an increase from $122.5 to $144.6 with real operation. On the other hand, the total LNG −20% fuel cost is $1645.1 million with agent-based simulation compared to $1866.5 million with real operation. The resulting average Megawatt hour price decreased from $120.8 to $96.7 with agent-based simulation compared to a decrease from $122.5 to $101.2 with real operation. These variations are mainly driven by the fact that agents shift power generation to renewable power plants to produce maximum power at peak hours. As a result, there is a positive economic impact in both cases, due to the fact that agents coordinate to better displace expensive thermal generation with renewable generation. It is evident that renewable resources compensate for power at peak times and provide economic benefits and savings.

In economic terms, an average saving of 0.0079% in energy costs is made when using agent technologies. Assuming an average daily generation capacity of 3500 MW, then the amount of yearly savings will be $35 million. It is clear that MAS technologies reduce fuel costs and energy prices as they drive renewable resources to compensate for power at peak time and consequently, provide economic benefits and savings. The results of this research are consistent with the results of MAS research in power grid systems [2] [3] [19] in support the use of agents as viable tools for power system management and energy efficiency.

CONCLUSION

A conceptual agent design has been presented for a smart grid system using agent technologies. The agents act as smart self-organizing elements, which inherent a distributed nature of adaptive interacting systems. It has been shown that agents are important technologies for setting up smart grids. The economic impact of agent integration with renewable resources is noticeable. The economic analyses are performed for the year 2016, both for MAS integration and real operation with respect to price of fuel gas and operating reserves. A day-ahead simulation has been used in order to simulate hourly generation costs. The results of the simulation show an average saving of 0.0079% in energy costs when using agent technologies. Assuming an average daily generation capacity of 3500 MW, then the amount of yearly savings will be $35 million. It should be noted that MAS technologies may be explored for different grid systems while taking into account the intermittent nature of renewable resources and unpredictable weather conditions. It is anticipated that MAS technologies can be useful for further research and would support the development of smart grid systems in a larger context.

REFERENCES

1. Al-Agtash, S. (2012) Electricity Agents in Smart Grid Markets. Computers in Industry, 64, 235-241. https://doi.org/10.1016/j.compind.2012.10.009
2. Khazaei, J. and Nguyen, D. (2019) Multi-Agent Consensus Design for Heterogeneous Energy Storage Devices with Droop Control in Smart Grids. IEEE Transactions on Smart Grid, 10, 1395-1404. https://doi.org/10.1109/TSG.2017.2765241
3. Lau, J., Huang, G., Mak, K. and Liang, L. (2006) Agent-Based Modeling of Supply Chains for Distributed Scheduling. IEEE Transactions on Systems, Man, and Cybernetics, 36, 847-861. https://doi.org/10.1109/TSMCA.2005.854231
4. Bose, A. (2010) Smart Transmission Grid Applications and Their Supporting Infrastructure. IEEE Transactions on Smart Grid, 1, 11-19. https://doi.org/10.1109/TSG.2010.2044899
5. Bintoudi, A., Zyglakis, L., Tsolakis, A., Ioannidis, D., Hadjidemetriou, L., Zacharia, L., Mutlaq, N., Hashem, M., Al-Agtash, S., Kyriakides, E., Demoulias, C. and Tzovaras, C. (2020) Hybrid Multi-Agent-Based Adaptive Control Scheme for AC Microgrids with Increased Fault-Tolerance Needs. IET Renewable Power Generation, 14, 1-6. https://doi.org/10.1049/iet-rpg.2019.0468
6. Lopez-Rodriguez, I., Hernandez-Tejera, M. and Lopez, A. (2016) Methods for the Management of Distributed Electricity Networks Using Software Agents and Market Mechanisms: A Survey. Electric Power Systems Research, 136, 362-369. https://doi.org/10.1016/j.epsr.2016.03.016
7. Molderink, A., Bakker, V., Bosman, M., Hurink, J. and Smit (2010) Management and Control of Domestic Smart Grid Technology. IEEE Transactions on Smart Grid, 1, 109-119. https://doi.org/10.1109/TSG.2010.2055904
8. Rohbogner, G., Fey, S., Hahnel, U., Benoit, P. and Wille-Haussmann, B. (2012) What the Term Agent Stands for in the Smart Grid Definition of Agents and Multi-Agent Systems from an Engineer's Perspective. Proceedings of the 2012 Federated Conference on Computer Science and Information Systems, Wroclaw, 9-12 September 2012, 1301-1305.
9. National Institute of Standards and Technology (2010) NIST

Framework and Roadmap for Smart Grid Interoperability Standards Release 1.0.

10. Spataru, C. and Barrett, M. (2013) Smart Consumers, Smart Controls, Smart Grid. In: Hakansson, A., Hojer, M., Howlett, R. and Jain, L., Eds., Sustainability in Energy and Buildings, Smart Innovation, Systems and Technologies, Vol. 22, Springer, Berlin, Heidelberg, 381-389. https://doi.org/10.1007/978-3-642-36645-1_36
11. Li, F., Qiao, W., Sun, H., Wan, H., Wang, J., Xia, Y., Xu, Z. and Zhang, P. (2010) Smart Transmission Grid: Vision and Framework. IEEE Transactions on Smart Grid, 1, 168-177. https://doi.org/10.1109/TSG.2010.2053726
12. Shawon, M., Muyeen, S., Ghosh, A., Islam, S. and Baptista, M. (2019) Multi-Agent Systems in ICT Enabled Smart Grid: A Status Update on Technology Framework and Applications. IEEE Access, 7, 97959-97973. https://doi.org/10.1109/ACCESS.2019.2929577
13. Rahman, S., Pipattanasomporn, M. and Teklu, Y. (2007) Intelligent Distributed Autonomous Power Systems. Proceedings of 2007 IEEE Power Engineering Society General Meeting, Tampa, 24-28 June 2007, 1-8. https://doi.org/10.1109/PES.2007.386043
14. Bellifemine, F., Caire, G. and Greenwood, D. (2008) Developing Multi-Agent Systems with JADE. John Wiley & Sons, Hoboken. https://doi.org/10.1002/9780470058411
15. Wooldridge, W. (2009) An Introduction to Multiagent Systems. John Wiley & Sons, Hoboken.
16. Jain, M., Gupta, S., Masand, D., Agnihotri, G. and Jain, S. (2016) Real-Time Implementation of Islanded Microgrid for Remote Areas. Journal of Control Science and Engineering, 2016, Article ID: 5710950. https://doi.org/10.1155/2016/5710950
17. Brown, R. (2008) Impact of Smart Grid on Distribution System Design. Proceedings of 2008 IEEE Power and Energy Society General Meeting—Conversion and Delivery of Electrical Energy in the 21st Century, Pittsburgh, 20-24 July 2008, 1-4. https://doi.org/10.1109/PES.2008.4596843
18. AbdelHafez, H. (2017) Multi Agent Systems for Smart Power Grids. Master Thesis, German Jordanian University, Amman.
19. Elsied, M., Oukaour, A., Gualous, A. and Ottavio, B. (2016) Optimal Economic and Environment Operation of Micro-Grid Power Systems.

Energy Conversion and Management, 122, 182-194. https://doi.org/10.1016/j.enconman.2016.05.074

20. Alzyoud, F., Alsharman, N. and Almofleh, A. (2019) Best Practice Fundamentals in Smart Grids For a Modern Energy System Development in Jordan. Proceedings of the 9th International Conference on Advances in Computing, Communication and Information Technology, Rome, 7-8 December 2019, 80-86.
21. Jordan National Electric Power Company NEPCO (2015) The 2015 Annual Report. http://www.nepco.com.jo/store/docs/web/2015_en.pdf

Chapter 5

Distributed Optimal Control of Transient Stability for a Power Information Physical System

Shiming Chen and Kaiqiang Li

East China Jiaotong University, Nanchang 330013, China

ABSTRACT

The development of power energy structures and information communication technology has promoted the renewal of smart grid information-physical structures. At the same time, the changes in the smart grid energy structure and the vulnerability of the information network threaten the stability of the power system and uses multiagent control theory to improve the transient stability of the power grid which has strong practicability. In this paper, an optimized distributed control scheme is proposed for application to the smart grid model so that the grid system can flexibly adapt to the external operating conditions and recover to stable operating conditions after being disturbed. In this paper, an intelligent power grid information-physical

Citation: Shiming Chen, Kaiqiang Li, "Distributed Optimal Control of Transient Stability for a Power Information Physical System", Mathematical Problems in Engineering, vol. 2020, Article ID 1393216, 11 pages, 2020. https://doi.org/10.1155/2020/1393216.

network simulation system is established. According to the information exchange within the multiagent system, groups of coherent generators in the disturbed power grid in different regions are identified and controlled. Distributed control is applied to maintain the exponential frequency synchronization and phase angle aggregation of the synchronous generators to achieve transient stability. Finally, the effectiveness and rapidity of the proposed distributed optimal control scheme are verified by simulation analysis of the IEEE 39 node model.

INTRODUCTION

With the reorganization and renewal of the energy structure in recent years, the standards of smart grid systems in terms of their transient stability have been raised. A smart grid itself is based on the ultra-high-voltage UHV backbone transmission network and deeply integrates more stable information network technology to establish a strong power system network, realizing close connection and integration of the power system and information system [1]. The development of CPSs promotes the deep integration of physical power systems and power information systems. It provides a new way to realize the goal of power grid intelligence [2]. In the smart grid model, the information network dispatching centre node collects and processes the information of each physical power station and sends corresponding control instructions. Each physical power station supplies power to the information node to ensure its normal operation. Through this connection, the mutual dependent relationship between the physical network and the information network is formed (Figure 1) [3]. With the intelligent development of smart grids, there is a dynamic interaction among the five "generation, transmission, supply, distribution, and use" terminals of smart grids with the two-way flow of energy and information [4], new energy, and other distributed energy grid connections and huge data processing between the information network and the physical network. Additionally, the basic characteristics of smart grid informatization, automation, and interaction are the operational belt of the great smart grid challenge [5, 6].

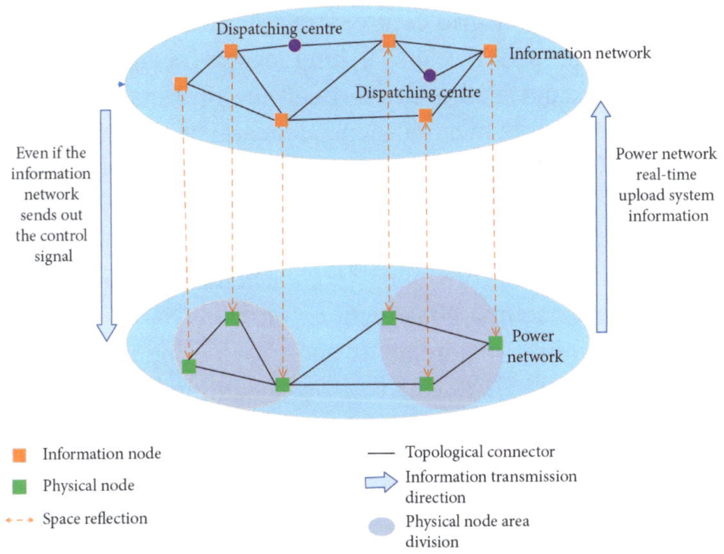

Figure 1: Model of a cyber-physical smart grid.

Information systems and information networks play an important role in enhancing the safe operation of smart grids. In actual operation, the dependence of the physical system on the information system will significantly affect the stable operation of the physical system in the case of information system failure. In actual operation, the cost of communication and information processing should also be considered. The use of redundant information and weakly correlated information in computing will cause a too high computational burden, resulting in poor performance of the system state. The stability and flexibility of smart grids can be effectively improved by determining an appropriate dependency on the information network and information flow. For phasor measurement units (PMUs), which are synchronous phasor measurement devices, a reasonable number of them at suitable locations can synchronously collect the corresponding parameters of the power system at the installation site and achieve the overall observability, economic and reliability goals of the system online state calculation, and system state prediction [7–9]. In this paper, PMUs are introduced into the optimal configuration of the power system network structure to measure the voltage vectors of some nodes, which can improve the dependence on information, providing strong initial conditions and a discrimination basis for system stability analysis, protection, and reconstruction. This is of great significance for system online state calculation and system state prediction.

The traditional power grid control strategy mostly adopts a centralized scheduling method, and the dispatching centre needs to communicate with all generators and load nodes through the information network, with a high degree of communication dependence [10]. The "plug and play" technology required by the grid connection and independent operation of new energy and power components make the topological structure of the information network and physical network of the smart grid changeable [11]. The centralized control strategy has difficulty avoiding the high cost of communication facility construction. Distributed control technology can meet the flexibility requirements and provide high-efficiency control constraints [12]. In [13], a smart grid scheduling strategy considering flexible load is designed, which have better convergence and application prospect compared with the centralized scheduling strategy.

In [14, 15], distributed control is used to reduce the cost of the communication network, and power distribution and frequency stability are achieved. Energy storage technology provides a new technology choice for the safe and economic operation of power grids, stabilization of power fluctuations, and distributed control [16–19]. In [16–18], the influence of flywheel energy storage on the power balance and stability of the power grid is studied. In [19], a fast control method for a battery energy storage system is adopted to improve the transient stability of the power grid. In [20], a distributed control method of energy storage system based on real-time wind power output regulation of energy sharing is proposed. In [21], a distributed intelligent power grid controller based on the Robust Consistency Algorithm of the second-order multiagent system is proposed. The controller is composed of a physical topology model of the network. The delay characteristics and anti-interference ability of the controller are verified by the IEEE39 node system.

In [22], the problem of cluster behavior and target consistent tracking of heterogeneous multi-inertial bodies with limited communication distance is studied. A distributed control protocol is designed to enable agents to achieve stable group behavior. In [23], a distributed observer is designed to estimate the relatively complete state of a general linear multiagent system that does not directly enter the real-time state, and it is used to track the consistency protocol, so as to achieve the overall consistency. In [24], a distributed multicluster method based on partial information exchange is designed by studying the multiobjective consistency pursuit of the multiagent system, and sufficient conditions are given to realize the dynamic goal consistency pursuit, so that the agent can adaptively select the target for tracking. In

[25], we study the problem of consistency tracking of heterogeneous interdependent group systems with fixed communication topology and propose a distributed consistency tracking control protocol, which uses local information to achieve the consistency tracking of heterogeneous interdependent group systems. In [26], a new scale consistency protocol is proposed to solve the problem of time-varying delay in nonlinear dynamics and communication networks. In this paper, based on the rules of speed consistency and group concentration of a multiagent system, the phase angle (rotor angle) aggregation and frequency consistency of power grid generators are realized by using the local and associated node information to inject power into or absorb power from generator nodes through an external energy storage device (such as a flywheel) to ensure synchronous operation of the generators and realize transient stability of the power system.

In the context of transient stability, this paper integrates information and physical nodes into an agent, adopts the distributed control strategy of a "leading assisting" agent, and utilizes the information of local and other regional physical nodes and the coupling relationship of the physical network. It also realizes global information flow and stable operation of the system through the interaction of the information and physical networks and external energy storage. In this paper, the IEEE 39 node grid model is used as an example to demonstrate the effectiveness of the application of optimized distributed control in the smart grid model to maintain the stability of the system [27].

TRANSIENT STABILITY CONTROL OF A SMART GRID

To ensure that the power grid can quickly respond after being disturbed by information and a physical disturbance and reduce the implementation time of transient stability zone control, necessary information network reliability optimization measures are taken to reduce information redundancy and unnecessary calculation and achieve the purpose of real-time control.

Reliability Optimization of a Smart Grid Information Network

Considering the reliability of the data processing of the information network under the influence of a single line fault in the physical network, the goal of a fully observable power network can be achieved by ensuring the optimal configuration of PMUs (synchronous vector measurement devices) with different channel capacities in the information network. Based on the configuration rule [28] of the minimum number of PMUs in the information network, the PMU optimization method to avoid scheme repetition is adopted, and the model is defined as follows:

$$\begin{cases} \min\left(x = \sum_{k}^{n} y_i x_k \right), \\ s.t\ \text{AX} \geq 1, \\ \max\left(f = \sum_{k}^{n} f_k \right), \end{cases} \tag{1}$$

where x_k is 1 or 0, which indicates whether physical grid node k has a PMU installed; $A = (a_{ij})$ is the node correlation matrix, where if the two nodes i, j are adjacent or the same, then $a_{ij} = 1$and otherwise, $a_{ij} = 0$; 1 is the column vector whose elements are all 1; $f_k = x_k + \sum_{l \in G} x_l$; and G indicates the set of all nodes connected to node k. Considering the 39-node grid model, the node correlation matrix is as follows:

$$A = \begin{bmatrix} A_1 \\ A_2 \\ A_3 \\ \vdots \\ A_{38} \\ A_{39} \end{bmatrix} = \begin{bmatrix} a_{1,1} & a_{1,2} & a_{1,3} & \cdots & a_{1,38} & a_{1,39} \\ a_{2,1} & a_{2,2} & a_{2,3} & \cdots & a_{2,38} & a_{2,39} \\ a_{3,1} & a_{3,2} & a_{3,3} & \cdots & a_{3,38} & a_{3,39} \\ \vdots & \vdots & \vdots & \vdots & \vdots & \vdots \\ a_{38,1} & a_{38,2} & a_{38,3} & \cdots & a_{38,38} & a_{38,39} \\ a_{39,1} & a_{39,2} & a_{39,3} & \cdots & a_{39,38} & a_{39,39} \end{bmatrix} = \begin{bmatrix} 1 & 1 & 0 & \cdots & 0 & 1 \\ 1 & 1 & 1 & \cdots & 0 & 0 \\ 0 & 1 & 1 & \cdots & 0 & 0 \\ \vdots & \vdots & \vdots & \vdots & \vdots & \vdots \\ 0 & 0 & 0 & \cdots & 1 & 0 \\ 1 & 0 & 0 & \cdots & 0 & 1 \end{bmatrix}. \tag{2}$$

The cost of installing PMU equipment at node i is as follows:

$$y_i = 1 + 0.1 n_{\text{ch}}. \tag{3}$$

The cost of equipment installation is the sum of a fixed cost and a variable cost, n_{ch} is the number of PMU channels, indicating the number of observable adjacent lines, generally 1–5, and the variable cost is increased by 0.1 for each additional line.

According to the above constraint rules, the following formula can be obtained to realize the fully observable IEEE 39 node power network topology:

$$\begin{cases} f_1 = x_1 + x_2 + x_{39} \geq 1, \\ f_2 = x_1 + x_2 + x_3 + x_{25} + x_{30} \geq 1, \\ f_3 = x_2 + x_3 + x_4 + x_{18} \geq 1, \\ \vdots \quad \vdots \\ f_{38} = x_{24} + x_{38} \geq 1, \\ f_{39} = x_1 + x_9 + x_{39} \geq 1. \end{cases} \tag{4}$$

To further explain the impact of the number of channels, the following variables are introduced:

$$w_{i,j} = \begin{cases} 1 & \text{PMU was installed on node } i \text{ and made node } j \text{observable,} \\ 0 & \text{Other.} \end{cases} \tag{5}$$

Then, the above formula can be rewritten to obtain the following inequalities:

$$\begin{cases} g_1 = x_1 + w_{2,1} + w_{39,1} \geq 1, \\ g_2 = x_2 + w_{1,2} + w_{3,2} + w_{25,2} + w_{30,2} \geq 1, \\ g_3 = x_3 + w_{2,3} + w_{4,3} + w_{18,3} \geq 1, \\ \vdots \quad \vdots \\ g_{38} = x_{38} + w_{24,38} \geq 1, \\ g_{39} = x_{39} + w_{1,39} + w_{9,39} \geq 1. \end{cases} \tag{6}$$

In the case of a single line fault in the physical network, the number of channels on the right side of the fault point at the end of the node line is accordingly changed (i.e., multiplied by 2 to ensure that each node is observed by at least two PMUs). The configuration with the lowest cost is 2, 6, 8, 15, 16, 17, 20, 23, 25, 26, 29, and 35. In the distributed control framework, the sensitivity of the information nodes is considered to select the dominant generator in the area.

Smart Grid Distributed Framework

According to the information network reliability optimization results and the multiagent system model, a distributed control framework is designed for cooperation between the generator and the external energy storage system in the smart grid. In this framework, the corresponding information and physical equipment at each generator node are combined to form an agent. Each agent represents its main properties by the physical quantities of the generator. The information equipment includes a measurement unit (PMU) and a controller, both of which perform calculations and applications related to information and data. The physical equipment also includes external energy storage equipment in addition to the generator. Only when external energy storage is needed to inject power or absorb a certain power, $P_{I,i}$ will the corresponding controller of the generator agent be activated.

The agents that describe the information-physical network interaction and the system model (Figure 2). The generator participating in stability control of the physical power grid and frequency adjustment without difference is the leading generator, which is realized by the information network through selective control; the power regulation of other generators in the same zone is controlled by the information network to proportionally change with the power of the leading generator to assist in frequency regulation of the leading generator.

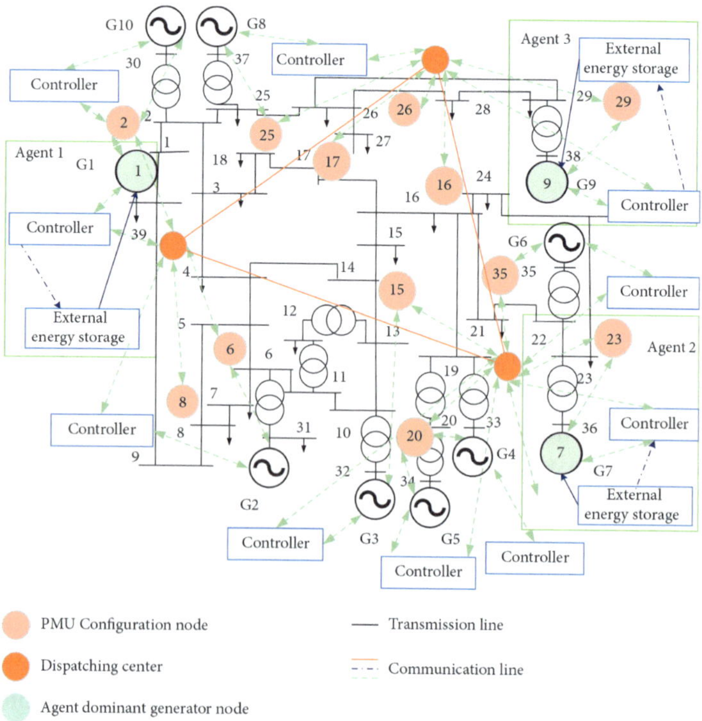

Figure 2: Model of the IEEE 39-bus cyber-physical smart grid with agents.

The synchronous generators play a leading role in the physical power grid, and thus the stability of the power grid is largely a problem of keeping the interconnected generators synchronized [19]. A smart grid maintains synchronous generator frequency synchronization and phase-angle aggregation by using a distributed control framework based on the multiagent model. According to the requirements of frequency synchronization, the normalized speed corresponding to the grid frequency is selected and denoted by ω, and the generator label in the physical grid is i ∈ V, V = {1, 2, . . . , Z}, with

$$\omega_i(t) \longrightarrow 1\,(t \longrightarrow \infty). \tag{7}$$

To avoid the rapid drop of the node voltage caused by the phase angle difference of the generator terminal voltage exceeding a certain range, thus affecting the safe operation of the unit, the phase angle aggregation requires that generators i and j satisfy the following:

$$\left|\theta_i - \theta_j\right| \leq \gamma, \tag{8}$$

where γ is a predefined threshold, usually 5π/9.

Smart Grid Control Target Model

The information network selection coefficient h_i is set as follows:

$$h_i = \left\{ \begin{array}{c} 1, \text{the } i-\text{th agent contains the leading generator,} \\ 0, \text{others.} \end{array} \right\}, \tag{9}$$

This coefficient is used to determine the selective control behaviour of the information network with respect to the physical network and ensure timely control of the main generator agent contained in each partition.

If there is a controller that connects the physical power grid with the information network, then the oscillating equation between motors can be obtained under the constant voltage condition [29]:

$$\begin{aligned} M_i\dot{\omega}_i &= -D_i\omega_i + P_{m,i} - |E_i|^2 G_{ii} \\ &\quad - \sum_{j=1, j\neq i}^{Z} P_{ij}\sin\left(\theta_i - \theta_j + \varphi_{ij}\right) + h_i u_i. \end{aligned} \tag{10}$$

The transient stability problem of a smart grid is described as the control task of multiagent swarm concentration with consistent speed, and then the second-order dynamics model of multiagents is used to describe each agent as follows [30]:

$$\begin{cases} \dot{\theta}_i = \omega_i, \\ D_i\dot{\omega}_i = -M_i\ddot{\omega}_i - l_{ij}\omega_i + h_i\dot{u}_i. \end{cases} \tag{11}$$

The controller signal is defined as $u_i = P_{L,i}$ for agent i = 1, ..., F, which corresponds to the agent containing the main generator. Since only one main generator is selected for each partition, F also represents the number of physical power grid zones. $h = diag[h_1, ..., h_i, h_{i+1}...h_Z]$. Generally, $h_i = 1$ is set for $i \le F$; otherwise, $h_i = 1$ corresponds to other agents in the partition that assist in the adjustment, $V_{mf} = \{F + i, ..., Z\}$. The auxiliary regulated power of the corresponding generator is $P_{F+i} = a_{F+i} \cdot P_{L,i}$, where a is a proportional coefficient that varies with the power of the leading generator and l_{ij} is an element of the physical relation matrix L, which is represented as

$$l_{ij} = \begin{cases} \sum_{j=1,j\neq i}^{Z} P_{ij}\cos\left(\theta_i - \theta_j + \varphi_{ij}\right), i = j. \\ -P_{ij}\cos\left(\theta_i - \theta_j + \varphi_{ij}\right), i \neq j. \end{cases} \tag{12}$$

In this case, a new control quantity $\widetilde{u}_i$ is introduced to define the controller signal u_i as

$$\dot{u}_i = \widetilde{u}_i + l_{ij}\omega_i - B_i\dot{\omega}_i, \tag{13}$$

where $B_i \ge (100 * D_i)$ is an element in information relation matrix B.

DISTRIBUTED AREA OPTIMAL CONTROL FOR THE TRANSIENT STABILITY OF A SMART GRID

Smart Grid Area Division

The main properties of each agent in the smart grid are described by the physical quantities of the generators in the physical grid. Let the state information carried by agent i at time t = k be

$$\begin{cases} x_i(k) = \theta(k), \\ v_i(k) = \omega_i(k). \end{cases} \tag{14}$$

When the time step is t = k, $\theta_i(k)$ and $\omega_i(k)$ are the rotor angle and speed of the i-th generator, respectively, which are physical quantities directly obtained from the PMU. $S_i(k) = [x_i(k), v(k)]^T$ describes the status information carried by individuals, where x_i (k) and $v_i(k)$ correspond to the position and speed of the individual at time step t = k. If the distance between two individuals is less than the threshold r, then they are considered as being in each other's neighbourhood. The neighbourhood set for the i-th individual is defined as follows:

$$N_i(k) = \left\{ j \in V \,\middle|\, \left\| x_i(k) - x_j(k) \right\| < r \right\}. \tag{15}$$

According to the similarity of the individual dynamic state quantities, the region is divided; for $j \in N_i(k)$, the relationship between individuals i and j in the same group meets the following requirement:

$$S_i(k) - S_j(k) < S_{th}(k), \tag{16}$$

Where $S_{th}(k)$ is the set threshold value.

The following dynamic model is used:

$$x_i(k+1) = x_i(k) + \Delta t v_i(k). \tag{17}$$

The physical dynamic quantity of the agent has been transformed into the individual state quantity in the information space, and multiple groups containing different individuals are simultaneously obtained, which are composed of agents with close physical coupling in the actual smart grid corresponding to the partition.

Smart Grid Distributed Controller

According to formula (13), letting $\underline{M}_i = D_i + B_i$, the second-order system of the set containing the leading generator agent is determined as

$$\begin{cases} \dot{\theta}_i = \omega_i, \\ \underline{M}_i \dot{\omega}_i = \tilde{u}_i. \end{cases} \tag{18}$$

To achieve transient stability of the smart grid, the distributed control signal applied to the agent (label i = 1, 2, · · · , F) with the leading generator in each partition is as follows:

$$\tilde{u}_i = -\nabla V(\theta_{ij}) - \tilde{l}_{ij}\omega_i - c(\omega_i - \omega^*). \tag{19}$$

where c is a constant representing the coefficient of the feedback term; ω is the expected normalized speed value. $\widetilde{l}_{ij}$ is an element in the velocity matching matrix $\widetilde{L}$, and the first term is the V(θ_{ij}) gradient, where the potential function V(θ_{ij}) is defined as follows:

$$V(\theta_{ij}) = \frac{1}{2}\sum_{i=1}^{F}\sum_{j=1, j\neq i}^{F}\chi(\theta_i - \theta_j). \tag{20}$$

The above functions ensure that the rotor angle difference between agents is bounded.

Analysis of the Distributed Control Stability of a Smart Grid

Theorem 1. *Consider a second-order system with expected values of the guidance feedback terms in kinetic equations (18) and (19). Suppose that the initial velocity does not match the expected value and that the initial energy H_0 is finite; under the action of control protocol equation (19), all agents asymptotically and uniformly converge to the corresponding reference speed, corresponding to the expected speed of the generator of ω*, and the group behaviour is consistent with global stability.*

Proof. Let $\widetilde{\theta}_i = \theta_i - \theta^*$ and $\widetilde{\omega}_i = \omega_i - \omega^*$ represent the differences between the physical quantities of the agent and the expected values, respectively, then:

$$\begin{cases} \dot{\widetilde{\theta}}_i = \widetilde{\omega}_i, \\ \underline{M}_i\dot{\widetilde{\omega}}_i = \widetilde{u}_i. \end{cases} \tag{21}$$

According to equations (18), (19), and (21), the control function can be obtained as follows:

$$\widetilde{u}_i = -\nabla V(\widetilde{\theta}_{ij}) - \widetilde{l}_{ij}\widetilde{\omega}_i - c\widetilde{\omega}_i. \tag{22}$$

The Lyapunov function is defined as the total energy of the system, that is, the sum of the total potential energy between agents and the relative potential energy and kinetic energy between the physical quantities of the agents and the expected reference quantities:

$$H(\widetilde{\theta}_i, \widetilde{\omega}_i) = \frac{1}{2}\sum_{i=1}^{F} V(\widetilde{\theta}_{ij}) + \frac{1}{2}\sum_{i=1}^{F}\widetilde{\omega}_i^{\,T}\underline{M}_i\widetilde{\omega}_i. \tag{23}$$

Since the potential function $V(\theta_{ij})$ is symmetric with respect to $\theta_{ij}(i, j = 1, \ldots, F)$,

$$\frac{\partial V(\tilde{\theta}_{ij})}{\partial \tilde{\theta}_{ij}} = \frac{\partial V(\tilde{\theta}_{ij})}{\partial \tilde{\theta}_{i}} = -\frac{\partial V(\tilde{\theta}_{ij})}{\partial \tilde{\theta}_{j}}. \tag{24}$$

The derivative of equation (23) can be obtained:

$$\dot{H} = \frac{1}{2}\sum_{i=1}^{F}\frac{\partial V(\tilde{\theta}_{ij})}{\partial \tilde{\theta}_{i}} + \sum_{i=1}^{F}\tilde{\omega}_i{}^{T}\underline{M}_i\dot{\tilde{\omega}}_i = \sum_{i=1}^{F}\tilde{\omega}_i{}^{T}\nabla V(\tilde{\theta}_{ij}) + \sum_{i=1}^{F}\tilde{\omega}_i{}^{T}\tilde{u}_i. \tag{25}$$

Equation (22) can be substituted into equation (25) to obtain

$$\begin{aligned}\dot{H} &= \sum_{l=1}^{F}\tilde{\omega}_i^{T}\nabla V(\tilde{\theta}_{ij}) - \sum_{i=1}^{F}\tilde{\omega}_i^{T}\nabla V(\tilde{\theta}_{ij}) - \sum_{i=1}^{F}\tilde{\omega}_i^{T}\tilde{l}_{ij}\tilde{\omega}_i - \sum_{i=1}^{F}\tilde{\omega}_i^{T}c\tilde{\omega}_i \\ &= -\sum_{i=1}^{F}\tilde{\omega}_i^{T}\tilde{l}_{ij}\tilde{\omega}_i - \sum_{i=1}^{F}\tilde{\omega}_i^{T}c\tilde{\omega}_i = -\sum_{i=1}^{F}\tilde{\omega}_i^{T}\left[\left(\tilde{l}_{ij} + c\right)\right]\tilde{\omega}_i \\ &= -\tilde{\omega}^{T}[(\tilde{L} + cI)]\tilde{\omega}.\end{aligned} \tag{26}$$

Since $\tilde{L}$ is a semipositive definite Laplace matrix and $c > 0$, $\tilde{L} + cI$ is a positive definite matrix, and thus, $\tilde{\omega}^{T}[(\tilde{L} + cI)]\tilde{\omega} > 0$ and $\dot{H} < 0$. Therefore, the agents asymptotically stably converge to their expected reference values under control input (19).

SIMULATION EXAMPLE

In this paper, the IEEE 39 node smart grid model is analysed as an example, which includes 19 loads and 46 lines, which are simulated by the MATLAB/Simulink platform. According to the classification of a short-circuit fault, centralized control after a fault and swarm control after a fault, the examples are divided into four scenarios.

Scenario 1. Suppose at t = 0 s, a three-phase short-circuit fault occurs on disturbed line 21–22 of the power grid. Disconnect the fault line and set $h_i = 0$ for $i \in V$, $V = \{1, 2, \cdots, Z\}$, that is, the controller connecting the information and physical networks is not activated and the information network traffic and calculation amount are minimal. The relationships between the speeds of each generator and the rotor angles are given in Figure 3. The system operation is obviously unstable, the rotor angle has an increasing difference trend, and the power network has a serious fault.

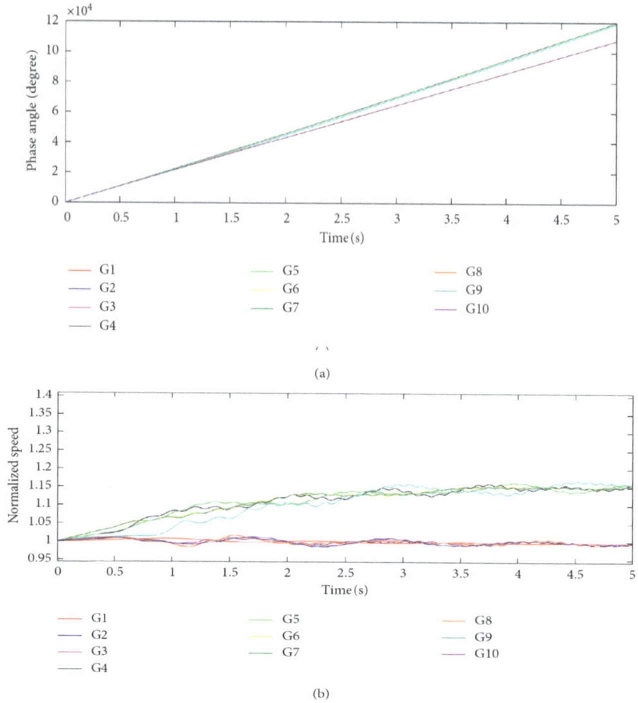

Figure 3: Relative rotor angles and rotor speeds of generators after the fault.

Scenario 2. Under centralized control, each node of the power grid needs to transmit information to the dispatching centre. After centralized processing, a large number of PMUs and communication lines are needed to ensure information collection and transmission. In the period before centralized control after fault removal, it is necessary to collect the state information of the whole power grid, assuming that the data can be collected and processed within the preset critical time after fault treatment under ideal conditions. When t = 0.15 s, set $h_i = 0$ for $i \in V$, $V = \{1, 2, \cdots, Z\}$, so that all generator nodes are controlled by no difference regulation. Under ideal conditions(Figure 4), the rotor angles and speeds of the generators change with time and the system achieves a stable state in a short time. However, the system is in the state of maximum information transmission and processing. The control equipment of each generator node frequently uses external energy devices to adjust the power of the corresponding nodes. The overall energy consumption of the whole regulation process is large, and the network topology and performance of the related equipment are not fully utilized.

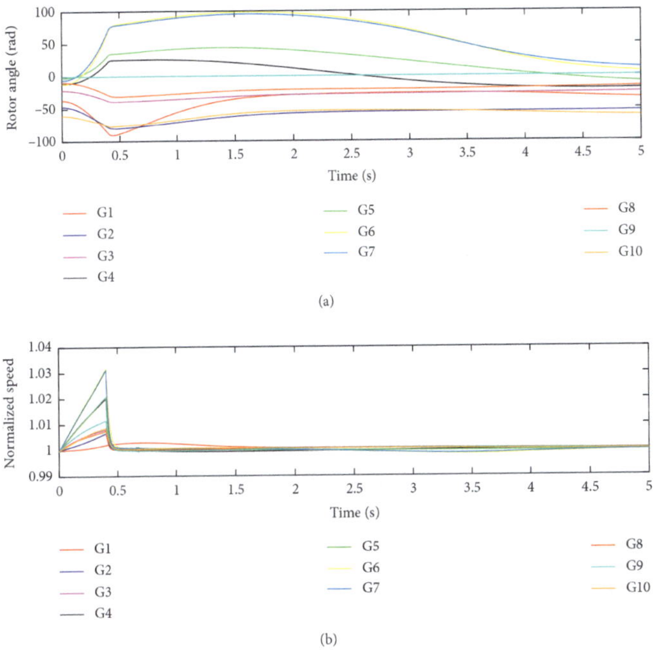

Figure 4: Relative rotor angles and rotor speeds of generators under centralized control after the fault.

Scenario 3. The flocking control method leads to a stable performance (Figure 5). Distributed flocking control can reduce the equipment cost and improve the centralized control of information processing, the energy consumption, and the critical stability range in the 5 s of the observation period, but the fluctuation range is larger. The smart grid after using each agent to facilitate control tends to be stable for a long time.

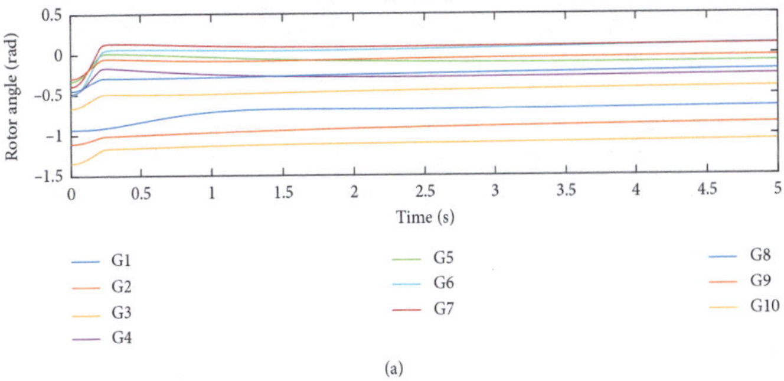

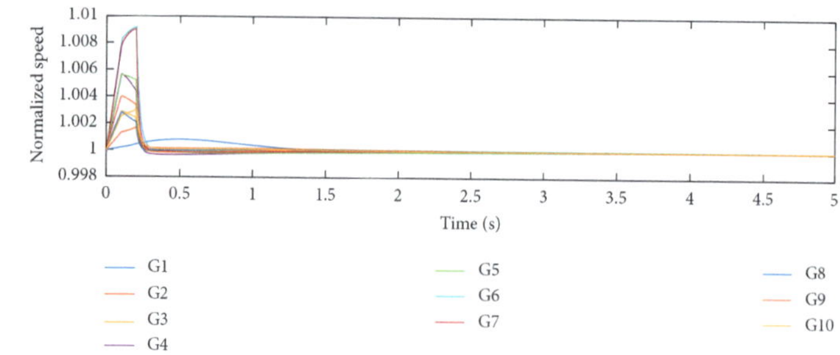

(b)

Figure 5: Relative rotor angles and rotor speeds of generators under flocking control after the fault.

Scenario 4. The distributed optimal control method in this paper is used to control the performance (Figure 6). The smart grid partition identification scheme should be enabled within the preset critical time point after fault treatment to avoid more damage and improve the ability of the smart grid to restore transient stability. The distributed optimal control is applied to the agents containing the main generators, that is, setting $h_i = 1$ for $i = 1, ..., F$.

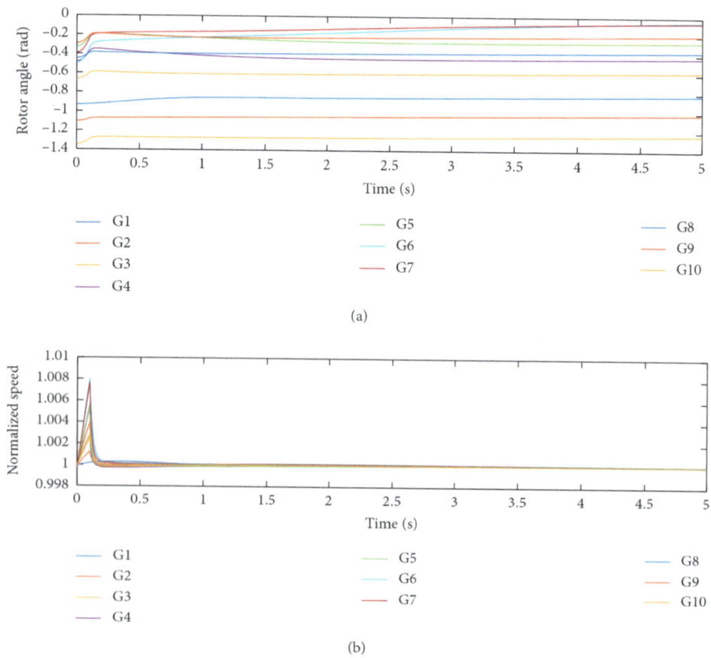

Figure 6: Relative rotor angles and rotor speeds of generators under distributed optimal control after the fault.

First, determine the smart grid partition based on the similarity of the intelligent agent dynamic characteristics; that is, the smart grid agent quantity changes corresponding to the movement trend of individuals in the information space. Next, obtain the state information of individual trajectories, described as individual motion curves in a two-dimensional space. From the trend of separation of individuals in the information space, that is, the change of the physical quantities of each agent after the failure, it can be found which individuals are obviously far from other individuals and which movement trends are the closest. The following partitions can be obtained from the partitioning scheme of dynamic recognition of each agent: {G1}, {G2 G3 G8 G9 G10}, and {G4 G5 G6 G7}.

Second, on the basis of information network reliability optimization combined with the actual active power output for each agent and sensitivity weight arrangement of the node data, reasonable leading generators G1, G7, and G9 are selected for each partition, and distributed optimal control is applied to the intelligent agent with the leading generator to achieve the goal of system transient stability.

Within the critical time after the fault is cut off, partition identification and leading generator selection are completed. At t = 0.15 s, the distributed control scheme combining the information and physical networks is applied to the smart grid. The state quantities of each agent change with time (Figure 6). The control objective is achieved for each agent in each partition, and the corresponding physical quantity converges to a certain value according to the partition. Since external energy injection could impact actual operation and lead to certain fluctuations, we quickly recover to the stable operation of the steady-state power system. The generator speeds are stable and gradually converge, and the rotor angles satisfy the requirements of the constraint and stable state. Under the conditions of information use and smaller external energy injection, the good effect of a shorter stable system recovery time, a smaller physical quantity fluctuation range, and an increased stability margin of the system are achieved.

Under the action of centralized control method, the system can resume stable operation in a short time, but the disadvantage is that the system is in the state of maximum information transmission and processing and the overall energy consumption is large. Compared with centralized control, distributed swarm control improves the disadvantages of large amount of information processing and energy consumption, but the time of system stability is prolonged. The distributed optimal control method proposed in this paper can quickly restore the stability of the system, reduce the energy consumption, and improve the stability margin of the smart grid.

CONCLUSIONS

In this paper on transient stability control of the smart grid model, the distributed multiagent control method adopted in the power grid model after reliability optimization of the information network is described. In the distributed control framework, the dynamic similarity of the physical parameters of the generators is used to form each zone, and external energy storage is used to adjust the agents in each zone. The leading generator realizes no difference adjustment, and other generators in the same zone adjust their power proportional to the adjusted power of the leading generator to achieve asymptotic stability of each zone. The simulation results in this paper verify the effectiveness of the proposed control scheme, which finally achieves transient stability and control of the smart grid model, and it has fast and stable performance compared with swarm control. The idea presented in this paper has a certain reference value for improving the transient stability of smart grids in practical projects.

ACKNOWLEDGMENTS

This work was supported in part by the National Natural Science Foundation of China under Grant no. 11662002.

REFERENCES

1. S. Y. Chen, S. F. Song, and L. X. Li, "Survey on smart grid technology," *Power System Technology*, vol. 33, no. 8, pp. 1–7, 2009.
2. P. Kundur, N. J. Balu, and M. G. Lauby, *Power System Stability and Control*, McGraw-Hill, New York, NY, USA, 1994.
3. H. Li, Z. Han, A. D. Dimitrovski, and Z. Zhang, "Data traffic scheduling for cyber physical systems with application in voltage control of distributed generations: a hybrid system framework," *IEEE Systems Journal*, vol. 8, no. 2, pp. 542–552, 2014.
4. E. Mengelkamp, B. Notheisen, C. Beer et al., "A blockchain-based smart grid: towards sustainable local energy markets," vol. 33, no. 1-2, Computer Science Research & Development, 2018.
5. K. Muralitharan, R. Sakthivel, and R. Vishnuvarthan, "Neural network based optimization approach for energy demand prediction in smart grid," *Neurocomputing*, vol. 273, pp. 199–208, 2018.
6. Q. Y. Sun, F. Teng, H. G. Zhang et al., "Construction of dynamic coordinated optimization control system for energy internet," *Proceedings of the CSEE*, vol. 14, pp. 3667–3677, 2015.
7. D. Carrion and J. W. Gonzalez, "Optimal PMU location in electrical power systems under N-1 contingency," in *Proceedings of the International Conference on Information Systems and Computer Science*, INCISCOS) IEEE Computer Society, Kuala Lumpur, Malaysia, March 2018.
8. E. Abiri, F. Rashidi, and T. Niknam, "An optimal PMU placement method for power system observability under various contingencies," *International Transactions on Electrical Energy Systems*, vol. 25, no. 4, pp. 589–606, 2015.
9. F. Rashidi, E. Abiri, T. Niknam, and M. R. Salehi, "Optimal placement of PMUs with limited number of channels for complete topological observability of power systems under various contingencies," *International Journal of Electrical Power & Energy Systems*, vol. 67, pp. 125–137, 2015.
10. Z. Wang and W. Jie, "A novel finite-time control scheme for enhancing smart grid frequency stability and resilience," *IEEE Transactions on Power Systems*, vol. 10, no. 6, pp. 6538–6551, 2019.
11. A. Q. Huang, M. L. Crow, G. T. Heydt, J. P. Zheng, and S. J. Dale, "The future renewable electric energy delivery and management (FREEDM)

system: the energy internet : the energy internet," *Proceedings of the IEEE*, vol. 99, no. 1, pp. 133–148, 2011.

12. Z. Li, W. Ren, X. Liu, and M. Fu, "Distributed containment control of multi-agent systems with general linear dynamics in the presence of multiple leaders," *International Journal of Robust and Nonlinear Control*, vol. 23, no. 5, pp. 534–547, 2013.
13. D. U. Jia-Bing, C. Yan-Peng, L. Man-Fa et al., "Design and implementation of smart grid control system based on distributed real-time scheduling strategy," *Electronic Design Engineering*, vol. 26, no. 7, pp. 119–127, 2018.
14. Y. Xu, W. Zhang, M.-Y. Chow, H. Sun, H. B. Gooi, and J. Peng, "A distributed model-free controller for enhancing power system transient frequency stability," *IEEE Transactions on Industrial Informatics*, vol. 15, no. 3, pp. 1361–1371, 2019.
15. Y. Zhou, K. Wang, and G. Li, "Distributed hierarchical control for microgrid based on multi-agent consensus Algorithm," *Automation of Electric Power Systems*, vol. 41, no. 11, pp. 142–149, 2017.
16. G. Dekker and J. Frunt, *Frequency Stability Contribution of Wartsila Combustion on Engines*, Wartsila Finland OY, Vaasa, Finland, 2012.
17. A. Farraj, E. Hammad, D. Kundur et al., "On the use of energy storage systems and linear feedback optimal control for transient stability," *IEEE Transactions on Industrial Informatics*, vol. 13, no. 4, pp. 1575–1585, 2017.
18. H. Silva-Saravia, H. Pulgar-Painemal, and J. Mauricio, "Flywheel energy storage model, control and location for improving stability: the Chilean case," *IEEE Transactions on Power Systems*, vol. 1, no. 1, p. 99, 2016.
19. K. Kawabe and A. Yokoyama, "Improvement of transient stability and short-term voltage stability by rapid control of batteries on EHV network in power systems," *Electrical Engineering in Japan*, vol. 188, no. 3, pp. 1–10, 2014.
20. S. Baros and M. D. Ili, "A consensus approach to real-time distributed control of energy storage systems in wind farms," *IEEE Transactions on Smart Grid*, vol. 10, no. 1, pp. 613–625, 2019.
21. W. Ziqiang and W. Jie, "A distributed control considering the cyber-physical characteristics of smart grid," *Proceedings of the CSEE*, vol. 39, no. 23, pp. 6921–6933, 2019.

22. S. Chen, H. Pei, Q. Lai et al., "Multitarget tracking control for coupled heterogeneous inertial agents systems based on flocking behavior," in *Proceedings of the IEEE Transactions on Systems Man & Cybernetics Systems*, pp. 1–7, January 2018.
23. C. Shiming, G. Junjie, G. Yanli et al., "Observer-based event-triggered tracking consensus of non-ideal general linear multi-agent systems," *Journal of the Franklin Institute*, vol. 356, no. 17, 2019.
24. H. Pei, S. Chen, and Q. Lai, "Multi-target consensus circle pursuit for multi-agent systems via a distributed multi-flocking method," *International Journal of Systems Science*, vol. 47, no. 16, 2016.
25. H. Pei, S. Chen, Q. Lai et al., "Consensus tracking for heterogeneous interdependent group systems," *IEEE Transactions on Cybernetics*, vol. 50, no. 4, 2020.
26. Z. Zhang, S. Chen, and H. Su, "Scaled consensus of second-order nonlinear multiagent systems with time-varying delays via aperiodically intermittent control," *IEEE Transactions on Cybernetics*, no. 99, pp. 1–14, 2019.
27. A. MOEINI, I. KAMWA, P. BRUNELLE et al., in *Proceedings of the 2015 50th International Universities Power Engineering Conference*, pp. 1–6, Stroke-on-Trent, UK, September 2015.
28. Z. Zhuoli, Y. Ping, X. U. Zhirong et al., "Review on large-disturbance transient stability research of microgrid with multiple energy resources and multiple conversions," *Power System Technology*, vol. 41, no. 7, pp. 2195–2204, 2017.
29. F. Dorfler, M. Chertkov, and F. Bullo, "Synchronization in complex oscillator networks and smart grids," *Proceedings of the National Academy of Sciences*, vol. 110, no. 6, pp. 2005–2010, 2013.
30. R. Olfati-Saber, "Flocking for multi-agent dynamic systems: algorithms and theory," *IEEE Transactions on Automatic Control*, vol. 51, no. 3, pp. 401–420, 2006.

SECTION 2:
SUSTAINABILITY AND ENERGY EFFICIENCY

Chapter 6

Energy Efficiency in Smart Grid: A Prospective Study on Energy Management Systems

Hermes José Loschi, Julio Leon, Yuzo Iano, Ernesto Ruppert Filho, Fabrizzio Daibert Conte, Telmo Cardoso Lustosa, Priscila Oliveira Freitas

Department of Communications, Faculty of Electrical and Computer Engineering, University of Campinas (UNICAMP), Campinas/SP, Brazil

ABSTRACT

The term Smart Grid has become a term to represent the benefits of a smart and sophisticated electrical grid, which can meet various social expectations related to sustainability and energy efficiency. The Smart Grid promises to enable a better power management for energy utilities and consumers, to provide the ability to integrate the power grid, to support the development of micro grids, and to involve citizens in energy management with higher levels of responsibility. However, this context comes with potential pitfalls, such as vulnerabilities to cyber-security and privacy risks. In this article, a prospective study about energy management, and exploring critical issues of

Citation: Loschi, H. , Leon, J. , Iano, Y. , Filho, E. , Conte, F. , Lustosa, T. and Freitas, P. (2015), "Energy Efficiency in Smart Grid: A Prospective Study on Energy Management Systems". *Smart Grid and Renewable Energy*, **6**, 250-259. doi: 10.4236/sgre.2015.68021.

modeling of energy management systems in a context Smart. Grid is presented along with background of energy management systems. An analysis of the demand response condition is also presented. Finally, the advantages and disadvantages of the implementation of energy management systems, and a comparison with the Brazilian electricity system are discussed.

Keywords: Smart Grid, Management, Cloud Computing, Energy, Efficiency

INTRODUCTION

Electric energy is essential to increase productivity and ensure a high quality of life; therefore, the relationship between electric power and economic growth is crucial. However, the consequence of the current worldwide economic growth and electricity demand is the depletion of energy resources. An essential and effective way to prevent the depletion of resources and promote economic growth at the same time is the application of the concept of energy efficiency through energy management systems, this is being the basic principle of the Smart Grid [1] .

With the development of the Smart Grid, especially in the distribution grid, and with the possibility of load modeling, control over the peaks of energy demand becomes vital. The peaks of demand are serious problems and present themselves in the electrical system. The demand management in residential, commercial and industrial sectors can play an important role in reducing peak demand, reducing stress, overhead transmission and distribution lines. In many countries, there are various demand response programs, implemented for industrial and commercial loads [2] .

There are few demand response (DR) programs in use for energy management in the residential sector. Direct load restriction is the most popular method used to reduce peak demand. However, when using direct control/restriction load, the consumer comfort can be compromised. In contrast, reduction of peak demand through load displacement can benefit consumers and energy utilities [2] .

Peak demand of energy has caused adverse effects to the reliability and stability of the power system during recent decades. Reducing peak demand can reduce the risk of faults on transmission and distribution grid, consequently, the risk of interruptions. Demand response is one way to deal with peak events and avoid overload on the grid by providing the necessary flexibility through load displacement [2] [3] .

Multiple energy management systems use the concepts of demand response; however, these DR systems are not broadly implemented due to complexity of the automation involved, particularly in industrial and commercial buildings. Although the residential sector is responsible for a significant amount of electrical energy demand, few DR programs are currently used in the residential field. Direct Load Control (DLC) is the method most used by energy utilities in Brazil to manage peak demand in the residential sector, where consumer loads are adjusted in time during peak network events [4] [5] .

An energy management system that can automatically switch the operation of appliances during peak hours, can be used to management of peak demand without compromising the quality of supply to the consumer [6] .

Up recognizing the very different perspectives and priorities of individuals and organizations involved in the electrical system, the authors propose a broad analysis through a prospective study to examine the different ways for the Smart Grid meeting the demands and developments of society. This study focuses on comparing the development of "Smart Grid" in different regions of the world and demonstrating the commitment of these countries to change the social and political contexts and expectations, which often are shaped by specific regions, goals and available resources.

SMART GRID TECHNOLOGIES

The term Smart Grid means more than a single technology or even a clear set of individual technologies. Is an "umbrella" term under which various technologies of electric power systems are considered, both in hardware and software. For some people, Smart Grid is characterized primarily as the addition of an information and communication technology (ICT), superimposed in a way on existing infrastructure. For others, Smart Grid represents the installation of new transmission lines, meters, and renewable energy generation [7] .

However, in order for both conditions to comply, first it is necessary to understand the legacy electric systems worldwide. The current dominant infrastructure of electric power systems involves four basic elements:

- Generation: Electric power is generated in large-scale power plants;

- Transmission: High-voltage electrical energy is transported from the plant to substations closer to consumers;
- Distribution: Low voltage energy is distributed from substations to residences and commercial buildings;
- Consumer: Electricity used for consumer devices such as refrigerators, computers, lights, pumps and other devices used by residential, commercial, and industrial devices.

The main mechanism in power generation by current systems depends on the heat produced by burning fossil fuels, division of atoms in nuclear energy, or from the hydroelectric stations water movement. Except for solar cells, almost all other forms of power generation, including the burning of fossil fuels, nuclear, biomass, hydro, wind, concentrated solar, cogeneration, and need driving a turbine to produce electricity [7] [8] .

The generation usually produces electricity with relatively low voltages ranging from 2 to 30 kilovolts (kV), depending on the size of the unit. Since electricity is generated, its tension is amplified before transmission. A critical step between the generation of electricity and long-distance transmission involves a transformer to increase the voltage. Often, the generation of electric energy occurs far from the places where the electricity is needed, making the long distances of high voltage transmission lines, a crucial part of the electrical system. The long-distance transmission voltage varies from 115 to 120 kilovolts (KV), so the transformer plays a crucial role in increasing the voltage for transmission [7] [8] .

The high-voltage transmission lines carry electricity from generating plants to local substations, where the energy is "left over" for a lower voltage and then sent through electric energy distribution networks for local users, including the industrial, commercial, and residential consumers. From the substation, the electrical energy is distributed locally within a community to individual buildings and houses. The voltage is usually reduced at the point of use, to the standard voltage of that region, which varies in different countries (with most consumers receiving 110 - 120 V in the United States and 220 - 240 V in Europe) and with the requirements of electric power use [7] [8] .

Main Technologies

The term Smart Grid represents the integration of digital technologies, sensors, and ICTs to empower and make the management on the use of electricity more reliable and efficient. Smart Grid includes technologies for the consumer (with which consumers interact) and the grid (transmission and distribution that are less visible to consumers). The Smart Grid technologies also include hardware and software [7] [8] .

One of the definitions of the term Smart Grid is the integration of various technologies, products and services, from the generation, transmission, and distribution; using advanced communication and control technologies. Figure 1 illustrates this concept.

Table 1 presents the main technologies proposals for Smart Grid and their definitions.

In addition to the above-mentioned Smart Grid technology, a more holistic approach to energy management systems is presented in the next section. A major challenge with current systems is the limited mechanisms for coordination and communication between the management of the different parts of the system. The management of the current systems' transmission and distribution are carried out through separate activities that occur in different parts of the network [8] .

ENERGY MANAGEMENT SYSTEMS IN SMART GRID

The vast majority of the energy management systems just consider the monitoring and data statistics of energy consumption of consumer electronics. For these systems, manual actuation is necessary in each device to reduce energy consumption. However, the Smart Grid technologies require management systems to be smarter and able to respond to demands related to the charge control, energy management, and timing systems with micro grids [9] .

Some architectural solutions for energy management systems are being integrated into a concept of Smart Grid; these are presented blow. Chang-Sci Choi develops an architecture using AMI solutions for an energy management system entitled EMM (Energy Monitoring and Management), interoperating this system with the Smart Grid. Among the assumptions adopted, the most important is the division of the stream of operation and settings of residences and buildings [9] .

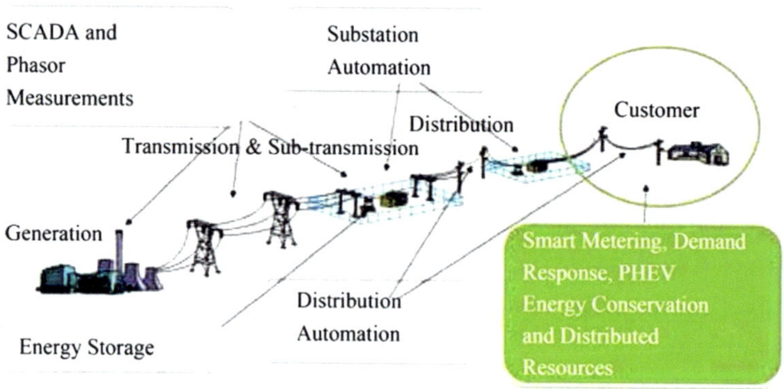

Figure 1: Concept for control and communication systems in Smart [9] .

Table 1: Main technologies proposals for Smart Grid [7]

Phases	Technologies	Definition
Generation	Inverters "Grid Tie"	Use of "Grid Tie" inverter, connecting to mains, e.g. photovoltaic systems
Transmission	Synchrophasors	Measurement of Synchronized Phasor, i.e. sinusoidal measurements of AC magnitudes, and synchronized in time expressed in phasors. To determine useful information about grid performance
	FACTS (Flexible Alternate Current Transmission Systems) para HVAC (High Voltage Alternating Current) e HVDC (High Voltage Direct Current)	AC or DC (High Voltage ou Low Voltage) voltage, transmission from generation to distribution
	Static Deviation/Compensator VAR (Volt-amper reactive)	Used for static VAR compensation, have mutual inductance lines, consuming reactive power
Test and modeling of transmission	Software for Analysis of Transmitted Power	Full suite of tools to create, configure, customize and manage power transmission systems
Transmission in cases of distribution of HVDC and generation by renewable sources	Inverters and rectifiers	Used for conversion AC-DC and DC-AC

Transmission and distribution	Substation automation	Applications of automation: Control, timing, voltage transfer, reduction and detection of loads
Transmission, distribution and substation	Relays and circuit breakers	The relay is actuated to detect any failure in the power system, disarming the breakers of the circuits, avoiding damage to the equipment
Distribution (Distribution management system-DMS)	Fault location for distribution System.	Fault locators are devices and software, usually installed in substations to identify possible fault events, calculating the distance from the point of the failure to the source monitored identified in distribution system
Distribution	AMI (Advanced metering Infraestrutura)	AMI provides the communication with the smart meters and other power management devices
Distribution, information management.	Distribution automation	The system consists of equipment's, communication infrastructure and information technology, which are used as intelligence distribution system
Information management	Management system of measurement data	Used for complex data collection processes measured for multiple data recording technologies
	RTU (Remote Terminal Unit)	The function of the RTU is the remote location of the system SCADA, for monitoring and control of the necessary equipment
	SCADA (Supervisory Control and Data Acquisition)	Computational system that monitors and controls infrastructure and industrial plant processes based installation
	EMS (Energy Management System)	The EMS are tools used for operation of electrical grid, to monitor, control and optimize the performance of systems for generation, transmission and distribution
Information management and consumer economics	Smart Meters	The smart meter registers the range of the electric energy consumption
Information management: HAN (Home are network) and LCM (Load Control Modules).	PCT (Program Communication Thermostat)	Components of the control system, to detect the temperature of a system, controlling the same in the desired set point.
Residential Consumption	Load control receiver	Are devices used for load control, directly or indirectly, through voltage circuits, such as air conditioning thermostat.

Consumption of the generation, distribution and transmission	Short circuit current limiter	Current limitation is the practice in electrical or electronic circuits, the imposition of an upper limit to the current that can be delivered, avoiding damage to the equipment
Transformers generation, distribution and transmission	OLTC (On-Load Tap-Changer)	The OLTC is used to change the transformer voltage ratio, without interrupting the load service

The sizing system EMM first considered complex apartments and residential environments popular in South Korea, with advanced infrastructure of home networks and several features, such as: electricity, gas, water, and cogeneration. Figure 2 shows the architecture that is installed via the internet providing accessibility, mobility, and interoperation with other systems. Each EMM installed performs communication with smart watches "Smart Meters", and interface with networked home appliances. To report the current status of the house, the energy measurement system sends data in real time via the Internet [9] .

For commercial buildings and industrial plants, the author proposes the BEMS (Building Energy Management System). However, these systems and its communication protocols depend on the market and corporate strategies at the time of construction, and its concepts of energy efficiency and energy management, which adopt generic and growing open communication standards and protocols, such as: BACnet, LonWorks, Modbus, KNX, WLAN, Zigbee, SNMP, IEC61850, DNP3, etc. To overcome these constraints, the author proposes the development of 2 CCL (Common Communication Layer) protocols. The first one is called BAS CCL, and the other one, EMM CCL. Figure 3 illustrates the configuration of EMM for heterogeneous integration BEMS in each building [9] .

In both systems, the server data is analyzed by software, providing information about weather and conditions of operation of the entire plant. This information can be broadly used and incorporated into a maintenance planning, among other applications [9] [10] .

Levels of data analyzed by the software can be broadly classified as:

- Level 1, Management: Supervision of computers, servers and services data management servers;
- Level 2, Automation: Smart G/W (gateway);
- Level 3, Installation: Sensors, actuators and controllers.

The software operates considering a data group, composed of service server and data management that are installed on the building control center, as illustrated in Figure 4. The Smart Gateway is installed inside the building, collecting the data of the electric power consumption through intelligent devices such as meters, sensors and actuators, etc., that are engaged in the construction of an operating system, including electricity control systems, HVAC/HVDC, lighting systems, etc. [10] .

This architecture allows the system to provide multiple functions for energy saving control, such as: control of maximum demand of energy, weather-based light and dimming control by means of sensors. The information generated by the system makes it possible to provide the function of energy efficiency through predictive analysis of the trend of energy consumption, comparing with the data in real time, through a comparative analysis with similar facilities. Another option is the provision of services, such as load shifting adapting to pricing/rates through Smart Grid function; connection with renewable energy sources; and energy exchange function. In addition, it allows a selection of ideal tariff systems appropriate to the construction standard and energy consumption of the installation [10] .

ANALYSIS OF THE CONDITION OF DEMAND RESPONSE

To meet peak demand, high-cost generating stations are required. Adding more generation was the strategy used in the past to meet the demand of electricity.

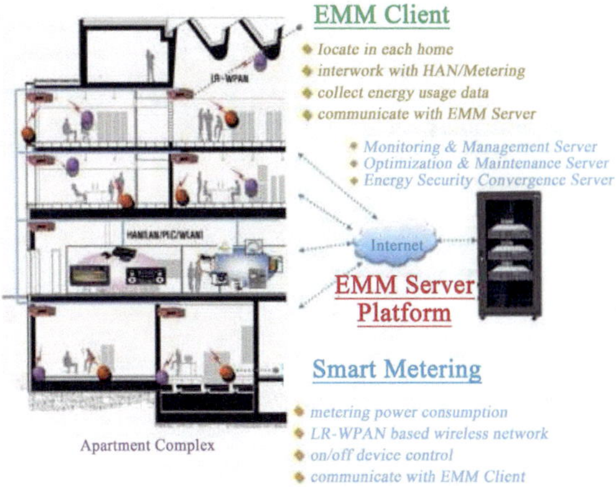

Figure 2: Systems configuration of EMM for residence [9] .

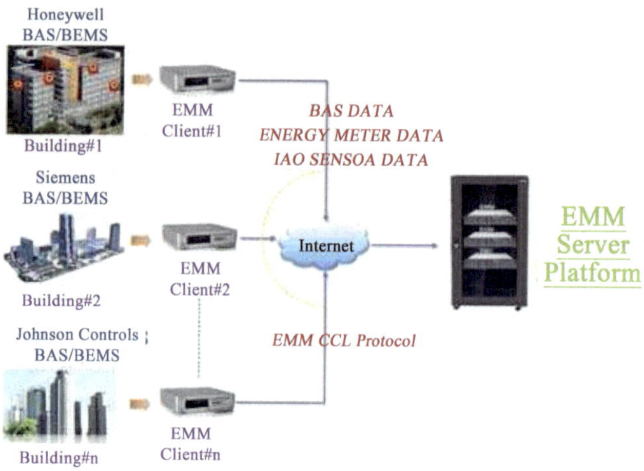

Figure 3: Systems configuration of EMM for integration with BEMS [9] .

Currently, the energy utilities have given more attention to demand management in order to reduce peak demand. Using the Smart Grid concept means more than a single technology or even a clear set of individual technologies for this [2] .

Demand response (DR) is a key concept in energy demand management, which helps to reduce peak demand in critical situations. DR is defined as the changes in the use of electricity, for end consumers, of their normal consumption patterns in response to changes in the price of electrical energy over time, or the incentive payments intended to induce a better use of electricity at peak hours [11] .

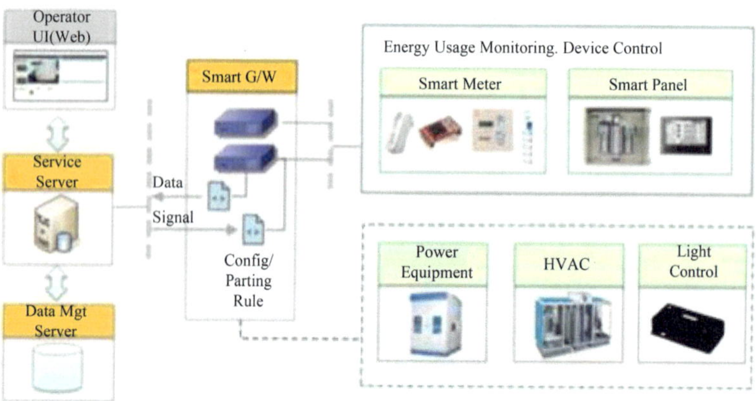

Figure 4: Performance software architecture [10] .

Load management is defined as a set of objectives which aim to control directly or indirectly and/or modify patterns of electricity consumption of various consumers, aiming to reduce peak demand. This control and modification enable the supply system to meet the demand, making better use of its available generation and transmission capacity [12] .

For leveling the peaks of demand, three common strategies for load management are used: "Peak Clipping", "Load Shifting" and "Valley Filling", as illustrated in Figure 5.

"Peak Clipping": Load reduction for short peaks and periods of use, usually performed by the direct load control. In this method, the energy utilities disconnect the consumer when there is a critical situation. This direct control can be used to reduce capacity requirements, operational costs and dependence on fossil fuel generation [12] .

"Valley Filling": Creation of loads during the peak period. This helps to reduce the average price of electricity. One of the methods used in industrial production, which uses the loads generated by fossil fuels [12] .

"Load Shifting": Moves the peak loads for other periods of time without necessarily changing the global consumption. This method combines the benefit of "Peak Clipping" and "Valley Filling" moving existing loads during off-peak hours [12] .

In programs of DR, electricity consumers play an important role in the reduction of peak demand during peak hours. Consumers can move their loads and thus help the energy utilities to prevent failures and blackouts in the electrical system, reducing the probability of stress conditions of the system. Improve energy security through the DR increases productivity and customer satisfaction. The DR also eliminates the need for high-cost generators and eventually reduces the cost of electricity [2] .

In order to inform consumers with real-time data, there must be a communication link between the energy utilities and consumers. Consumers must be able to measure their electrical energy demand, in real time, in order to act for demand response events. Advanced metering infrastructure implementations (AMI) and other technologies allows the user to measure the real-time energy demand and further enhance the use of resources of DR in daily operation [13] .

Therefore, it is evident that there is a need for an automatic energy management system in DR programs, which will provide more flexibility consumers.

CLOUD COMPUTING SOLUTIONS IN SMART GRID

The term cloud computing has many definitions; in scenarios as Smart Grid, it is defined as a model for enabling convenient, on-demand network access to a shared pool of configurable computing resources (e.g. networks, servers, storage, applications, and services) that can be rapidly provisioned and released with minimal management effort or service provider interaction. The characteristics of cloud computing include on-demand service, ubiquitous network access, location independent resource pooling, rapid elasticity, and measured service.

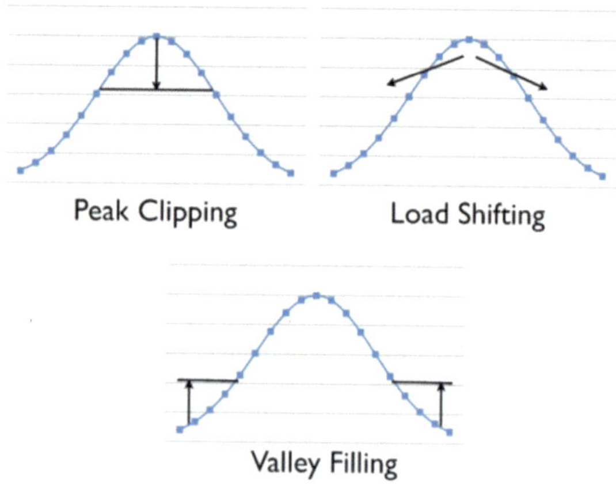

Figure 5: Strategies for load management [12] .

Cloud computing is not a universal solution. It has strengths and weaknesses, and understanding them is the key to making a decision about whether it is right for a particular application [14] . The main advantages of cloud computing are: Economy of scale: sharing of computing resources between different customers; Pay per use: customers pay for the service instead of buying software licenses and hardware; On-demand usage/flexibility: cloud services can be used almost instantly and can easily be scaled up and down; External data storage: customers' data is stored externally at the location of the cloud computing provider; Highly reliable services: clouds manage themselves in case of failures or the performance degradation [15] .

One of the most common applications for real-time data in manufacturing and process industries is SCADA, supervising remote processes over a network. With the growing popularity of cloud computing, many engineers and managers in the automation sector are looking at the possibility of using the cloud for SCADA. SCADA systems have evolved over time and have followed the progress of computing in general. As many view cloud computing as the next logical step in this evolution, enthusiastic visionaries foresee a fourth, "cloud" generation of SCADA, where an entire control system would be running in the cloud [16] . The cloud is creating a revolution in SCADA system architecture because it provides very high redundancy, virtually unlimited data storage, and worldwide data access all at a very low cost [17] .

The cloud computing can support SCADA applications in two ways: The SCADA application is running on-site, directly connected to the control network and delivering information to the cloud where it can be stored and disseminated. The control functions of the SCADA application are entirely isolated to the control network. However, the SCADA application is connected to a service in the cloud that provides visualization, reporting, and access to remote users. These applications are commonly implemented using public cloud infrastructures (PaaS cloud service). The SCADA application is running entirely in the cloud and remotely connected to the control network. The controllers are connected via WAN links to the SCADA application running entirely in the cloud. These applications are commonly implemented using private or hybrid cloud architectures (IaaS cloud service) [18] [19] .

Because of wide ranging variability of the entities in Smart Grids, there is a very high level of potential complexity in finding the optimal solution for each different Smart Grid. Smart Grid will eventually be deployed across all types of infrastructure using widespread Internet of Services, connecting all smart objects worldwide. It will become the major application domain of the Internet of Things, perhaps even referred to as the Internet of Energy [19] .

BARRIERS OF BRAZILIAN ELECTRICAL GRID

Energy management systems in residential, commercial, and industrial sectors can play an important role in reducing peak demand of electric network. Eventually, it can help in reducing overhead and stress on transmission and distribution lines. In many countries, there are various demand response programs implemented for the industrial and commercial sector. With the installation of energy management systems, it becomes possible to perform load control, mainly through the models of RTP (Real Time Pricing) and TOU (Time-of-Use) [20] .

Few demand response programs are used for management energy in the residential sector in Brazil. Direct restriction of loads is the most popular method used to reduce peak demand. But by direct control of the load, the consumer comfort can be compromised. In contrast, by the method of load displacement, the loads that have less impact to the consumer lifestyle are displaced outside of peak hours, ensuring a better stability of the network.

To analyze and consider the use of the concept of demand response is important to understand the load models "Hung" on the grid, for such function Smart Grid technologies play a fundamental role. This identification in conjunction with a proper communication between the consumer and the energy utilities and demand management of domestic charges are indispensable factors for high efficiency power management systems.

FINAL CONSIDERATIONS

Energy management systems, when developed in a context Smart Grid, have their functions enhanced, in regards of the technologies making up the system. Among the main features of an energy management system, the following stand out as the most enhanced:

- Real-time performance monitoring;
- Information to compose predictive maintenance planning;
- Energy management;
- The efficiency of the power system;
- Financial economics, avoiding any sanction from the energy utilities;
- To keep the tolerance of voltage and current to the extent allowed;
- Continuous monitoring for power quality;
- Load control and management with appropriate methodology;
- To keep consumption with the limit signed;
- Database of information that can assist future decision-making;
- Identify and correct the causes of energy disorder to avoid recurrences.

Among the main Smart Grid technologies, the most important and that directly impacts the design of an energy management system are the solutions on cloud computing. Cloud computing has established itself as an adequate means to provide resources to customers, primarily in energy management systems, with access to a large amount of information and computer storage. With cloud computing, customers do not have to manage and maintain their own information technology (IT), and are not bound to its local resources which often are limited. However, for customers and energy utilities, making sure that your cloud services are usable, an appropriate level of guarantees of Quality of Service (QoS) is needed.

In recent years, the creation of solutions in Data Center Networks (DCNS) came with rapid growth in scale and complexity, making possible hosting large applications, known as cloud hosting. This growth imposes enormous challenges to update the current datacenter infrastructure, especially considering a scenario of Smart Grid with cloud computing solutions, broadly used in energy management systems.

The proliferation of the adoption of cloud computing solutions in recent years is driven by the potential for obtaining benefits such as reduced costs, greater agility, and better use of resources. However, there are many challenges to ensure the success of these cloud-based services, and these need to be understood and managed before the major use in concepts such as Smart Grid.

However, the major current infrastructures are owned by a large number of Internet Service Providers (ISPs); and it is difficult to adopt new architectures without the agreement of all parties concerned. This includes the standardization of communication protocols and creating regulations for wide use of cloud computing solutions.

ACKNOWLEDGEMENTS

The authors would like to thank the Coordenação de Aperfeiçoamento de Pessoal de Nível Superior (CAPES), the Concelho Naional de Desenvolvimento Científico e Tecnológico (CNPq), the Fundação de Amparo à Pesquisa do Estado de São Paulo (FAPESP), the Departamento de Comunicações (DECOM), the Faculdade de Engenharia Elétrica e de Computação (FEEC), and the Universidade Estadual de Campinas (UNICAMP), for their support in the development of this research.

REFERENCES

1. Eissa, M. (2012) Energy Efficiency—The Innovative Ways for Smart Energy, the Future towards Modern Utilities.
2. Ireshika, M.A.S.T. (2014) Home Energy Management System. Universitetet I Agder, Kristiansand & Grimstad.
3. Khodaei, A., Shahidehpour, M. and Bahramirad, S. (2011) SCUC with Hourly Demand Response Considering Inter-temporal Load Characteristics. IEEE Transactions on Smart Grid, 2, 564-571. http://dx.doi.org/10.1109/TSG.2011.2157181
4. Du, P. and Lu, N. (2011) Appliance Commitment for Household Load Scheduling. IEEE Transactions on Smart Grid, 2, 411-419. http://dx.doi.org/10.1109/TSG.2011.2140344
5. Gatsis, N. and Giannakis, G.B. (2012) Residential Load Control: Distributed Scheduling and Convergence with Lost AMI Messages. IEEE Transactions on Smart Grid, 3, 770-786. http://dx.doi.org/10.1109/TSG.2011.2176518
6. Li, J., Chung, J.Y., Xiao, J., Hong, J.W.-K. and Boutaba, R. (2011) On the Design and Implementation of a Home Energy Management System. Proceedings of the 6th International Symposium on Wireless and Pervasive Computing, Hong Kong, 23-25 February 2011, 1-6. http://dx.doi.org/10.1109/iswpc.2011.5751338
7. Stephens, J., Wilson, E.J. and Peterson, T.R. (2015) Smart Grid (R) Evolution. Cambridge University Press, Cambri-dge.
8. Jiang, T., Yu, L. and Cao, Y. (2015) Energy Management of Internet Data Centers in Smart Grid.
9. Choi, C.-S., Ian, J.I., Park, W.-K., Jeong, Y.-K. and Lee, I.-W. (2011) Proactive Energy Management System Archi-tecture Interworing with Smart Grid. Proceedings of the IEEE 15th International Symposium on Consumer Electronics, Singapore, 14-17 June 2011, 1-4.
10. Park, K., Kim, Y., Kim, S., Kim, K., Lee, W. and Park, H. (2011) Building Energy Management System based on Smart Grid. Proceedings of the IEEE 33rd International Telecommunications Energy Conference, Amsterdam, 9-13 October 2011, 1-4.
11. Balijepalli, V.S.K.M., Pradhan, V., Khaparde, S.A. and Shereef, R.M. (2011) Review of Demand Response under Smart Grid Paradigm. Proceedings of the 2011 IEEE PES International Conference on Innovative Smart Grid Techno-logies-India, Kollam, 1-3 December

2011, 236-243. http://dx.doi.org/10.1109/iset-india.2011.6145388

12. Paracha, Z.J. and Doulai, P. (1998) Load Management: Techniques and Methods in Electric Power System. Proceedings of the International Conference on Energy Management and Power Delivery, Singapore, 3-5 March 1998, 213-217.
13. Medina, J., Muller, N. and Roytelman, I. (2010) Demand Response and Distribution Grid Operations: Opportunities and Challenges. IEEE Transactions on Smart Grid, 1, 193-198. http://dx.doi.org/10.1109/TSG.2010.2050156
14. Bashir, A.K., Ohsita, Y. and Murata, M. (2015) Abstraction Layer Based Distributed Architecture for Virtualized Data Centers. Proceedings of the Sixth International Conference on Cloud Computing, GRIDs, and Virtualization, Nice, 22-27 March 2015, 62-67.
15. Frey, S., Disch, S., Reich, C., Knahl, M. and Clarke, N. (2015) Cloud Storage Prediction with Neural Networks. Proceedings of the Sixth International Conference on Cloud Computing, GRIDs, and Virtualization, Nice, 22-27 March 2015, 68-72.
16. Mcilvride, B. (2012) Will SCADA Envolve to the Cloud? http://real-timecloud.com/
17. Combs, L. (2011) Cloud Computing for SCADA. http://www.controleng.com/single-ar
18. Conway, G., Carcary, M. and Doherty, E. (2015) A Conceptual Framework to Implement and Manage a Cloud Computing Environment. Proceedings of the Sixth International Conference on Cloud Computing, GRIDs, and Vir-tualization, Nice, 22-27 March 2015, 138-142.
19. Markovic, D.S., Zivkovic, D., Branovic, I., Popovic, R. and Cvetkovic, D. (2013) Smart Power Grid and Cloud Com-puting. Renewable & Sustainable Energy Reviews, 24, 566-577. http://dx.doi.org/10.1016/j.rser.2013.03.068
20. Albadi, M.H. and El-Saadany, E.F. (2007) Demand Response in Electricity Markets: An Overview. Proceedings of the 2007 IEEE Power Engineering Society General Meeting, Tampa, 24-28 June 2007, 1-5.

Chapter 7

Energy Efficiency and Renewable Energy Technologies Using Smart Grids: Study Case on NIPE Building at UNICAMP Campus

M. D. Berni, P. C. Manduca, S. V. Bajay, J. T. V. Pereira, J. T. Fantinelli

Interdisciplinary Center on Energy Planning (NIPE), State University of Campinas (UNICAMP), Campinas, Brazil

ABSTRACT

In its broadest interpretation, the smart grid vision sees the future of power industry transformed by the introduction of intelligent two-way communications, ubiquitous metering and measurement. This enables much finer control of energy flows and the integration and efficient use of renewable forms of energy, energy efficiency methodologies and technologies, as well as many other advanced technologies, techniques and processes that wouldn't have been practicable until present. The smart grid vision also enables the creation of more reliable, more robust and more secure power supply infrastructure, and helps optimize the enormous investments

Citation: Berni, M. , Manduca, P. , Bajay, S. , Pereira, J. and Fantinelli, J. (2014), "Energy Efficiency and Renewable Energy Technologies Using Smart Grids: Study Case on NIPE Building at UNICAMP Campus". *Smart Grid and Renewable Energy*, **5**, 193-197. doi: 10.4236/sgre.2014.58018.

required to build and operate the physical infrastructure required. The smart grid promises to revolutionize the electric power business that has been in place for the past 75 years. This work discusses the efficiency, targeted at the consumer units of electricity, with a view to sustainability and potential for technological innovation. The issue is addressed from two perspectives: the systems for generation and power distribution, and the design of a building "smart energy". Because of the novelty of the subject in our country, the concepts presented and treated throughout this work come from material obtained at events and specialized sites on electric power system in Brazil and worldwide, being accompanied by information and data from NIPE's building at University of Campinas's campus case study in which it exemplifies the applicability of the techniques and recommended technologies.

Keywords: Smart Grids, Energy Efficiency, Renewable Energy, Smart Building, Generation and Distribution System, Decentralized Generation

INTRODUCTION

The ongoing structural and technological changes make it possible to expand the renewable energy use, decentralized generation and energy efficiency in order to supply the demand for electricity. In this way, besides the fundamental concepts of operation, transmission and distribution knowledge on electrical systems, it requires to include topics such as information and communications technologies, and signal processing to improve energy used by smart grid (SG). SG is becoming a reality in many developed countries like United States, Japan and Germany with the implementation of pilot projects countries and improvement actions in their supplying networks [1] . Brazil's Federal Government establishes the guidelines of the Brazilian Program of Smart Grids (PBREI) through the National Electric Energy Agency (ANEEL) [2] and the 482/2012 Resolution paves the way to replace 67 million conventional electricity meters for smart meters. Universities, corporations, NGOs, government and society worldwide had been concerned with energy issues. The energy field involves many aspects of social life including the background on engineering, environment, logistics, sources, and, more recently the use of information technology (IT) "intelligence" to optimize and manage power systems, adjusting supply, demand and efficiency in the use of electricity. Modern life has made electricity an increasingly vital product. In any segment such as production of goods or services for the public safety, health or simply for the comfort of homes, electricity is an indispensable element. However, its intense use stresses the system and request for more power production. Whereas the world is generating hydro, an equally increasing impact on the environment has been seen. Environmental issues have reached their limit and have the most importance for the mankind survival and that is the reason why to think of solutions that reconcile energy production and environmental preservation. In this context, the point of view of the power distribution system and the future of the consumer unit, covered by SG concept, will be prioritized. Utilities and consumers will converge their interests and make use of new technologies to achieve energy efficiency, reliability on electricity distribution systems, decrease cost and reducing environment and natural resources. This work discusses the efficiency, targeted at the consumer units of electricity, with a view to sustainability and potential for technological innovation. The issue is addressed from two perspectives: the systems for generation and power distribution, and the design of a building "smart energy". Furthermore, this work discusses a case study on NIPE Building at UNICAMP campus, which has good values to the scientific community and contributed new information to the related field. Finally, Table 1 provides an example project approach for analysis energy efficiency and smart grid for this case study.

RENEWABLE ENERGY AND ENERGY EFFICIENCY ON BRAZILIAN CONTEXT

In Brazil energy efficiency is clearly less important than the addition of "new energy" to the grid, despite the great potential of reducing energy intensity of the Brazilian GDP and the recent successful experience in increase energy conservation during the blackout in 2000. Making energy efficiency the key topic in the whole society should be priority in the government agenda. The little importance given to the subject in the Ten Year Energy Plan in 2019 [3] , neglects the fact that investments in the area are smaller and faster return and therefore should be highlighted in the policy and government plans, especially when the country should observe growth of energy demand of 54 GW over the next ten years.

Renewable energy together with energy efficiency alternatives should be considered because the wide social and environmental benefits that often result to generate electricity. Brazil has several options for generating clean and competitive energy for its expansion: hydropower, biomass, wind and solar power. In less developed countries, there are also ocean energy and geothermal energy. But as important as increasing the supply of renewable energy is to increase the efficiency of energy consumption generated from renewable sources or not in the economic fields. Moreover, the improving of energy efficiency can occur in more narrow term to transition to a more renewable energy sources, which will not happen abruptly, since the dynamics that sustain the current non-renewable energy model cant not be easily reversed for many reasons such as: (i) the high level of material consumption and energy in an emerging country, (ii) the non-renewable energy infrastructure already established, (iii) the growing demand for electricity services by applicants, and (iv) population growth.

SMART GRID AND ENERGY EFFICIENCY

SG represents the application of IT in the integrated communication electrical system. This technology involves installing sensors on the lines of the electric power grid, embedded with chips that detect data on the operation and performance of the network like voltage and current.

Table 1: Phases in a sustainable building retrofit

Phase 1	Phase 4
Project setup	I & C
definescopeofwork	implementation
availableresources	comissioning
prediagnostic	Phase 5
Phase 2	Validation
Energy auditing	verification M & V
selectindicators	
measures	
finaldiagnostic	SUSTAINABLE
Phase 3	BUILDING
Retrofitoptions	
energysaving	energyefficiency
economicanalysis	and
energygenerationon site	smart grid

The devices analyse those information to determine what is significant. For example, if the voltage is too high or too low. When the sensors detect significant information communication data to a central analytical system where it will analyse them and determine what is wrong and what should be done to improve network performance occurs. In the case of very high voltage, the software will instruct one of the devices already installed in the network to reduce the voltage, thus saving energy.

SG technology has been benefits with fuel efficiency, which means less energy the utility company to provide equal or better quality of service to its customers; increases the reliability of the electric system, reducing costs and carbon emissions. The SG detects when the assets of a network fail or are with declining performance, will identify them to the concessionary can repair them or replace them before there is a real power outage, but also allows you to isolate the impact of a failure to customers, so that fewer customers are affected when there is a power failure. Lastly is the integration of cutting edge , ranging from reading a smart meter system to interact with the client's management at home, solar panels , which require interaction with the network to achieve success [4] . Furthermore, the SG lever distributed generation from renewable sources, in that power generation can be carried out at or near the independent power consumer, technology and energy source. The main advantage of distributed generation systems is the savings in investments, transmission and reduction in losses in these systems, enhancing the stability of electric power service and increases energy efficiency.

With the SG tool it can relax the retrofits. In the energy context, retrofit is used to define changes and upgrades in systems generators and consumers of electricity aimed at their conservation. This type of application occurs in power plants with reform or adding equipment to increase efficiency, production and life. A common case is the retrofit of boilers in power plants. Another important example of retrofit is the object of the case study presented below in the NIPE's building at University of Campinas's campus. Retrofit buildings is normally associated with the change in lighting systems, electrical and plumbing installations through energy efficient communication technologies and advanced quality. In addition to improving energy efficiency, retrofit helps to reduce emissions of greenhouse gases over the life cycle of buildings.

RETROFIT BUILDINGS: BRIEF DESCRIPTION OF THE CASE STUDY PROGRESS IN BUILDING NIPE-UNICAMP

Campus at UNICAMP as same micro representatives of society contains buildings with a variety of functions and uses, each with their own characteristics in the consumption of energy, water and waste. As a task to become "good examples" for society in general, universities need to develop "best practices" to make sustainable buildings, both in the rehabilitation of buildings and in new buildings. Sustainable buildings on campus may have educational goals for college students and the academic community, reflected in potential "best practices" for society at large gains.

Energy efficiency and renewable energy in the built environment stand out as one of the main alternatives to minimize problems arising from climate change. The retrofit project of NIPE-UNICAMP will demonstrate concrete gains on rehabilitation of a building live "post-occupation" by proposing actions to saving energy, maximizing the educational impact of these actions and awareness through interaction with students and the university's community, under the supervision of experts. Participatory nature of the project will allow the perception of specific possibilities of a green building among users of the Campus, increasing local knowledge to be adapted to the "perfecting" of the sustainability process extra-campus. The project will share know-how to process a wider spread in the society. Against this background, the NIPE-UNICAMP project deals with the theme "Smart Grid and Sustainable Buildings" with the main objective of the study and evaluation of opportunities for energy efficiency and comfort environmental post-occupancy building for research, considering the surroundings, air conditioning system, lighting type and acoustic quality, based on the evaluation of sustainability of the built environment through possibilities INMETRO nota certification. Therefore, it is a survey of constructive characteristics, architectural and the thermal NIPE-UNICAMP, followed by the determination of energy needs through software simulation, and finally evaluating the results and proposing alternatives for energy optimization and environmental comfort NIPE through the retrofit of the building interventions from the sustainability approach, involving the energy performance, thermal quality , lighting quality and acoustics quality, should integrate the agendas of pilot projects and retrofits that begin to be stimulated and developed by various national and international organizations that study the impacts of the sector building and construction for society and the environment [5] .

There are many softwares available for the study and evaluation of opportunities for energy efficiency and environmental comfort, and mention: i) DOE2 (EQUEST), ii) Design Builder, iii) ECOTEC, iv) TRNSYS, v) and vi) Energy Plus and ESP-r, among others [6] . Besides these, there RET Screen software support for decision making opportunities for energy efficiency with a focus on renewable energy to a building. This is free software, provided by the Government of Canada, which helps the decision maker to identify and access potential energy projects that bring reduction in energy intensity of a building, including its technical and economical viability [7] . Regarding the thermal comfort of the environment, their qualification and quantification requires making measurements, for example if the air temperature and the Globe. From measurements of the thermal evaluation can be performed by the Predicted Mean Vote (PMV), while simulations with 2:03 Comfort software method. Regarding the study and analysis of the luminal comfort, it has been alternatively windsurfing and Reluxcad and finally, evaluation of acoustic comfort in the Brazilian case can be according to NBR 10152 [8] .

FINAL CONSIDERATIONS

There is a strong correlation between energy consumption, environmental comfort and the life cycle of a building. Indicators of energy intensity throughout the life cycle analysis of a building provide a way to understand the evolution of that correlation. The energy intensity can be reduced in two ways. First, greater energy efficiency can reduce the energy consumed to produce the same level of energy services (for example, a more efficient light bulb produces the same light with less energy consumption). Second, the issues surrounding sustainability, markets and peer pressure, end up imposing changes in energy intensive activities, such as the search for the lowest energy consumption activities, greater comfort and environmental activities and/or less carbon intensive process. Energy efficiency assisted by replacing fossil energy by renewable and sustainability criteria in the lifecycle is the key to driving incremental reduction in energy intensity and can offer solutions as diverse as climate change, energy security, competitiveness, and human being and socio-economic development.

M. D. Berni, P. C. Manduca, S. V. Bajay, J. T. V. Pereira, J. T. Fantinelli (GHG) is 30% [9] . The final energy consumption in Brazil, only for residential household, more public trade, represents 47.1% of the total consumption of energy sources [10] . The evaluation of energy efficiency and environmental comfort in a built environment, post-occupancy, as NIPE-UNICAMP, is requesting a careful study to obtain data that allow integrate simulations and diagnostics to make decisions.

REFERENCES

1. IEEE PES (2013) http://smartgrid.ieee.org/resources
2. Silveira, P.M. and Ribeiro, P.F. (2012) Introdução do Conceito de Redes Elétricas Inteligentes no Currículo do Engenheiro Eletricista Brasileiro. XL Congresso Brasileiro de Educação em Engenharia, Belém, Pará, Brazil.
3. EPE (2013) Empresa de Pesquisa Energética, Plano Decenal de Energia 2019, Ministério de Minas e Energia. http://www.epe.gov.br/imprensa/PressReleases/20100504_2.pdf
4. GVces, Centro de Estudos em Sustentabilidade, Fundação Getulio Vargas (FGV) (2011) Fontes de Energia e Eficiência Energética, Plataforma Empresas pelo Clima, Oficina de Trabalho IV, 17 p.
5. UNEP-SBCI (2012) United Nations Environment Programme—Common Carbon Metric: Protocol for Measuring Energy Use and Reporting Greenhouse Gas Emissions fro Building Operations. www.unepsbci.org
6. EERE (2013) Energy Efficiency and Renewable Energy, Building Energy Software Tools Directory. http://apps1.eere.energy.gov/buildings/tools_directory/alpha_list.cfm
7. (2012) http://www.retscreen.net
8. Kowaltowski, D.C.C.K., Fávero, E., Borges Filho, F., Gouveia, A.P., Ruschel, R.C., Pina, S.A.G. and Gomez, V.S. (2001) Ensino do Projeto Arquitetônico: A Teoria Traduzida em Exercícios no Ensino do Projeto Arquitetônico. Revista da Escola de Minas, Ouro Preto, Vol. 54, 1-6.
9. UNEP-SBCI (2013) United Nations Environment Programme—Sustainable Building & Climate Iniciative. www.unepsbci.org
10. EPE-BEN (2012) Balanço Energético Nacional ano Base 2011, Empresa de Pesquisa Energética (EPE), Ministério de Minas e Energia (MME), Brasília, DF.

Chapter 8

Towards Attaining Reliable and Efficient Green Cloud Computing Using Micro-Smart Grids to Power Internet Data Center

Mohammed Mansur Ibrahim[1], Anas Ahmad Danbala[2], Mustapha Ismail[3]

[1]Department of Mathematics & Computer Science, Federal University of Kashere, Kashere, Gombe State, Nigeria
[2]Trans. Access Planning Engineer, Mobile Telecom. Network (MTN) Abuja Switch Office, Abuja FCT, Nigeria
[3]Department of Mathematics & Comp. Sci., Gombe State University, Gombe, Gombe State, Nigeria

ABSTRACT

Energy generation and consumption are the main aspects of social life due to the fact that modern people's necessity for energy is a crucial ingredient for existence. Therefore, energy efficiency is regarded as the best economical approach to provide safer and affordable energy for both utilities and consumers, through the enhancement of energy security and reduction of energy emissions. One of the problems of cloud computing service

Citation: Ibrahim, M.M., Danbala, A.A. and Ismail, M. (2019), "Towards Attaining Reliable and Efficient Green Cloud Computing Using Micro-Smart Grids to Power Internet Data Center". *Journal of Computer and Communications*, **7**, 195-205. doi: 10.4236/jcc.2019.77016.

providers is the high rise in the cost of energy, efficiency together with carbon emission with regards to the running of their internet data centres (IDCs). In order to mitigate these issues, smart micro-grid was found to be suitable in increasing the energy efficiency, sustainability together with the reliability of electrical services for the IDCs. Therefore, this paper presents idea on how smart micro-grids can bring down the disturbing cost of energy, carbon emission by the IDCs with some level of energy efficiency all in an effort to attain green cloud computing services from the service providers. In specific term, we aim at achieving green information and communication technology (ICT) in the field of cloud computing in relations to energy efficiency, cost-effectiveness and carbon emission reduction from cloud data center's perspective.

Keywords: Cloud Computing, Internet Data Center, Green IT, Energy Efficiency, Mi-cro-Smart Grids

INTRODUCTION

Cloud computing is an area of interest in the field of information and communication technologies (ICT) due to the minimisation of operational cost associated with it. Many organisations such as business, government and academic institutions utilised cloud services nowadays from software, a platform to infrastructure to have proper and flexible management of multiple IT resources [1] . However, due to extensive internet coverage across the globe, it became imperative to directly focus on green technologies to achieve minimum energy consumption concerning IT resources towards environmental protection. Cloud computing appeared to be the solution towards achieving green ICT without compromising the quality of internet services. The technology offers tremendous benefits to the users in terms of having their considerable resources stored and managed by the service providers instead of managing their systems [2] . Additionally, economic and environmental issues related to cloud computing with respect to power consumption can be solved through micro grid application, thus, through providing information exchange that is extensive intra- and inter-utility together with numerous real-time information in a cost-effective manner [3] . The smart grid is an enhancement of the traditional power grid whereby it utilises two-way communication of the flow of electricity to produce automated and distributed advanced network for energy distribution [4] . This paper focuses on how to mitigate issues of high energy consumption, cost and environmental pollution related to activities of IDCs to achieve green cloud computing service.

The paper first introduces the paradigm of green IT together with concern areas, and that is followed by cloud computing, models and services offered by the cloud. The paper then introduced smart micro-grids in brief and most importantly, how the concept of green IT in the area of IDC for cloud computing could be achieved with the smart micro-grids' application. The paper also carefully studied the existing practice based on the perspectives of energy and cost efficiency together with a carbon emission of the IDCs and projected a convincing solution to IDC issues highlighted.

GREEN IT

Green IT is a new computing model that transformed IT resources into energy consumption efficiency concerning cost and power. The IT resources in interest here are the IDCs and other IT facilities. The application areas for green IT includes proper power management, server visualisation, design of data centers plus eco-labelling for IT products. Other important areas include environmental sustainability plus the design and energy efficient resources [5] . While [6] defined green IT as a "study area that deals with designing, manufacturing, using and disposing of computers, servers and associated sub-systems such as monitors, printers, storage devices and networking and communication systems efficiently and effectively with minimal or no impact on the environment". The idea of green IT aims at realising the cost-effectiveness of IT infrastructure by improving systems performance and at the same time respecting social and ethical responsibilities. Additionally, green IT definition suggests the inclusion of green IS to have a complete description. Based on that, green IT is defined as a combination of people, processes together with software and information technologies to support different institution or societal set goals. This definition emphasised that both green IT and green IS can bring about tremendous and realistic initiatives to promote the sustainability of business processes and add to a different aspect of organisational responsibilities together with private green IT initiatives.

However, in general term green IT is regarded based on the perception of "environmental sustainability, the economics of energy efficiency in computing, power management, data centre design and development. Likewise, virtualisation of server, proper recycling method and disposal, complying with regulatory bodies in terms of green metrics; assessment tools and methodology, environmental-related risk mitigation use of renewable energy resources and eco-labelling of IT products" [7] . Therefore, it is imperative to have a good understanding of these green IT concern areas in order to mitigate the issues. Below subsection present a foundation for a good understanding of these green IT concern areas.

1) Concern Areas of Green IT

There are numerous concern areas in the field of green IT that attracted much attention, and a lot of research was carried out in these areas to achieve an environmentally friendly and cost-effective IT resource management. The most concern areas that were given much attention recently are power management and energy efficiency. While areas like sustaining the natural resources by providing a green scenario with cost-efficiency, carbon emissions reduction, virtualization of servers and management of servers were less focused in the past and hence require much attention in order to overcome the challenges associated with these areas [5] [6] Figure 1. below presents the major green IT concern areas.

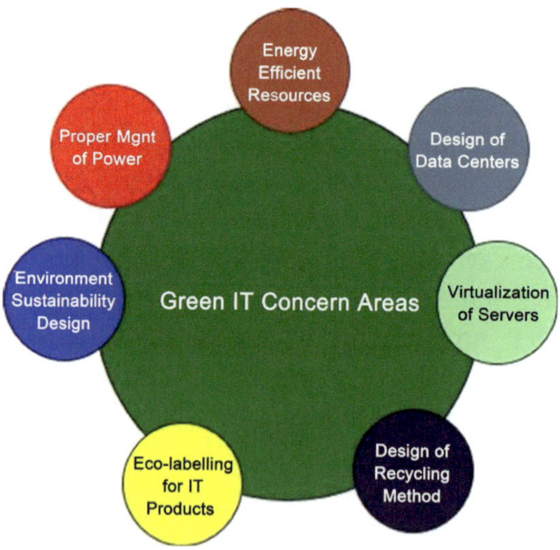

Figure 1: Concern areas of green IT.

2) Green Computing

More so, have explained the concept of green IT; it is important to have a good understanding of the term green computing as well. The idea of green computing means the way and manner computers and other IT resources are used with high regard of the environment to achieve energy efficiency, the proper energy consumption of resources with minimal electronic waste. Technically, green computing involves two aspects which are as follows;

- In terms of software technology with a particular focus on enhancing storage, program and energy efficiency.
- In terms of hardware with interest in technologies with lower energy consumption together with economic efficiency in respect of recycling [5] . However, have gone through the field of green IT, it's now imperative and ideal to discuss the concept of cloud computing as a whole.

CLOUD COMPUTING

It means accessing computing product and services via the cloud powered by external parties through a remote server [8] . In another perspective cloud computing referred to applications deployed as a service via the internal network with a physical data centre responsible for implementing and managing these services [9] . While [10] [11] defined cloud computing as a subsidiary of grid computing and the model is based on pay-per-use. It provides efficient use of energy, affordable price technologies that help when accessing, sharing of services together with storage and management of resources. Of course, the idea has numerous advantages which among them are; network access, infrastructure sharing, reduction of cost of operations, maintainability, reliability, flexibility together with different service, and so on. Figure 2 depicts the concept of cloud computing model or services.

Figure 2: Cloud computing scenario [8] .

Additionally, it is a storage, network and another vital aspect of computing services that can be deployed and run arbitrary software which includes in both operating system and applications. Also, here the user has no business with management and control of cloud infrastructure rather the end user can have control over the operating systems applications, storage and some limited control over specific networking components like host firewalls [1] [5] [11] [12] .

1) Deployment Models of Cloud Computing

The whole idea of cloud computing is to offer on-demand services to end-consumers. There are three different deployment models namely; Public cloud, Private cloud and hybrid cloud [2] below details on the explanation of these three deployment models.

2) Public Cloud

The model is designed in such a way that any user can access it based on pay-as-you-go manner and of course hosted by the internet. The famous public clouds are as follows; Amazon Web Services (AWS), Google App Engine, Microsoft Aure and all of them support the three cloud computing services namely; IaaS, SaaS and PaaS. For instant, Google App Engine is a public cloud offering application development platforms, and Salesforce.com is also a public cloud that provides software as a service while Amazon EC2 is a public cloud that gives infrastructure as a service. Public cloud is basically for commercial purposes [2] [5] .

3) Private Cloud

The private cloud serves a particular enterprise, and the responsibilities of hosting the network lie in that specific enterprise. Cloud infrastructure is managed by the enterprise together with the network design to serve one or fewer more organisations. Part of its benefits is that it allows greater access and control over the infrastructure with much security since its access is limited to one party [2] [5] .

4) Hybrid Cloud

However, hybrid cloud emerged from public and private clouds diffusion. The model allows organisations to outsource less priority information processing to the public cloud while keeping classified services and data under their control. It takes into account three computing services which are stored as a service, processing as a service and software as a service. Storage service permits the user to save their data safely on the cloud, processing as service allows the user to outsource certain computational services while software as a service gives the users the ability to merge the above two services and outsource them all to the cloud with little utilisation of processing power terminal home. Hybrid cloud basically can move data across the two mentioned clouds through their interfaces [2] [5] .

Cloud Computing Services

Services of the cloud mean a chunk of services powered by the data centres via the cloud with the use of the internet. These services are categorized into three major categories, namely; Platform as a Service–PaaS, Software as a Service–SaaS and Infrastructure as a Services–IaaS. Although there is a fourth category communication as a service but is being regarded as a subset of software as a service and it's mainly related to industries that used hosted services of IP telephony [5] .

a) Platform as a Service—PaaS: The ability for the end user to access or acquired applications platforms which are being developed using programming languages libraries, services and tools powered by the cloud owner. The user has no business with the maintenance of cloud infrastructures such as servers, network, storage and operating systems but can manage the powered applications with the ability to configure settings for the applications with hosting environment. In a simple term, PaaS comprises programming environment and execution environment, e.g. commercially Google Application Engine [1] [11] [12] .

b) Software as a Services—SaaS: SaaS deals with different applications required by many end users across a different geographical location for their daily routine work. Example of such services is words processors and spreadsheets. End-User doesn't manage or control the cloud in terms of network servers, operating systems, storage and individual application access but can have some limited user-specific application configuration settings [1] [5] [11] [12] .

c) Infrastructure as a Service—IaaS: Cloud services developed on top of bare hardware which is achieved by the use of virtualization technologies, the model offered a user with services such as processing,

Subsequently, this paper would focus on how to mitigate issues related IDCs that power these cloud computing services in terms of high energy consumption and related environmental pollution for an environmentally friendly IDC.

A PROPOSED SOLUTION TO ALARMING HIGH ENERGY CONSUMPTION AND RELATED ISSUES OF IDCS

1) Micro Smart Grids

Micro-Smart grids evolved out of microgrids environment as a smaller version of today's big centralised power systems. It is capable of generating, distributing and regulating the flow of electricity to consumers. Smart micro-grids can form a network with one another as well as a central grid to add to capacity, efficiency as well as reliability. Due to the fact that micro-grids are not often owned and operate by a utility company, that allows micro-grids to be built, owned and run by an entity, be it community, neighborhood, university, corporation, hospital, individual or a corporate body with absolute legal right over their power infrastructure (i.e. power transmission lines, meters, generation source etc.) [13] . There are two principal operational modes in smart micro-grids, i.e. islanded mode and grid-connected mode. Operation is said to be in islanded mode when SMGs supply their loads utilising different energy resource, e.g. energy storage devices, back up with renewable generators. SMG can transact with the primary grid in terms of buying or selling of energy. Other essential components of SMG include generation systems, load, energy storage systems (ESS) and of course, the energy management system (EMS). The generation part comprises of many generators and a conventional one serving as back up in case of failure. While EMS has to do with energy scheduling for another component in SMG, and the overall load in the SMG is required by the data centre to process the service request distributed from the front-end servers [14] .

2) Efficient and Reliable Cloud Computing Application embedded with Smart grid

Generally, energy has been considered to be playing a vital role in the shaping of the human condition. Energy production and consumption are the main activities of social life due to of 21st-century people's necessity for energy as a fundamental for existence. Hence, the standard of living and quality of a civilisation is directly dependent on the quantity of energy a society uses. Energy efficiency is regarded as the best conservative approach to provide safer and affordable energy for both utilities and consumers through the enhancement of energy security and the reduction of energy emissions. With the exponential growth in the power industry, there is a need for continuously vast and real-time computing and storage capacity. The amount of these resources will exhibit a uniformly distributed growth at all levels of the grip; as such, the importance and significance of the cloud model come into play. The scalable and flexible characteristics of Cloud computing, with its capability to handle large amounts of data, make it the most straight forward and best-suited way to this kind of smart grid applications. The development of a smart grid requires large-scale real-time computing capabilities for the processing of communication, the transportation and also storage of huge amount of transferable data. Cloud computing helps to unbundle the smart grid by providing automatic updates, remote data storage and also reduced maintenance of IT systems, thereby saving money, manpower and energy [15] . Several problems have been encountered in the conventional smart grid architecture which does not have cloud integration. Some of these issues include the following below;

- Cyber-attacks such as Distributed Denial of Service could arise from the Master-Slave architected grid (without cloud).
- If there is any failure in the Master-Slave architecture, the system could also fail.
- The limited server capacity can only serve for several users (customers).
- Limited memory and storage will pose a significant problem when serving a large number of customers.
- There is a need for management and stability issues [16] .

However, scholars have proposed several solution concepts and approaches regarding demand response and micro-grid management and efficiency. The implementation of dynamic pricing is a basic approach that can be used to address energy management. Additionally, the integration of cloud into the microgrid system will make it possible to schedule incoming jobs to be carried out by the availability of resources, their order of priority and other applicable constraints. In the course of peak hours, there are more messages from the smart meters than those in the non-peak hours. Nonetheless, at this moment, scheduling of incoming jobs from users takes place according to their order of priority, availability of resources and applicable constraints. These issues can be addressed conveniently by the integration of dynamic bandwidth allotment mechanism using cloud application. The allotted bandwidth during peak-hour is higher than that in the non-peak hour, which serves all the incoming jobs simultaneously [17] .

3) Energy Consumption, Cost-Efficient and Carbon Emissions Reductions of Cloud Computing IDCs with Smart Microgrids.

In cloud computing, internet data centres are the backbone of the operation, which is also responsible for high energy consumption together with cost and carbon emission at the same time. To realise a green concept in the area of data centres, Smart-Micro-grids application would support lower electricity distribution loss, higher utilisation of ratio concerning energy cost and free carbon emission to the environment. One of the concerned areas of green IT in cloud computing is internet data centres, due to it high power consumption and bills associated with it, for example, Google annual power consumption is about 6.3 * 105 MWh which is about $38 million respectively, in the year 2010 google consumed 2260 MWh translating into more than 1.35 billion dollar electricity bill. Another concern area of green IT concerning cloud computing is the alarming environmental impact associated with internet data centres, for example in the year 2008 carbon emission of data centres was 0.6% of the global carbon emission, and this proportion of 0.6% was projected to reach 2.6% by 2020. However, for a data centre to achieve social responsibility, these mentioned problems need to be addressed together. More so, in a quest to mitigate these issues associated with internet data centres IDCs, smart micro-grids evolved from smart grid environment. Smart micro-grids have the potential of providing fault isolation and easy kind of distributed generation handling. And without a doubt there are many benefits of running internet data centres with smart micro-grids which are as follows;

- Cooling power consumption can be achieved through direct current microgrids by reducing the distribution loss, which in turn would lower the cost of energy.
- Transaction of power between smart microgrids and the primary grid can reduce waste of renewable energy and lower down carbon emission.
- In case of power failure from the central grid, smart microgrids would pick up the power operation in islanded mode. Hence the rate of carbon emission can also be exploited to minimise energy cost and bring down the carbon emission.

However, in the quest to move towards achieving a socially responsible IDC, operations must focus on cutting energy cost and carbon emission reduction simultaneously and that we believed can be realised through the use of smart microgrids. On the real scenario of green IDC, the operator on that regards would have in place different IDCs in self-owned SMGs with independent electric power regions (ERs) [14] [15] [18] [19] [20] .

Another way of addressing energy management and energy efficiency is by adopting the concept of Cloud-Based Demand Response (CDR) for getting fast response times in large scale deployment. The CDR presents a master/slave demand response model whereby both the smart meters and the Home Energy Management System (EMS) act slaves, while the utility acts as the master. In this situation, the CDR controls data-centric communication, subscriber/publisher and topic-based group management, rather than IP-centric communication. Furthermore, two cloud-based demand response models are to be established, namely as follows:

a) data-centric communication.

b) topic-based group communication.

The CDR approach can render a secure, reliable and scalable demand response. However, the demand response has an overhead problem in the cause of implementing a private cloud for a small-sized network, which may include implementation cost and the choosing of an appropriate strategy [21] . Therefore, this study is proposing a solution to the said issues via model design with a mathematical algorithm to support the model in order to achieve a practical solution to the issues.

CONCLUSION

The paper presents an idea of how green cloud computing can be attained with regards to alarming energy consumption and carbon emission of cloud data centres. Smart micro-grid was found to be suitable and reliable in mitigating such issues with a high sense of optimality. Although achieving green IT in cloud computing doesn't stop at addressing IDC issues alone. Instead, it requires full attention in other green IT concern areas such as virtualisation of servers, proper power management and environmental sustainability design concerning IT products. And to realise a comprehensive green cloud computing service, more research should focus on the areas mentioned above. However, to make this idea possible, we recommend a model design with a possible mathematical algorithm to support and validate the model.

ACKNOWLEDGEMENTS

Much gratitude to our lecturer Assoc. Professor Mehmet Toycan, Director Telecommunication Research Center Cyprus International University via Mersin 10 Turkey, for his valuable comments and guidance while developing this paper. And also, to all the reviewers, we say thank you for finding the time to review and recommend some improvement on this research paper.

REFERENCES

1. Da Cunha Rodrigues, G., Calheiros, R.N., Guimaraes, V.T., Santos, G.L., de Carvalho, M.B., Granville, L.Z., Tarouco, L.M. and Buyya, R. (2016) Monitoring of Cloud Computing Environments: Concepts, Solutions, Trends, and Future Directions. In: Proceedings of the 31st Annual ACM Symposium on Applied Computing, ACM, New York, 378-383. https://doi.org/10.1145/2851613.2851619
2. Pandya, S.S. (2014) Green Cloud Computing. International Journal of Information and Computation Technology, 4, 431-436. http://www.irphouse.com/ijict.htm
3. Vaccaro, A., Popov, M., Villacci, D. and Terzija, V. (2011) An Integrated Framework for Smart Microgrids Modeling, Monitoring, Control, Communication, and Verification. Proceedings of the IEEE, 99, 119-132. https://doi.org/10.1109/JPROC.2010.2081651
4. Xi, F., Misra, S., Xue, G. and Yang, D. (2012) Smart Grid—The New and Improved Power Grid: A Survey. IEEE Communications Surveys & Tutorials, 14, 944-980. https://doi.org/10.1109/SURV.2011.101911.00087
5. Patel, Y.S., Mehrotra, N. and Soner, S. (2015) Green Cloud Computing: A Review on Green IT Areas for Cloud Computing Environment. 2015 International Conference on Futuristic Trend in Computational Analysis and Knowledge Management, Noida, India, 25-27 February 2015, 327-332.
6. Mithas, S., Khuntia, J. and Roy, P.K. (2010) Green Information Technology, Energy Efficiency, and Profits: Evidence from an Emerging Economy. ICIS 2010 Proceedings. Paper No. 11.
7. Murugesan, S. (2008) Harnessing Green IT: Principles and Practices. IT Professional, 10, 24-33. https://doi.org/10.1109/MITP.2008.10
8. Computing, C. (2011) Cloud Computing Privacy Concerns on Our Doorstep. Communications of the ACM, 54, 36-38. https://doi.org/10.1145/1866739.1866751
9. Armbrust, M., Fox, A., Griffith, R., Joseph, A.D., Katz, R.H., Konwinski, A., Lee, G., Patterson, D.A., Rabkin, A., Stoica, I. and Zaharia, M. (2009) Above the Clouds: A Berkeley View of Cloud Computing. Technical Report No. UCB/EECS-2009-28, University of California at Berkeley, Berkeley, CA.

10. Buyya, R., Yeo, C.S., Venugopal, S., Broberg, J. and Brandic, I. (2009) Cloud Computing and Emerging IT Platforms: Vision, Hype, and Reality for Delivering Computing as the 5th Utility. Future Generation Computer Systems, 25, 599-616. https://doi.org/10.1016/j.future.2008.12.001
11. Zhang, L., Zhang, L., Ge, M.-H. and Bi, X.-L. (2011) Research and Application on the Cloud-Computing-Based Power Dispatching IT Architecture. 2011 Asia-Pacific Power and Energy Engineering Conference, Wuhan, 25-28 March 2011, 1-4. https://doi.org/10.1109/APPEEC.2011.5748813
12. Mell, P. and Grance, T. (2011) The NIST Definition of Cloud Computing: Recommendations of the National Institute of Standards and Technology. Computer Security Division Information Technology Laboratory National Institute of Standards and Technology, Gaithersburg, MD, 20899-8930.
13. Liang, Y., Jiang, T. and Zou, Y. (2016) Real-Time Energy Management for Cloud Data Centers in Smart-Microgrids. IEEE Access, 4, 941-950. https://doi.org/10.1109/ACCESS.2016.2539369
14. Popeang, J. (2012) Cloud Computing and Smart Grids. Database Systems Journal, 3, 57-66.
15. Liang, Y., Zou, Y. and Wang, K. (2016) Online Energy Management for Data Centers and Electric Vehicles in Smart Grid Environment. 42nd Annual Conference of the IEEE Industrial Electronics Society, Florence, Italy, 23-26 October 2016, 5415-5419.
16. Naveen, P., Wong, K.I., Danquah, M.K., Sidhu, A.S. and Abu-Siada, A. (2016) Cloud Computing for Energy Management in Smart Grid-An Application Survey. IOP Conference Series: Materials Science and Engineering, 121, Article ID: 012010. https://doi.org/10.1088/1757-899X/121/1/012010
17. Li, X. and Lo, J.-C. (2012) Pricing and Peak Aware Scheduling Algorithm for Cloud Computing. 2012 IEEE PES Innovative Smart Grid Technologies, Washington DC, 16-20 January 2012, 1-7. https://doi.org/10.1109/ISGT.2012.6175636
18. Jiang, T., Liang, Y. and Yang, C. (2015) Energy Management of Internet Data Centers in Smart Grid. Springer, Berlin, Heidelberg. https://doi.org/10.1007/978-3-662-45676-7

19. Kim, H., Kim, Y.J., Yang, K. and Thottan, M. (2011) Cloud-Based Demand Response for Smart Grid-Architecture and Distributed Algorithms. 2011 IEEE International Conference on Smart Grid Communications, Brussels, 17-20 October 2011, 398-403.
20. Aliyu, A., Tayyab, M., Abdullah, A.H., Mohammed Joda, U. and Kaiwartya, O. (2018) Mobile Cloud Computing: Layered Architecture. 2018 Seventh ICT International Student Project Conference, Nakhonpathom, Thailand, 11-13 July 2018, 1-6. https://doi.org/10.1109/ICT-ISPC.2018.8523997
21. Mohammed Joda, U., et al. (2019) Energy-Efficient Virtual Machine Allocation Technique Using Flower Pollination Algorithm in Cloud Datacenter: A Panacea to Green Computing. Journal of Bionic Engineering, 16, 354-366. https://doi.org/10.1007/s42235-019-0030-7

Chapter 9

The Development of Electricity Grid, Smart Grid and Renewable Energy in Taiwan

Hwa Meei Liou

Graduate Institute of Technology Management, National Taiwan University of Science and Technology, Taipei, Taiwan

ABSTRACT

The grid has played a vital role in the evolution of the electricity market; from traditional to smart grids; from fossil fuel power generated electricity grid connections to the integration of other renewable energy forms such as solar and wind power; the grid has played a key role in each step in Taiwan's move towards energy transition. This study includes Taiwan's construction of its transmission and distribution network, the recently passed newly revised version of the Electricity Act with its revisions to its transmission and distribution related content, and policies promoting the smart grid as well as issues that the renewable energy grid has raised in both the technical and

Citation: Liou, H. (2017), "The Development of Electricity Grid, Smart Grid and Renewable Energy in Taiwan". *Smart Grid and Renewable Energy*, **8**, 163-177. doi: 10.4236/sgre.2017.86011.

legal aspects. Taiwan's electricity supply system is made up of the northern, central and southern systems. The Transmission and distribution grid have been defined as a common carrier, maintaining state-owned monopoly. The smart grid has 6 main facets to promote, including smart generation and dispatch, smart transmission, smart consumers, smart grid electricity grid industry and the establishment of a smart grid environment. Due to the possible effects of the integration of renewable energy generated electricity, there is a vital need for the regulation of the grid's management and skills.

Keywords: Electricity Grid, Smart Grid, Renewable Energy

INTRODUCTION

From traditional electricity grids to smart grids, from power grids based on fossil fuels such as coal and petroleum to grids integrating renewable energy sources such as solar and wind power, the grid system itself plays a key role in the operation of the electricity market, revealing the evolution process of this market, in particular in light of the problems which climate change and global warming raise for us now and for countries around the world; these issues are ones that will have to be addressed as we face the future. Energy saving methods can be achieved through advancements in technology and the development of new forms of power grids, while improving the production of renewable energy and reducing emission levels of greenhouse gases; those are all key links in enabling Taiwan to move towards energy transition.

A trend towards liberalization of the power industry took place around the world in the 1980s, and this reformation of the power industry included the transmission and distribution line within the electricity industry. With power grids aging and the demand for power increasingly rising, the importance of investing in the both the broadening of the scope of current power grids as well as the maintenance becomes more obvious day by day. The installation of smart meters is able to manage demands and is also an important part of power grids modernization process.

From a legal perspective, the need for legislative amendments to support this process has been shown to be a vital factor in power industry reforms in past research. This includes updating legislation with amendments to both the management and technological aspects supporting new developments, while revising outdated legislation to keep up with the progress of leading countries worldwide. The purpose of this paper is in discussing the development of Taiwan's distribution and transmission power grid, the current status of attempts to promote a smart grid, and the affects of the newest revisions to power industry legislation on the scope and content of transmission and distribution as part of the electricity market's liberalization policy plans, as well as the effect of renewable energy integrated grid on traditional power grids and the legislative response needed. Finally, this paper will consider the development of power grid related energy policy and legislation in the US and Europe and discuss possible ways that such developments could help Taiwan as it continues to promote renewable energy development.

TRANSMISSION AND DISTRIBUTION NETWORK

A country's power grid is part of its basic infrastructure, and the development of such power grids has a history of more than a hundred years [1] , power systems include electric power generation, transmission, distribution and sale systems, and within this system, transmission and distribution are formed through a combination of transmission and distribution grid paths and power substations. Taiwan's power grid system is run by state owned Tai-Power, and can be separated into three main systems, the north, central and southern systems, respectively based in Xinzhu County, Fengshan River and Choshui River. Figure 1 and Figure 2 show the northern, central and southern electricity systems' power supply capacity and peak load of power utilization in 2015. We can see from these figures that both power utilization in the north and the northern system's supply level are higher than the central and southern systems, with utilization accounting for 39% while supply levels dropped to 34% meaning that the north is already reliant on the economic dispatch of power from the central and southern systems to the north.

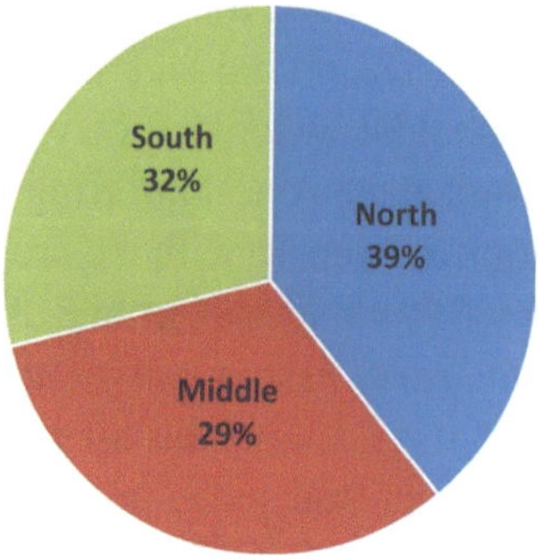

Figure 1: Peak load of northern, central and southern Taiwan.

Source: Taipower, available at: http://www.taipower.com.tw/content/new_info/new_info-c21.aspx?LinkID=12

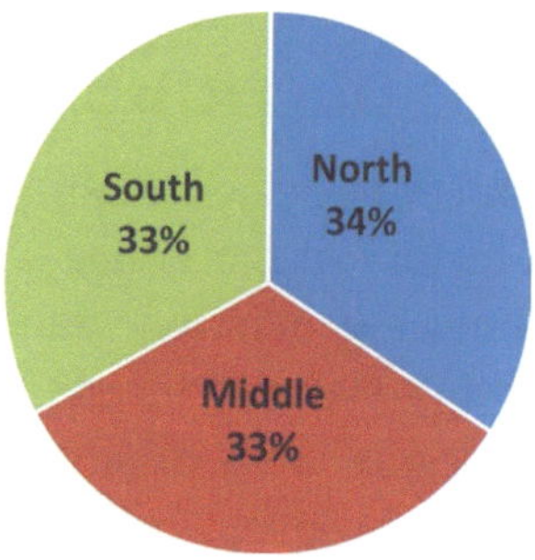

Figure 2: Load capability of northern, central and southern Taiwan.

Source: Taipower, available at: http://www.taipower.com.tw/content/new_info/new_info-c21.aspx?LinkID=12

For optimum results both the generation of power and power use should occur identically within the same system; the central system's generated power accounts for 33% of all power generated nationally, a figure higher than the 29% peak load the central system consumes. While the southern system's generated power also stands at 33%, slightly more than the system's peak load of 32%. Only the northern system produces less power than its peak load consumes, leading to a situation where the northern system lacks power. As for why the northern system lacks enough power, the main reason is the densely populated nature of the area, leading to higher demands for power than in central and southern Taiwan. Taipower's Power Development

Plan continues to emphasize the regional balance of power in order to decrease the trend of power being transmitted from the south to the north [2] .

The process of power transmission is dependent on PSA levels, with the power system being dividable into three types, as Figure 3 shows: 1) Nuclear power, large scale hydropower and thermal power plants, the power produce from this type of plant having gone through a transformer produces 345 KV, making use of transmission lines to transmit energy and having passed through the power substation voltage is reduced to 161 KV, 69 KV, providing large scale users in science parks, before going through a power distribution substation to reduce voltage again for use in normal homes and civilian electricity; 2) Medium scale hydropower, thermal power and large scale renewable energy power plants; 3) Small scale hydropower and medium scale renewable power plants [3] .

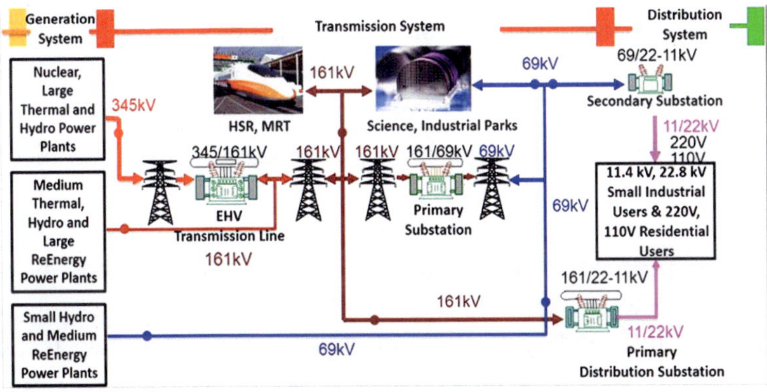

Figure 3: Introduction of Taiwan electricity supply system.

Source: Taipower, available at: http://www.taipower.com.tw/content/new_info/new_info-c21.aspx?LinkID=12

The power transmission and distribution line can be divided into three types: 1) 345 KV voltage-level extra high voltage (EHV) transmission power line, Taiwan has three such lines, using large quantity of electric charge, long distance transmission and 4004 KM loop; 2) 161 KV level primary transition line; 3) 69 KV level secondary transition line. The latter two use regional medium to small sized systems as a backbone with large power systems secondary networks, with a total of 13,281 KM loops [2] [4] . Distribution systems are the broadest link in the electricity grid, taking into account the size of the area and distinct nature of the distribution area, the density of load, system voltage and other factors bring out different patterns [5] .

Power grids can be divided into Super high Voltage (SHV), High voltage (HV), Low Voltage (LV) power grids [4] . Power transmission systems include 31 SHV substations (345/161KV), 45 primary transmission substations (161/69KV) and a distribution system including 234 distribution substations (161/22, 11 KV) and 293 secondary substations [3] . (Figure 3)

Taiwan's first power grid with the first transmission line was completed in 1951, connecting Hualian to the West coast's electricity system; in 1962 the Hualian Taitung secondary transmission line was completed, connecting Taitung to the west coast electricity system; in 1974 the first 345 KV SHV transmission system line between North and South was completed, at the time Taiwan was the first country in Asia to have a 345 KV transmission system [6].

Taipower's organizational structure includes a power generation division, transmission system division and distribution and service division, in charge of generating, transmitting and distribution of power [7] .

There are two ways to construct power transmission and distribution lines: the first is overhead lines, constructing transmission towers, cement poles, with one tower connected to the next tower covering the whole of Taiwan. The problem with this method is that it makes use of more land, making it easy to give rise to dispute. Another method is ground cables, building pipelines underground, along with cable conducting installation. In 1974 Taiwan began building Taiwan's first HV underground cables [6] .

REGULATIONS ON TRANSMISSION AND DISTRIBUTION NETWORKS FROM THE PERSPECTIVE OF THE NEWEST REVISION OF ELECTRICITY ACT

The liberalization of Taiwan's Electricity Industry planned to begin in 1995 when the Executive Yuan first sent the draft revision of the Electricity Act to the Legislative Yuan for deliberation, then after in 1999, 2002, 2007, 2008, 2015 and 2016 new amendments to the Electricity Act were sent to the Legislative Yuan. Currently, the newest version of the Electricity Act was revised in January of 2017, in particular with Taiwan having a new ruling party, the Bureau of Energy, Ministry of Economic Affairs once again proposed a new version in 2016. Revisions to the Act were accessible on line, and then gradually convened to discuss, consulting related administrative agencies, civil groups, electricity industry and labor groups to understand various opinions.

The latest Revision of the Electricity Act has 9 chapters, and 97 articles and the revisions made regarding regulations on Transmission and distribution networks are as follows: (please refer to Revision of the Electricity Act)

1) In the future the power industry will be divided into three distinct industries: power generation industry, transmission and distribution industry and electricity selling industry (article 2 item 1), each with its individual management.

Taiwan's current electricity market structure is based on one state-run integrated power company combining the business of the generation, transmission distribution and selling of electricity. While the electricity generation department is open to independent power producers (IPP), combined heat and power station, renewable energy power generator, yet besides Taipower, the electricity produced by other sources, legally speaking are only permitted to sell wholesale to Taipower, and are not allowed to sell directly to the consumer. That is to say that Taipower currently holds the exclusive right to the operation of the electricity business (old electricity act article 3). In the future users will not be limited to purchasing electricity from Taipower alone. Electricity industry organizations, other than renewable energy generator plants, will be limited to Company Limited by Shares.

2) The transmission and distribution sector will still remain a monopoly

"The transmission and distribution sector is defined as" points to the national installation power grid, providing power for public use (article 2 item 4), "the electricity grid is defined as refers to the demarcation point between the main electricity power generating equipment, the transmission and distribution industry and the user, all belonging to the same electricity transmission system with its supporting installations and substation installation." (article 2 item 14) Since the transmission and distribution network possesses both public and natural exclusive properties, by defining the transmission and distribution power industry equipment as "public access" or "common carrier", does the Act maintain the only one state-owned enterprise. The transmission and distribution industry's wheeling of renewable energy power generator's electricity, or renewable energy power generator's can customize an in-house route to directly transmit electricity to users.

3) Transmission and distribution power industry to operate power transmission and distribution network

The transmission and distribution power industry will be responsible for the national power transmission and distribution network, including the planning, construction and maintenance, with an obligation to ensure users are connected, while fulfilling their obligation to ensure interconnecting grid systems for other power industries. It is vital to ensure the fairness of the power grid, equal treatment for all and public provision for all power industry users. This version of the legislation has yet to require the establishing of an independent systems operator (ISO) for the transmission and distribution sector.

4) New organizational design

To coordinate with the development of the transmission and distribution industry, the new organization will have follow qualities. First, having referred to the experience of the UK, US and EU, an Electricity Industry regulatory agency would be assigned by the central competent authority in charge of the state- owned power related business. Secondly, a Platform for Power Trading would be set up by the Transmission and Distribution industry.

5) Concurrent limitations

There are two main limitations: 1) The operation of a transmission and distribution company, in principle, can't operate concurrently in terms of the power generating industry and the electricity selling industry; 2) The transmission and distribution industry supplementing other industries besides the power industry, must not influence its own business operations, nor should it limit competition or cause unfair competition, moreover it should be regulated by the Electricity Industry Regulatory Agency.

As for the power generation industry and normal Power Sales industry, since they count as non-public businesses, they can work in a free competitive market and diversify their businesses, unlike the old Electricity Act with its concurrent limitations.

6) Terms of license permits

The terms of license permits for the electricity industry are all the same, 20 years, once the license permit term comes to an end, the business can apply to the Electricity Industry Regulatory Agency for an extension.

7) The establishment of a power development foundation

This revision of the Electricity Act aims to set up a dry fund, in relation to the transmission and distribution industry: the transmission and distribution industry and power generating industry should invest a lump sum to establish a power development foundation, with the legal purpose of taking into account businesses' social responsibilities and feeding back to the community and residents of areas around power plant installations.

8) 2025 achieve a nuclear-free homeland

Taiwan currently has four nuclear power plants, the building of the 4th has already been brought to a halt, and there will not be an extension to the life of power plants' 1st to 3rd. This revision sets 2025 as a goal for the end of operations at all nuclear plants. Within this context, in order to make up for the lack in the power supply which the end of nuclear power in Taiwan will lead to, it is of vital importance to be proactive in developing forms of renewable energy and strengthening the construction of smart grids to enable the country to cope with the changes that will occur in the energy resources.

In Taiwan's promotion of the liberalization of the power industry, there should be an increase in competition for the power generation market; increasing business operations efficiency and bringing about growth in power production industry. At the same time the renewable energy industry would be developed, increasing Taiwan's image as a country that promotes green energy and low carbon society within an international community. Fitting this in with the long term plan for promoting the smart grid we can see that it would be contributions to the economy and society as a whole.

SMART GRID

The concept of developing Smart Grids was one of the main aims of the Executive Yuan's 2010 National Master Plan on Energy Conservation and Carbon Reduction. Beginning in 2009 with the National Energy Conference held by the Ministry of Economic Affairs, the developing of smart power meters and smart grids was set as a principle item for forthcoming promotion, including electrical terminal of the electricity transmission grid, power distribution all the way along to the user clients smart meter, smart grid and this has been written in as one of the principle axis of the National Energy Program for which the Ministry of Science and Technology is responsible, smart grids are also an important bridge to be established to enable greater interaction between the energy industry and information and communication industry. In 2010 the Executive Yuan passed the "Advanced Metering Infrastructure", AMI and in 2011 the Ministry of Economic Affairs Energy Board established a Working Group for the Smart Grid Principle Central Plan, as well as holding a Smart Grid Development Strategies Forum. In 2012 the Executive Yuan passed a Master Plan of Smart Grid in Taiwan, becoming an important policy in the promotion of smart grids, moreover establishing a cross-sector Interdepartmental Promotion Team of Smart Grid, while Tai Power also established a smart grid working group.

According to the definition used in the Master Plan of Smart Grid in Taiwan: "Through data, information and automatic technology, construct an integrated electricity grid integrating power generating, transmission and distribution, enabling automatic, secure close coordination between clients and the supply side, promoting greater efficiency in the electricity system's integrated transportation, providing quality and reliable power grid, while promoting the expansion the use of renewable energy and carbon reduction policy goals". And this smart grid can be divided structurally into 6 different aspects, including: Smart Generation and dispatch, smart transmission, smart distribution, smart consumers, smart grid industry and constriction of a smart grid friendly environment. It was estimated in 2011 that the smart grid's development would be separated into three stages, over a 20 year promotion period, these three stages are as follows: Progressively Extending Stage (2011-2015); Promoting Stage (2016-2020), Expansion Use Stage (2021-2030) [8] [9] , as shown in Figure 4. An estimated 139.9 billion NTD will be invested into this project; this will be invested correlating with each different stage as shown in Figure 5.

The definition of a smart meter, according to Brendan Cook et al.: A smart meter is a device which monitors a household's electricity consumption in real-time, and has the ability to display real-time pricing in each household [10] . A smart meter is capable of distinguishing between high voltage AMI and low voltage AMI. Between 2010-2012 the Ministry of Economic Affairs Energy Bureau completed the installation of a smart meter model system (Taipei, Xinzhu, Tainan), by the end of 2015, all 24,000 high voltage users, which accounts for 60% of power utilization in Taiwan, and 10,000 low voltage users, had already installed smart meters [11] .

Taiwan's first Smart Grid Demo Site and AMI is currently installed in Penghu [12] , There are other demo sites such as smart user energy management system demo sites in a number of places including Tainan,

Academia Sinica, convenience stores, science parks, China oil company, along with smart power distribution system technology and virtual power plant technology in various areas and institutions including Tai Power, the Business sector and Jinmen, altogether there are 18 smart grid technology demo sites, installed by the Ministry of Economic Affairs, as well as the technology and business sectors [4] [13] .

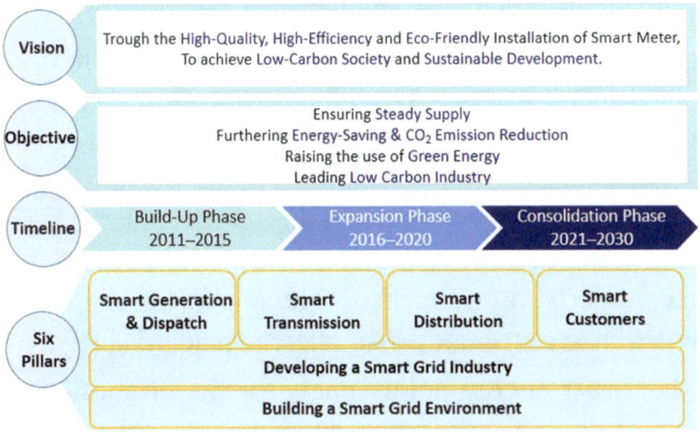

Source: Bureau of Energy, Ministry of Economic Affairs, R.O.C., General Program of Smart Grid

Figure 4: *General Framework of Smart Meter.*

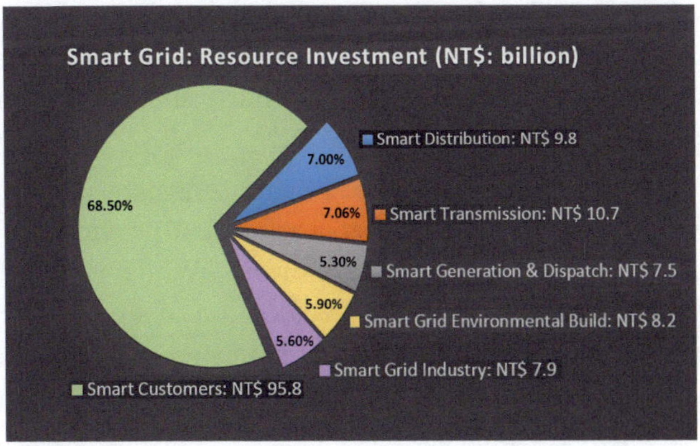

Source: Bureau of Energy, Ministry of Economic Affairs, R.O.C., General Program of Smart Grid.

Figure 5: *Smart Grid: Resource Investment.*

In terms of important plans for developing smart grid use, the National Energy Program—Phase II, included a "smart grid major central plan"[1] [14] . In terms of the smart grid industry there are currently two important associations: the Taiwan Smart Grid Industry Association and Taiwan Smart Energy Industry Association. In terms of the market, the market for building grid foundations can be divided into three main kinds: electricity wiring, substations and control facilities and terminal facilities [4] .

In terms of regulations regulating the construction of smart grids, including the Electricity Act, Grid Interconnection Code of Renewable Energy, Regulation of Exterior/Interior Circuit Installed [15] , and others such as the Energy Management Act, Renewable Energy Development Act are all related in some way [16] , besides this there are also related issues such as information security, privacy rights that have related regulations [17] [18] , intellectual property rights related regulations, building standardization and authentication systems and so on related standards [19] .

Under the larger energy policy goals for the promotion of renewable energy, smart grids play an irreplaceable role, as smart grids enable the promotion of large scale generation of renewable energy into the grid. Smart grids combine smart meter, enabling management of demands, reducing CO_2 emissions [4] . Through implementing smart grids, electricity systems will be able to ensure high level operations, improving the quality of electricity provision and systems security [20] . As for the development and improvement of smart grids nationally, this is affected by the promotion policies of the government that are dependent on a number of factors including the progression of power electronics, smart grid information and smart meter [21] .

As for South Korea's legislation related to promoting smart grids, in 2011 the Smart Grid Promotion Act was passed providing a sustainable smart grid development planning framework. While in the aftermath of Japan's Fukushima nuclear disaster the adoption of smart meters helped to manage demand. Under this context the Ministry of Economy, Trade and Industry (METI) continued to promote the construction of smart grids [22] . While in China smart grids are listed as part of the 12th Five-Year Plan as an important industry in the strategic development of new industries, the planning for which is separated into seven main areas: generated power, transmission of power, substations, allocation of power, power utilization, scheduling and communication platform [8] . In Taiwan smart grids were included in the energy policy as a priority for promoting in 2009 and now have already entered the second stage, though as yet no special law has been formulated.

INTEGRATION OF RENEWABLE ENERGY INTO THE GRID

Due to the market penetration rate and intermittent characteristic of renewable energy, once large amounts of renewable energy is incorporated into the grid, it will have a direct affect on the quality of electricity provision and reliability [23] . Many countries when allowing renewable energy to be connected to the grid have provided standards for the management and technology of such challenges including variation in voltage, frequency etc.; in order to ensure the flexibility of grid transmission once renewable energy is connected to the grid, standards were also put into place : voltage regulations, information data etc.; as well as grid operation, real-time pricing are all aspects which can be regulated through establishing standards to respond to the challenges that the connection of renewable energy to the grid could give rise to [24] . In terms of the extent to which legislation covers these issues, amendments have been made to current legislation to respond to the newest development trends in the electricity grid, as well as making new laws to regulate the future of electricity grid's modernization, and the challenges that smart technology could give rise to. Take for example Germany's Power Grid Expansion Act, EnLAG.

The key legislation for Taiwan's implementation of renewable energy being incorporated into the grid is the Renewable Energy Development Act Item 8: "Renewable energy power facilities and the generation of electricity power shall have the stability of their power grids evaluated by local electric power grid enterprises and have them paralleled and bought wholesale at the locations where existing power grids are closest to renewable energy power assembly sites and provide electricity required by such power facilities during maintenance shutdown period; electricity enterprises shall not reject the aforesaid request without proper reason and approval of the central competent authority." In particular this emphasizes the sharing of costs: other than the existing circuits, the cost for power grid enhancement will be shared between both the electricity enterprises and renewable energy power facility installers. The rising cost for the enhancement of the power grid is for example the expansion of current lines, increasing related transformers, or user line being changed to larger scale [25] . Moreover, in terms of line costs: The installer shall install and maintain the circuits connecting renewable energy power facilities and power grids. The renewable energy power generation facility installers shall bear the costs incurred.

Besides the regulations mentioned above, other legislation related to grid- connection also include standards related to electricity: for example: regulations related to electric utility circuit installations) (originally known as interior electricity supply power lines installation regulations, Ministry of Economic Affairs Standards); regulation of interior circuit installations) (Ministry of Economic Affairs Standards), along with business regulations as set out by Taipower: for example Taipower's Guide to Renewable Energy Electricity Generated System Grid Connection Integration Technique, Taipower's Guideline for energy distribution system planning, Third type (installation capacity less than 500 KW) Renewable energy self-use Electricity generating equipment parallel connection framework [25] . Renewable energy electricity generating system grid-connection technique framework regulating renewable energy grid-connection power distribution system electricity voltage level and capacity limits, stipulating renewable energy equipment grid-connection protection support plan, design and installation standards along with grid-connection transmission demands [26] ; Besides this, Taipower established the renewable energy electricity generation grid- connection capacity inquiry system, enabling users to check grid-connection capacity for their area with the intention of applying for grid-connection.

In terms of the constructing of the electricity grid, research points out that legislation's inability to coordinate with developments is the largest barrier against policies attempting to promote the extension of the electricity grid, with an influence far greater than that of the development maturity of electricity grid technology. This Legislation barrier includes: permission procedure delays, the procedure itself being too complicated and lacking transparency; nationally, there are a lack of well planned, long-term guidelines for the expansion of grid; internationally, there is also a lack of planned cooperation. The issues mentioned above show that there is a failure on the part of both the standard system content and implementation process. Sometimes the current standards themselves might be the greatest obstruction to development. In order to eliminate this legislation barrier to development, rather than pointing out small adjustments to be made within the narrow range of current legislation in terms of revisions to specific content or improvements to the legal body at an implementation level, there is a need for significant changes to be made to the regulatory framework itself [27] .

US AND EU DEVELOPMENT

The US legal system and the legal process in relation to smart grids, is closely related to the development some bills related to US electricity market [28] : In 1965, the northeast experienced its first power failure, in 1977 the Federal Power Commission (FRC) reformed as the Federal Energy Regulatory Commission (FERC), then in 1978 the National Energy Act was passed, which then led to the Public Utilities Regulatory Policies Act (PURPA). In 1992 the Energy Policy Act was passed and then in 2003 Northeast once again experienced a power failure, following this in the years 2005, 2007 and 2009 the 2005 Energy Policy Act, Energy Independence and Security Act (EISA07) and the American Recovery and Reinvestment Act (ARRA09) were each passed respectively. Of these Acts the Energy Independence and Security Act directly dealt with Smart Grids and added related regulations, in Title XIII specified the "Statement of policy on modernization of electricity grid", a "Smart grid advisory committee" and "Smart grid task force", "Smart grid technology research, development, and demonstration" and so on. Then the American Recovery and Reinvestment Act set a budget to invest in the development of Smart Grid, including Smart Grid Investment Grant Program, Smart Grid Demonstration Program [29] . In terms of policy, in 2003 the U.S. Department of Energy proposed the "'Grid 2030': A National Vision for Electricity's Second 100 Years", in 2012 it proposed the "2010 Smart Grid Report" [30] .

As for EU, their smart grid legislation is related to climate change and energy policy: the EU's 20-20-20 strategic goals and the Strategic Energy Technology Plan (Set-Plan) and other related initiatives such as the European Electricity Grid Initiative (EEGI) the plan of which is part of the Set-Plan. In 2005 the EU established the Smart Grids European Technology Platform; in 2009 the Third Energy Package goal was to complete the Internal Energy Market. The EU's R&D Framework Programme (FP6, FP7) all support smart grid development, according to the newest R&D Framework Programme (Horizon 2020), between 2014-2020 the EU will invest an estimated 80 billion euros into research innovation plans, on a theme called "Societal Challenge" directly related to smart grid and referred to as Secure, clean and efficient energy [31] . The EU's Electricity Directive Appendix I.2 demands that member states develop a smart grid system plan in compliance with the directive and establish a time frame for its development by the 3rd September 2012 [32] . Other related legislation include the Energy Infrastructure Package, the Regulation on Energy Market Integrity and Transparency and the Energy Efficiency Directive. In

2011 the EU Committee proposed the Connecting Europe Facility (CEF) legislation package, with goals for improving energy, transport, information these three connected infrastructures [33] . The future development of the EU's electricity grid will be closely related to whether or not they achieve the 20-20-20 goals.

CONCLUSIONS

As a result of the rapid development of renewable energy, advancements in digital technology and the growing connection between energy and information technology, the electricity grid is currently in a period of reform. And while this could benefit aims to ensure energy saving and improvement of energy efficiency, reliability and security are still the electricity market's major principles within the electricity provision process.

Taiwan's electricity supply system is actually made up of three systems: the north, central and south systems, of these systems the northern system's supply and demand are the highest. The Transmission system includes high voltage substations and primary substations; power distribution systems include power distribution substation and secondary substations; altogether Taiwan has three high voltage transmission electrical lines. Currently the electricity market is dominated by the state-owned Taipower; however the liberalization of the power industry is currently under way and is already in full spring. In January of 2017 the Revision of Electricity Act is passed. Transmission and distribution grid have been defined as a common carrier, maintaining state-owned monopoly, not opening up to competition. The Amendment plans for Taiwan to become a nuclear free country by 2025.

As for the Smart meter, in 2010 the Executive Yuan passed the Advanced Metering Infrastructure, (AMI) plan; currently high voltage users' installations have already fulfilled the quota set in the AMI plan, while low voltage users installations are on track for reaching targets. As for the Smart grid, policy is mainly based on the "Smart Grid Principle Central Plan" passed by the Executive Yuan in 2012, and is separated into six facets, including Smart Generation and dispatch, Smart Transmission, Smart distribution, Smart Consumers, Smart Grid Industry and Smart Grid Environment Construction. Beginning in 2011 the promotion of Smart Grid has been separated into three stages; and currently development has reached the stage of having built demo sites, accumulating experience and persisting in pushing forward, in terms of national plans; the development of smart grid is being pushed through the Smart Grid Major Central Plan. Renewable energy grid-connection is mainly based on Renewable Energy Development Act article 8, related to strengthening electricity grid pricing, as the current legislation states that renewable energy generators and electricity generating enterprises should share the burden. Other legislation influencing the grid-connection issue, is the Ministry for Economic Affairs, the electricity related executive directive and Taipower company's business regulations, Taipower has also set up a renewable energy electricity generator grid-connection capacity inquiry system, which enables one to directly confirm the actual grid- connection capacity. In order to respond to the intermittent characteristic of renewable energy, when connecting to the grid, it will have a direct affect on the stability and reliability of the grid; therefore there is a need to learn from the US, EU and other countries experience and methods in developing grid-connection techniques, management methods and general standards.

NOTES

[1]Taiwan is currently executing a two phased National Energy Program, each stage taking five years, with the first phase running between 2009-2013 and the second going from 2014-2018. The second phase of the National Energy Program focus on 6 main areas: energy conservation, alternative energy, smart grid, offshore wind power and marine energy, geothermal and gas hydrates, carbon reduction and clean coal. See the introduction of National Energy Program-Phase II, National Energy Program Phase II office, 2014.

REFERENCES

1. Chang, C.A., Lan, B.R., Lin, Y.J., et al. (2014) Challenges in the Development of Smart Grid and Renewable Energy. Journal of Taiwan Energy, 1, 259-281.
2. Zeng, J.-R. (2015) The Transmission of Electricity from the South to the North. Energy Monthly.
3. Taipower Website (2016) http://www.taipower.com.tw/content/new_info/new_info-c21.aspx?LinkID=12
4. Taiwan Smart Grid Industry Association Website (2016) Understanding the Smart Grid. http://www.smart-grid.org.tw/
5. Wang, Y.C. and Huang, Q.T. (2014) The Development and Challenge of Smart Distribution Grid. Journal of Electrical, 168, 76.
6. Zhu, R.Y. (2010) The Transmission Line System of Taiwan Power Company Which Covers the Whole of Taiwan. Yuan Magazine, 79, 5-6.
7. Taiwan Power Company (2015) Taiwan Power Company Sustainability Report. Report, Taiwan Power Company, Taiwan.
8. Bureau of Energy, Ministry of Economic Affairs (2012) The Smart Grid Master Plan. http://web3.moeaboe.gov.tw/ecw/populace/content/SubMenu.aspx?menu_id=1946
9. Bureau of Energy, Ministry of Economic Affairs (2012) Program of Smart Grid Promotion Workforce. http://web3.moeaboe.gov.tw/ecw/populace/content/SubMenu.aspx?menu_id=1946
10. Cook, B., et al. (2012) The Smart Meter and a Smart Consumer: Quantifying the Benefits of Smart Meter Implementation in the United States. Chemistry Central Journal, 6, S5. https://doi.org/10.1186/1752-153X-6-S1-S5
11. Bureau of Energy, Ministry of Economic Affairs (2015) Energy Statistics Year Book. http://web3.moeaboe.gov.tw/ecw/populace/content/SubMenu.aspx?menu_id=1946
12. Chen, X.X. (2015) Plan for Smart Grid Technology Applications. Energy Monthly, 11, 34.
13. Chen, Y.H. (2014) The Current Status and Prospects of Taiwan's Smart Grid Industry. Taiwan Economic Research Monthly, 37, 17-18.

14. National Energy Program Phase II office (2014) The Introduction of National Energy Program-Phase II. http://www.nepii.tw/language/en/about-nep-ii/introduction/
15. Bureau of Energy, Ministry of Economic Affairs (2012) Briefing on the Smart Grid Master Plan. http://web3.moeaboe.gov.tw/ecw/populace/content/SubMenu.aspx?menu_id=1946
16. Lee, K.Y. (2009) A Discussion and Response to Research Analysis on the Issue of Key Legislation Related to the Smart Grid. Science & Technology Law Review, 12.
17. Su, W.X. and Lee, K.Y. (2012) Industry R&D Invests in Intelligent Energy for Privacy Protection and Information Security in Response to Suggestions—A Case Study of European Union Promoting Legal Policy. Science & Technology Law Review, 49.
18. Zeng, Y.X. (2016) A Brief Discussion of the Importance of My Country's Key Energy Related Infrastructure Information and Communication Technology (ICT) Security Legal Structure—An Example from the Progressive Development of the EU and Germany's Smart Meter. Science & Technology Law Review, 18.
19. Chen, Z.X. (2015) Countermeasure Research on the Trend of Smart Grid Technology Development and Renewable Energy Integration into the Grid. Journal of Taiwan Energy, 2, 148.
20. Brown, M.A. (2014) Enhancing Efficiency and Renewable with smart Grid Technologies and Policies. Futures, 58, 30. https://doi.org/10.1016/j.futures.2014.01.001
21. Hossain, M.S., et al. (2016) Role of Smart Grid in Renewable Energy: An Overview. Renewable and Sustainable Energy Reviews, 60, 1182. https://doi.org/10.1016/j.rser.2015.09.098
22. Tuballa, M.L. and Abundo, M.L. (2016) A Review of the Development of Smart Grid Technologies. Renewable and Sustainable Energy Reviews, 59, 710-725. https://doi.org/10.1016/j.rser.2016.01.011
23. Taiwan Power Company (2013) The Impact and Influence of Japan's Fukushima Nuclear Disaster on the Energy Policy and Direction of Reform in Taiwan's Power Industry as Well as the World's Major Countries. Report, Taiwan Power Company, Taiwan.
24. Chen, Z.X. (2015) Countermeasure Research on the Trend of Smart Grid Technology Development and High Renewable Energy Grid Operation. Journal of Taiwan Energy, 2, 146-147.

25. Chen, Y.C. (2015) Renewable Energy Integration into the Grid Operation Practices. Taiwan Power Company Tainan District Office, Taiwan.
26. Wang, Y.C. and Huang, Q.T. (2014) The Development and Challenge of Smart Distribution Grid. Journal of Electrical, 168, 77.
27. Battaglini, A., et al. (2012) Perception of Barriers for Expansion of Electricity Grids in the European Union. Energy Policy, 47, 254-259. https://doi.org/10.1016/j.enpol.2012.04.065
28. Sim?es, M.G., et al. (2012) A Comparison of Smart Grid Technologies and Progresses in Europe and the U.S. IEEE Transactions on Industry Applications, 48, 1154-1162. https://doi.org/10.1109/TIA.2012.2199730
29. Huang, Y.S. (2013) Smart Grid Promotion Policy in the United States (Part 1). Electricity Monthly, 23, 167.
30. Huang, Y.S. (2013) Smart Grid Promotion Policy in the United States (Part 1). Electricity Monthly, 23, 164-165.
31. Crispim, J., et al. (2014) Smart Grids in the EU with Smart Regulation: Experiences from the UK, Italy and Portugal. Utilities Policy, 31, 88. https://doi.org/10.1016/j.jup.2014.09.006
32. Chen, Y.H., et al. (2012) An Introduction to the Promotion of Smart Grid in Key Countries. Electricity Monthly, 22, 8.
33. Crispim, J., et al. (2014) Smart Grids in the EU with Smart Regulation: Experiences from the UK, Italy and Portugal. Utilities Policy, 31, 87-89. https://doi.org/10.1016/j.jup.2014.09.006
34. Chang, C.A., Lan, B.R., Lin, Y.J., et al. (2014) Challenges in the Development of Smart Grid and Renewable Energy. Journal of Taiwan Energy, 1, 259-281.
35. Zeng, J.-R. （2015） The Transmission of Electricity from the South to the North. Energy Monthly.
36. Taipower Website (2016) http://www.taipower.com.tw/content/new_info/new_info-c21.aspx?LinkID=12
37. Taiwan Smart Grid Industry Association Website (2016) Understanding the Smart Grid. http://www.smart-grid.org.tw/
38. Wang, Y.C. and Huang, Q.T. (2014) The Development and Challenge of Smart Distribution Grid. Journal of Electrical, 168, 76.

39. Zhu, R.Y. (2010) The Transmission Line System of Taiwan Power Company Which Covers the Whole of Taiwan. Yuan Magazine, 79, 5-6.
40. Taiwan Power Company (2015) Taiwan Power Company Sustainability Report. Report, Taiwan Power Company, Taiwan.
41. Bureau of Energy, Ministry of Economic Affairs (2012) The Smart Grid Master Plan. http://web3.moeaboe.gov.tw/ecw/populace/content/SubMenu.aspx?menu_id=1946
42. Bureau of Energy, Ministry of Economic Affairs (2012) Program of Smart Grid Promotion Workforce. http://web3.moeaboe.gov.tw/ecw/populace/content/SubMenu.aspx?menu_id=1946
43. Cook, B., et al. (2012) The Smart Meter and a Smart Consumer: Quantifying the Benefits of Smart Meter Implementation in the United States. Chemistry Central Journal, 6, S5. https://doi.org/10.1186/1752-153X-6-S1-S5
44. Bureau of Energy, Ministry of Economic Affairs (2015) Energy Statistics Year Book. http://web3.moeaboe.gov.tw/ecw/populace/content/SubMenu.aspx?menu_id=1946
45. Chen, X.X. (2015) Plan for Smart Grid Technology Applications. Energy Monthly, 11, 34.
46. Chen, Y.H. (2014) The Current Status and Prospects of Taiwan's Smart Grid Industry. Taiwan Economic Research Monthly, 37, 17-18.
47. National Energy Program Phase II office (2014) The Introduction of National Energy Program-Phase II. http://www.nepii.tw/language/en/about-nep-ii/introduction/
48. Bureau of Energy, Ministry of Economic Affairs (2012) Briefing on the Smart Grid Master Plan. http://web3.moeaboe.gov.tw/ecw/populace/content/SubMenu.aspx?menu_id=1946
49. Lee, K.Y. (2009) A Discussion and Response to Research Analysis on the Issue of Key Legislation Related to the Smart Grid. Science & Technology Law Review, 12.
50. Su, W.X. and Lee, K.Y. (2012) Industry R&D Invests in Intelligent Energy for Privacy Protection and Information Security in Response to Suggestions—A Case Study of European Union Promoting Legal Policy. Science & Technology Law Review, 49.
51. Zeng, Y.X. (2016) A Brief Discussion of the Importance of My Country's Key Energy Related Infrastructure Information and Communication

Technology (ICT) Security Legal Structure—An Example from the Progressive Development of the EU and Germany's Smart Meter. Science & Technology Law Review, 18.

52. Chen, Z.X. (2015) Countermeasure Research on the Trend of Smart Grid Technology Development and Renewable Energy Integration into the Grid. Journal of Taiwan Energy, 2, 148.
53. Brown, M.A. (2014) Enhancing Efficiency and Renewable with smart Grid Technologies and Policies. Futures, 58, 30. https://doi.org/10.1016/j.futures.2014.01.001
54. Hossain, M.S., et al. (2016) Role of Smart Grid in Renewable Energy: An Overview. Renewable and Sustainable Energy Reviews, 60, 1182. https://doi.org/10.1016/j.rser.2015.09.098
55. Tuballa, M.L. and Abundo, M.L. (2016) A Review of the Development of Smart Grid Technologies. Renewable and Sustainable Energy Reviews, 59, 710-725. https://doi.org/10.1016/j.rser.2016.01.011
56. Taiwan Power Company (2013) The Impact and Influence of Japan's Fukushima Nuclear Disaster on the Energy Policy and Direction of Reform in Taiwan's Power Industry as Well as the World's Major Countries. Report, Taiwan Power Company, Taiwan.
57. Chen, Z.X. (2015) Countermeasure Research on the Trend of Smart Grid Technology Development and High Renewable Energy Grid Operation. Journal of Taiwan Energy, 2, 146-147.
58. Chen, Y.C. (2015) Renewable Energy Integration into the Grid Operation Practices. Taiwan Power Company Tainan District Office, Taiwan.
59. Wang, Y.C. and Huang, Q.T. (2014) The Development and Challenge of Smart Distribution Grid. Journal of Electrical, 168, 77.
60. Battaglini, A., et al. (2012) Perception of Barriers for Expansion of Electricity Grids in the European Union. Energy Policy, 47, 254-259. https://doi.org/10.1016/j.enpol.2012.04.065
61. Sim?es, M.G., et al. (2012) A Comparison of Smart Grid Technologies and Progresses in Europe and the U.S. IEEE Transactions on Industry Applications, 48, 1154-1162. https://doi.org/10.1109/TIA.2012.2199730
62. Huang, Y.S. (2013) Smart Grid Promotion Policy in the United States (Part 1). Electricity Monthly, 23, 167.

63. Huang, Y.S. (2013) Smart Grid Promotion Policy in the United States (Part 1). Electricity Monthly, 23, 164-165.
64. Crispim, J., et al. (2014) Smart Grids in the EU with Smart Regulation: Experiences from the UK, Italy and Portugal. Utilities Policy, 31, 88. https://doi.org/10.1016/j.jup.2014.09.006
65. Chen, Y.H., et al. (2012) An Introduction to the Promotion of Smart Grid in Key Countries. Electricity Monthly, 22, 8.
66. Crispim, J., et al. (2014) Smart Grids in the EU with Smart Regulation: Experiences from the UK, Italy and Portugal. Utilities Policy, 31, 87-89. https://doi.org/10.1016/j.jup.2014.09.006

Chapter 10

Optimal Power Flow Approach for Cognitive and Reliable Operation of Distributed Generation as Smart Grid

Vilas S. Bugade

Dr. Babasaheb Ambedkar Technological University, Lonere, India

ABSTRACT

Smart grid expertise emphasises on the compound connections of the electricity to the grid, along with computing, control and communication interface. It will bring together in future smart infrastructure for power system. Investigating these complex dynamic interactions is crucial for the efficiency and robustness of the emerging smart grid. In particular, it is one of the key elements for smart-grids to establish the dynamics among sources of grid. This paper proposes the vital operation of renewable energy sources (RES) like Solar Photovoltaic (PV), wind energy with existing grid of a. c. network of power system in view of cognitive reliable operation of RES as add-on source of power. The research presents sequence of operation of these sources by optimal power flow based on power flow

---

Citation: Bugade, V. (2017), "Optimal Power Flow Approach for Cognitive and Reliable Operation of Distributed Generation as Smart Grid". *Smart Grid and Renewable Energy*, 8, 87-98. doi: 10.4236/sgre.2017.83006.

chart for demand side management as a smart grid of power system. The system fulfils realistic operation for power system, based on fundamentals of power system, therefore a necessary research topology is developed, for well-regarded schemes of RES for setting up a pilot model so that demand side load should not be hampered and same is verified for linear and non-linear loads of electrical networks.

Keywords: Smart Grid, Micro Grid, RES, Distributed Generation

INTRODUCTION

Renewable energy resources (RES), such as wind and solar PV, are considered highly auspicious in the face of emergent anxieties for the surroundings, energy conservation, and sustainable development. As seasonal variability are inherent to both wind and solar PV resources. Conventionally, the indecision of an independent solar PV panel or wind turbine installation is managed using a storage system. However, this results in an increased overall cost of the output energy, and therefore limits the benefits of using renewable energy. A hybrid system uses two renewable energy sources as solar PV and wind energy, thus improving the system efficiency, power reliability and reducing the energy storage obligation. However, accumulating integrally stochastic power sources such as wind and solar PV to achieve reliable electricity supply is a non-inconsequential problem. In the modern power system planning, cohort is dispatched to tie the load. Power system frequency directive, voltage mechanism and other key ancillary services are provided by controlling large-scale generators, transformers and other bulk-power devices [1] .

Smart grid skills and facilities are being developed from an assortment of perceptions. Whereas the innovative smart grid infrastructure is expected to increase the efficiency and reliability of grid operations, the increased complexity of the new system has to be addressed. Obtaining huge volumes of statistics from loads of advanced smart topologies self does not indicate that a grid becomes intelligent. Unlike traditional power grid systems, the future grid will invariably feature rapid integration of alternative forms of energy generation; thus, power flows bi-directionally as shown in Figure 1.

However it has been discussed in the literature for eras, the perception of monitoring end-use loads to match generation has only recently become gorgeous to scholars due, partly, to improvements in work out and communication technologies, but similarly due to tenacious anxieties regarding the integration of alternating and variable RES to the grid. The Central Energy Regulatory Commission (CERC) of India in recent times issued a report on the benefits of Demand Response (DR), the extensive concept under which this awareness falls, showing a possible 12% reduction of the forecasted peak demand for 2017 compared to the case without DR. This report, however, does not take into version sovereign, distributed device approaches. Though, study by imitations and a minor experimental program has established the benefits of using hundreds or even thousands of small controllable loads to balance the cumulative load and generation.

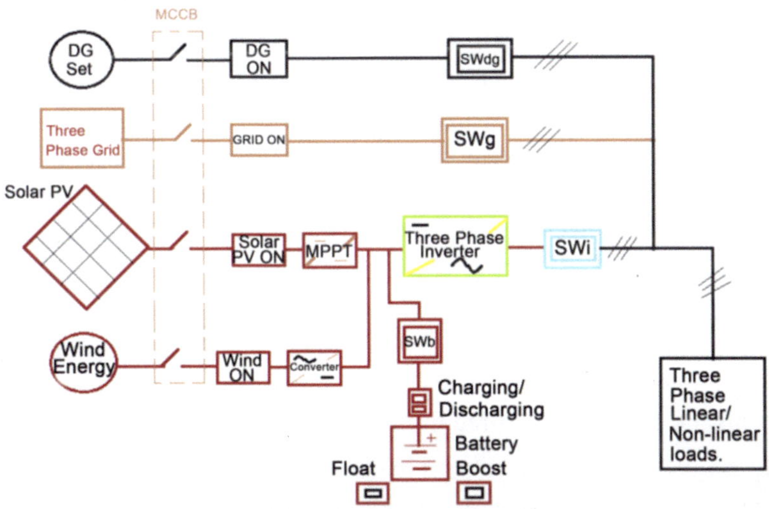

Figure 1: Block diagram of three phase SMARTGRID system with RES.

In recent times, there has been an enlarged interest in using thermostatically- controlled loads (TCLs) to leverage demand response [2] . Their capability to be swapped ON/OFF deprived of substantial impact to their end-use function makes them attractive for providing a short-time measure comeback to adverse frequency variations initiated by intermittent generation sources [3] .

In this paper, a demand side management approach is anticipated by Heuristic Optimization method for smart grid [4] . The strategy is based on load shifting technique, which can handle a large number of devices of several types [5] . A heuristic based evolutionary flowchart that can easily adapt its heuristics has been developed for solving the problem in section 1. Section 2 briefly describes the optimal power flow technique for future smart grid. The simulation for power flow analysis were carried out on a smart grid which has different types of generation sources with a variety of loads in section 3 and section 4 gives hardware results of each source and combination with existing grid. Section 5 concludes the paper.

EVOLUTIONARY FLOWCHART TOPOGRAPHIES FOR A SUITABLE DEMAND SIDE MANAGEMENT TECHNIQUE FOR FUTURE SMART GRID

This section gives conceptual demand side management with relative generation status of the system as shown in Figure 1.

The block diagram consists of solar PV, wind energy, battery, Diesel generator and exiting grid whose power rating is specified in simulation model. The three phase inverter is for converting unconditioned d. c. power in to conditioned a. c. power of solar PV, wind energy and battery source. At the outset the power is delivered to three phase load through point of common coupling (PCC) [6] .

Eventually each generation is represented as a source and at PCC the gathered parameters gives intimation for operating the switches at node of generating sources as Switch (SW) in auto mode. The subscript for each SW represents the generating source in Figure 2.

The sequential power flow chart in the form of exchange of power by RES to load is given in the flow chart. It represents the operation of SMARTGRID as RES-GRID-Battery-DG or RES-Battery-GRID-DG. The power flow is based on the generation priority of the source and its availability Figure 3 i.e. the fulfillment of load. An approach is to make system convenient with significant use of generation source by either way [7] .

OPTIMAL POWER FLOW STRATEGY FOR SMART GRID

The demand side is basically have two commands in the way the load connected and disconnected. For controlling the generation power dispatching flow at PCC these parameters need to be considered with real time. From the data available of the connected load in the system, the demand to be supplied or otherwise will be known from (i) the magnitude of load connected (demanded) (ii) the magnitude of load disconnected (Switched OFF) where in a balance is required maintained.

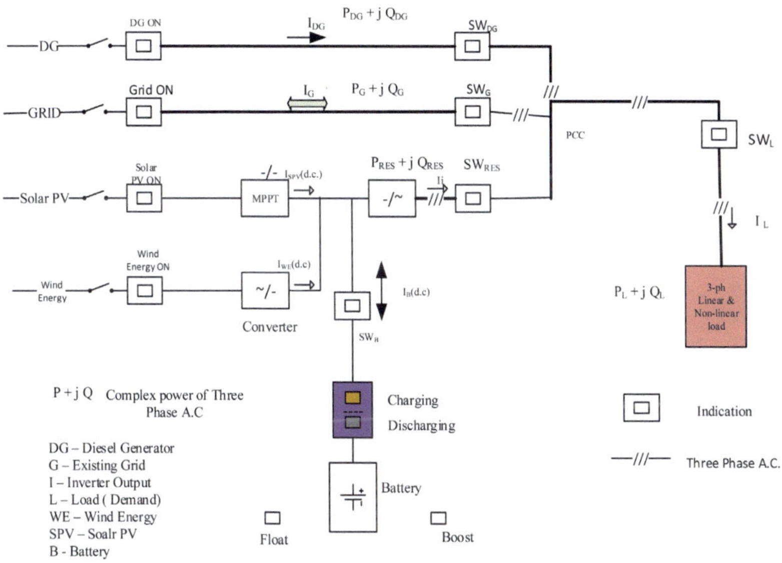

Figure 2: Power flow diagram of SMARTGRID with RES.

The proposed demand side strategy schedules the switching moments of each source of power for the system in a way that brings the load consumption as the objective of system. Proposed load fulfillment is mathematically formulated based on the power flow analysis as presented in flowchart shown in Figure 3,

$$\sum_{t=1}^{N} P_D(t) = \sum_{t=1}^{N} P_L(t) - \sum_{t=1}^{N} P_L(t) \quad \text{disconnected} \tag{1}$$

where $P_L(t)$ is actual consumption at time "t" sec.

The $P_i(t)$ the inverter output three phase a.c. power as shown in Figure 2 then, if

$$\sum_{t=1}^{N} P_D(t) = \sum_{t=1}^{N} P_i(t) \tag{2}$$

$$\sum_{t=1}^{N} P_D(t) < \sum_{t=1}^{N} P_i(t) \tag{3}$$

For the Equations (2) and (3) the SW_g, SW_b and SW_{dg} are at OFF position of Figure 2 and at Equation (3) the SW_g acts like bidirectional switch which will?ve sign returns the power to grid as Equation (4), the same will be displayed and reflected in power returns $P_{ir}(t)$ as,

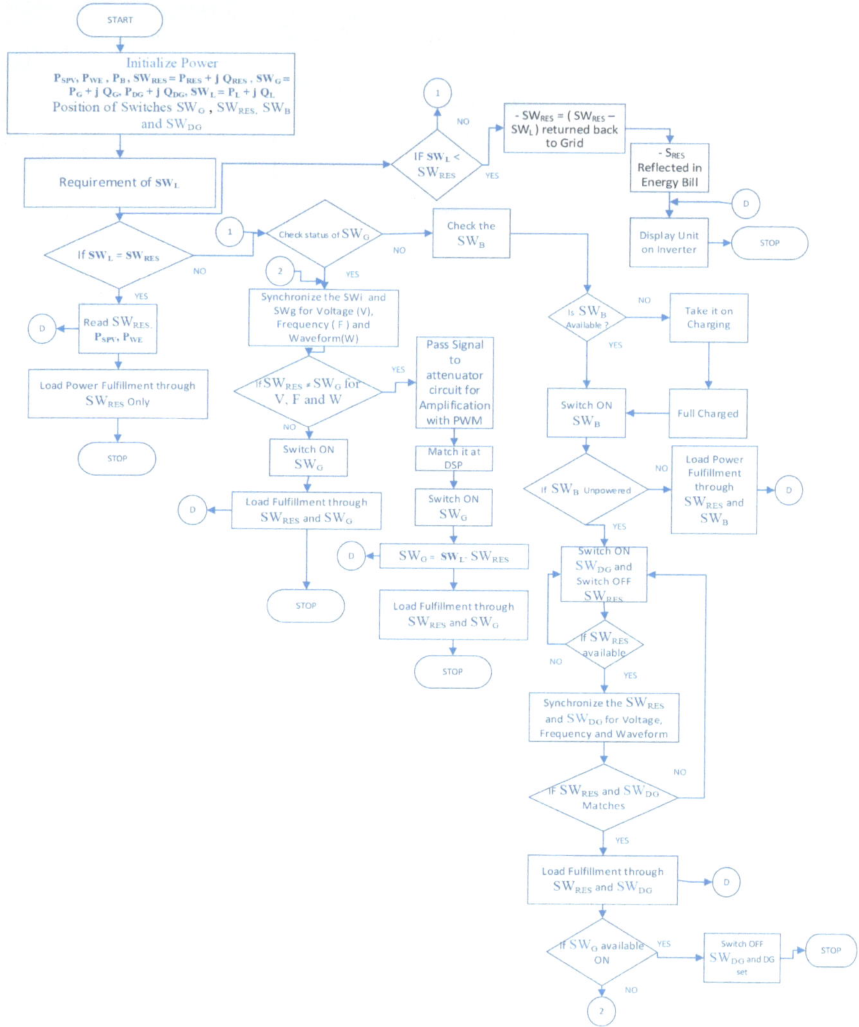

Figure 3: Power flow chart of SMARTGRID.

$$-\sum_{t=1}^{N} P_{ir}(t) = \sum_{t=1}^{N} P_i(t) - \sum_{t=1}^{N} P_D(t) \tag{4}$$

If,

$$\sum_{t=1}^{N} P_D(t) > \sum_{t=1}^{N} P_i(t) \tag{5}$$

Then

$$\sum_{t=1}^{N} P_D(t) = \sum_{t=1}^{N} P_i(t) + \sum_{t=1}^{N} P_G(t) \qquad (6)$$

For the Equation (6) the SW_b and SW_{dg} are at OFF position of Figure 2, Where $P_G(t)$ is the three phase grid power from Figure 2 and the power taken from grid will be,

$$\sum_{t=1}^{N} P_G(t) = \sum_{t=1}^{N} P_D(t) - \sum_{t=1}^{N} P_i(t) \qquad (7)$$

Hence the power flow analysis will be achieved from above but the limitation is that all the time $P_i(t)$ may or may not be available and 15% of rated capacity of RES needs to reserve the power. If $P_G(t)$ is not available then in Equation (6) and (7) the $P_G(t)$ will be replaced by $P_{dG}(t)$ of diesel generator for this SW_g and SW_b is in OFF position of Figure 2. Also if $P_G(t)$ and $P_i(t)$ both are unpowered then load fulfillment through P_{dg} for this SW_g, SW_i and SW_b is in OFF position of Figure 2.

The source battery in the Figure 2 is used as standby power source or as a back up to the network as well as conditioned power source for power drive and control circuit. Charging of battery is through RES only and during discharging the total d.c. input power to the $P_i(t)$ becomes $P_{spv} + P_{we} + P_b$.

To create an optimal power flow for equation (6) and equation (7) from first order transfer function, which eliminates the steady state error at the output for active (P_i) and reactive (Q_i) power of the inverter [8] .

$$\frac{dP_i}{dt} = \frac{P_{max}}{\tau_p}\left(U_p - \frac{P_i}{P_{max}}\right) \qquad (8)$$

$$\frac{dQ_i}{dt} = \frac{Q_{max}}{\tau_q}\left(U_q - \frac{Q_i}{Q_{max}}\right) \qquad (9)$$

where *Up, Uq* is output signal from inverter respectively, P_{max}, Q_{max} maximum output of inverter, τ_p, τ_q time constant w.r.t. active and reactive power respectively in (8), (9).

Then for battery as storage device [8] [9] ,

$$\frac{dP_b}{dt} = \frac{P_{max\,b}}{\tau_b}\left(U_b - \frac{P_b}{P_{max}}\right) \qquad (10)$$

$$\frac{d_{soc}}{dt} = b \cdot P_b \qquad (11)$$

U_b is the signal for power that the battery generates P_{max} b maximum output power of battery.

Based on this optimal flow power the problem formulation can be created to know power requirement at PCC on P_D (t) and P_G (t) as follows [8] :

Using Kirchhoff voltage law in Figure 4, the voltage of the terminal can be expressed as,

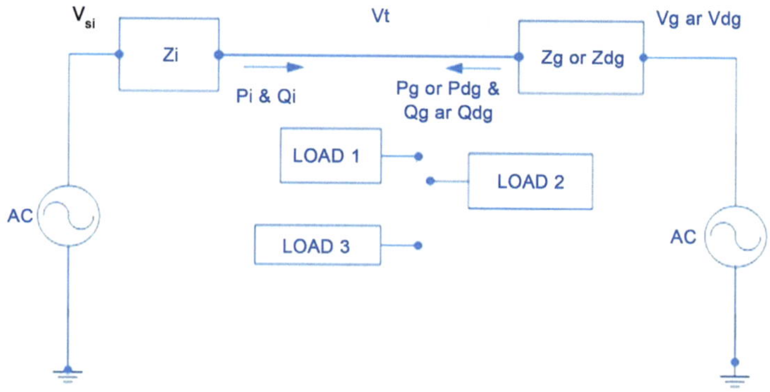

Figure 4: Terminal bus at PCC for voltage and frequency constraints.

$$V_t = V_{si} - \frac{|Z_i| S_i}{V_{S1}} = V_g \text{ or } V_{dg} - \frac{|Z_g \text{ or } Z_{dg}| S_g \text{ or } S_{dg}}{V_g \text{ or } V_{dg}} \qquad (12)$$

In which V_{si} and V_g or V_{dg} is the output voltage at inverter and grid or DG, Z_i and Z_g or Z_{dg} impedances of lines of inverter and grid or DG. The output voltage at respective nodes depends on frequency and current, therefore difference between on $P_L(t)$ Connected and $P(t)$ Generated could cause the variation in the frequency [9] [10] ,

$$\frac{d\omega}{dt} = \lambda\left(P_{\text{Generated}} - P_D\right) + \left(\omega_{\max} - \alpha P - \omega\right) \qquad (13)$$

For islanded distributed generation it is important to maintain the frequency and voltage regulation. The aim of control frequency and voltage a proper effective function could be defined as [8] :

$$J = \int_{t_0}^{t_f} \left(\xi_\omega \left|\omega - \omega_{ref}\right|^2 + \xi_V \left|V_t - V_{ref}\right|^2 \right) dt \qquad (14)$$

Using such equation the voltage and frequency remain unchanged around their reference values, if by adjusting ξ_ω, ξ_V the frequency and voltage regulation could be changed by providing such facility in the controller [8] . The power balance in terminal bus should be considered with system constraints as below:

$$P_G = P_{\text{Consumed}} - P_i \qquad (15)$$

The control action and related constraints for optimal power flow should be taken into account w.r.t. to minimum to maximum generating capacity of source. [11] .

SIMULATION RESULTS FOR OPTIMAL POWER FLOW

The simulation is based on power block diagram and equations related to it, the results are represented for its respective capacity of source which caters to power delivered by each source of the network [12] . The electricity generation by each source is for expected output of it, which will satisfy the load. The capacity of each is as specified below with variation in the load from 100 watt to 500 watt.

Solar PV	Wind Energy	Diesel Generator	Battery
500 watt, 70 Volt	200 watt, 230 Volt	5 KVA	12 V, 100 AH

For optimal power analysis the simulation model is simulated with specified parameters, for $P_D(t) \geq P_i(t)$ and $P_D(t) \leq P_i(t)$, the results gives satisfactory operation of all sources as shown in Figure 5, which represents contribution of each source with variation in load as per their electricity generation strategy [5] .

Distributed Generation (DG) Optimization Method

As seen from Figure 2 the process of optimization is carried out in following steps for both ON and OFF control of system:

a) Data and input parameters are specified which they include electricity generation, the load curve, the percentage of loads and the ecological characteristics of the site.

b) DG and switching ON/OFF of sources is as per consumption of power by loads datasheet.

c) Initial values for the capacity of DGs and loads are randomly identified based on optimal power flow.

d) Optimal instantaneous power flow, based on utilization over the entire process of the system, is determined by as load follows. Given the capacity for each DG, determined from the previous step, economic-environmental dispatch can be calculated by minimizing the power. Next, the power available from RES is determined based on (1) and (2) and the remaining required power is generated by other DGs and grid according the economic environmental dispatch.

At the same time, as a source during peak times and is subject to deliver the power. During off-peak hours as a load and its duration.

e) Given the capacity for each DG, as determined from the third step, power flow optimizes and switching ON/OFF based on reliability therefore, consumer outage are minimized.

f) Given the power generated by each unit, determined from the previous step of each source.

g) Using the capacity of each DG, as determined by power flow optimization. Using obtained optimum capacities, the type and size of the sources, forming Micro Grid's, are identified.

Also, Operational Strategy is determined on an hour-by-hour basis, using steps 4 - 7.

RESULTS

The optimal power flow for its power flow analysis with each source of generation of electricity is deployed for hardware implementation based on available Solar PV (500 W, 100 V), Wind energy (200 W, 230 V) and battery 12 V, 100 AH with existing grid of 230 V for nonlinear or variable loads shown in Figure 6.

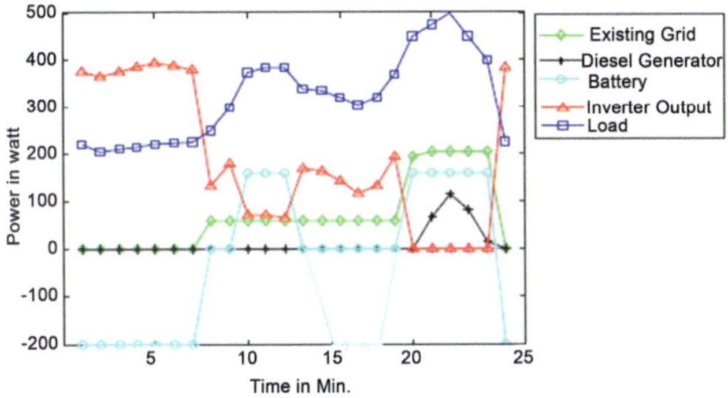

Figure 5: Optimal power flow analysis of simulation model for power system.

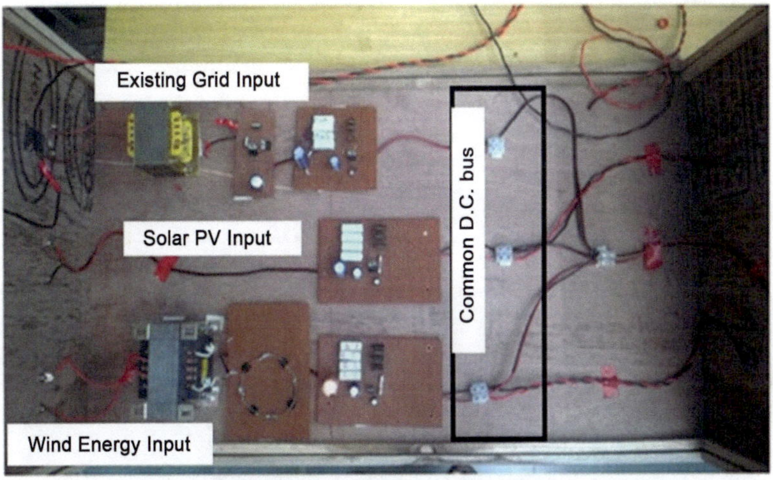

Figure 6: Power flow model of solar PV, wind energy and existing grid.

Each source is synchronised manually with existing grid for the parameters like voltage, frequency and waveform by power analyser and same is reported.

The proposed hardware consists of integration of distributed generation such as solar PV, wind energy system, battery with the conventional grid. From the various topologies available for integration, the common DC bus topology is selected and a common DC bus of 96 volt is formed. Figure 6 shows the interconnection of various equipment in the hardware setup. At the output of d. c. bus a single phase inverter connected which gives the supply to non-linear load.

Results and Discussion

From Figure 4 the simulation results for the system in Figure 5 have been obtained the same have explored for a small capacity system consisting of Solar PV, Wind Energy, Battery, Diesel Generator and Existing Grid of total capacity 1 KW. The results shown in Figures 7-10 with following sources are seen consistent with simulation results at Figure 5.

Sr. No.	Source	Capacity
1	Solar PV	500 watt, 70 Volt
2	Wind Energy	200 watt, 230 Volt
3	Battery	12 V, 100 AH
4	Existing Grid	230 V, 1 KW

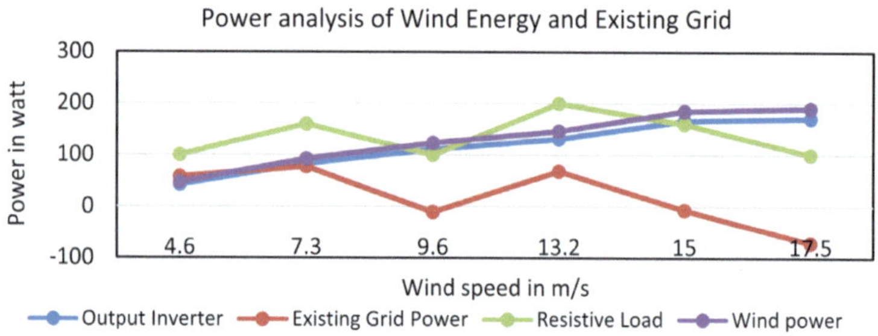

Figure 7: Power flow analysis of wind energy. (Synchronizing parameters at the instant of switching with grid f = 49.7 Hz, V = 217.5 volt, identical waveform).

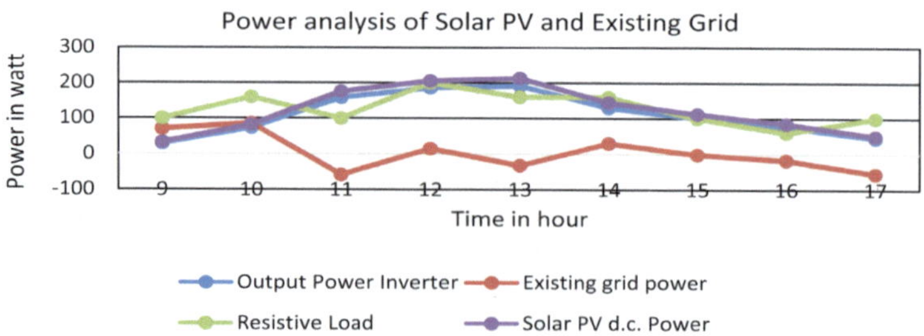

Figure 8: Power flow analysis of Solar PV. (Synchronizing parameters at the instant of switching with grid f = 49.9 Hz, V = 220.7 volt, identical waveform).

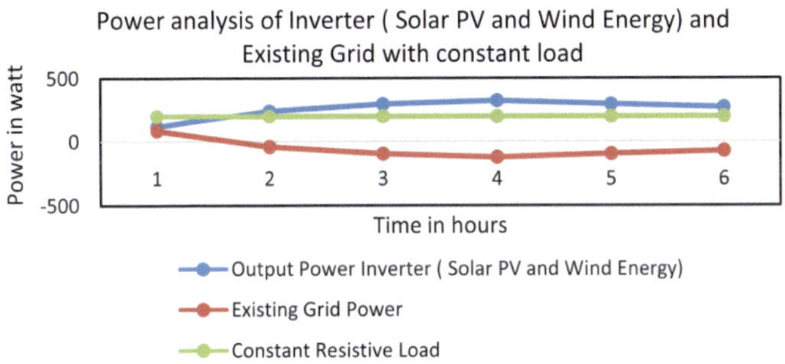

Figure 9: Power flow analysis with constant load. (Synchronizing parameters at the instant of switching with grid f = 50 Hz, V = 219.5 volt, identical waveform).

The results in Figures 7-10 are obtained for real time pilot system in the laboratory, which clearly depict the effectiveness of RES with conventional grid leading to so called Smart Grid concept.

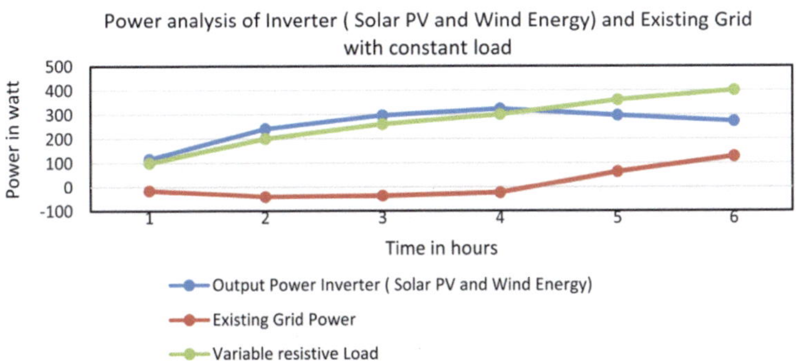

Figure 10: Power flow analysis with variable load. (Synchronizing parameters at the instant of switching with grid f = 49.9 Hz, V = 225.4 volt, identical waveform).

CONCLUSION

The adaptive power management to uncertain RES generation in single-feeder distribution networks was formulated as a power flow optimization. From Figure 5, Figures 7-10 the simulation and hardware results are consistent. It includes real power consumption for variable loads with respect to synchronising the RES and existing grid parameters at PCC. An optimal power flow solution was developed based on the RES generation sources leading to closed-form updates per node and per scenario. Tests were conducted on distribution networks with Solar PV, Wind Energy and existing grid by using a scenarios to illustrate the effectiveness of the formulation. The superior performance of the proposed system was also highlighted through comparisons with an individual generation scheme in terms of synchronising the voltage, frequency and waveform of these generation sources with manual mode of operation. In future it is entitled to develop an AUTO operation system for the same as Smart grid.

REFERENCES

1. Bolognaise, S., Cavraro, G., Cerruti, F. and Costabeber, A. (2011) A Linear Dynamic Model for Micro Grid Voltages in Presence of Distributed Generation. 2011 IEEE First International Workshop on Smart Grid Modeling, Simulation (SGMS), at IEEE SmartGridComm, 31-36.
2. Balenciaga, F. and Puleston, P.F. (2005) Supervisor Control for a Stand-Alone Hybrid Generation System Using Wind and Photovoltaic Energy. IEEE Transaction Energy Conversion, 20, 398-405.
3. Jafari, H., Mahmodi, M. and Rastegar, H. (2012) Frequency Control of Micro-Grid in Autonoums Mode Using Predictive Control. 2nd Iranian IEEE Conference on Smart Grid, 1, 231-236.
4. Logenthiran, T., Srinivasan, D. and Shun, T.Z. (2012) Demand Side Management in Smart Grid Using Heuristic Optimization. IEEE Transactions on Smart Grid, 3, 1244-1252.
5. Erseghe, T. (2014) A Distributed Algorithm for Fast Optimal Power Flow Regulation in Smart Grids. 2014 IEEE International Conference on Smart Grid Communications, 31-36.
6. Bugade, V.S. and Katti, P.K. (2015) Dynamic Modelling of Micro grid with Distributed Generation for Grid Integration. 2015 IEEE International Conference on Energy Systems and Applications (ICESA 2015), 30 October-1 November 2015, 102-106.
7. Chen, Y.-M., et al. (2007) Multi-Input Inverter for Grid-Connected Hybrid PV/Wind Power System. IEEE Transactions on Power Electronics, 22, 1070-1077.
8. Shabestary, S.M.A., Saeedmanesh, M., Rahimi-Kian, A. and Jalalabadi, E. (2012) Real Time Frequency and Voltage Control of Islanded Micro Grid. 2nd Iranian IEEE conference on Smart Grid, 1, 310-315.
9. Liu, Y., Zhang, Y., Chen, K.R., Chen, S.Z. and Tang, B. (2016) Equivalence of Multi-Time Scale Optimization for Home Energy Management Considering User Discomfort Preference. IEEE Transactions on Smart Grid, 1, 1-12.
10. Wu, D., Tang, F., Dragicevic, T., Vasquez, J.C. and Guerrero, J.M. (2015) A Control Architecture to Coordinate Renewable Energy Sources and Energy Storage Systems in Islanded Micro Grids. IEEE Transactions on Smart Grid, 6, 1156-1166.

11. Makhe, A., Bugade, V.S., Matkar, S. and Mothe, P. Digital Protection of LVDC and Integration of Distributed Generation. IEEE International Conference on Energy Efficient Technologies for Sustainability (ICEETS'16).
12. Liu, B.Q., Zhou, F., Zhu, Y.X. and Yi, H. (2015) System Operation and Energy Management of a Renewable Energy-Based DC Micro-Grid for High Penetration Depth Application. IEEE Transactions on Smart Grid, 6, 1147-1155.

SECTION 3: SECURITY AND STABILITY OF SMART GRIDS

Chapter 11

A Secure and Scalable Data Communication Scheme in Smart Grids

Chunqiang Hu[1,2,3], Hang Liu[3], Liran Ma[4], Yan Huo[5], Arwa Alrawais[6] , Xiuhua Li[7], Hong Li[8], and Qingyu Xiong[1,2]

[1]Key Laboratory of Dependable Service Computing in Cyber Physical Society, Chongqing University, Ministry of Education, Chongqing, China

[2]School of Software Engineering, Chongqing University, Chongqing, China

[3]Department of Electrical Engineering & Computer Science, The Catholic University of America, Washington, DC, USA

[4]Department of Computer Science, Texas Christian University, Fort Worth, TX, USA

[5]School of Electronics and Information Engineering, Beijing Jiaotong University, Beijing, China

[6]Department of Computer Science, The George Washington University, Washington DC, USA

[7]Department Electrical and Computer Engineering, University of British Columbia, Vancouver, BC, Canada

[8]Beijing Key Laboratory of IOT Information Security Technology, Institute of Information Engineering, CAS, Beijing, China

Citation: Bugade, V. (2017), "Optimal Power Flow Approach for Cognitive and Reliable Operation of Distributed Generation as Smart Grid". *Smart Grid and Renewable Energy*, 8, 87-98. doi: 10.4236/sgre.2017.83006.

ABSTRACT

The concept of smart grid gained tremendous attention among researchers and utility providers in recent years. How to establish a secure communication among smart meters, utility companies, and the service providers is a challenging issue. In this paper, we present a communication architecture for smart grids and propose a scheme to guarantee the security and privacy of data communications among smart meters, utility companies, and data repositories by employing decentralized attribute based encryption. The architecture is highly scalable, which employs an access control Linear Secret Sharing Scheme (LSSS) matrix to achieve a role-based access control. The security analysis demonstrated that the scheme ensures security and privacy. The performance analysis shows that the scheme is efficient in terms of computational cost.

INTRODUCTION

The concept of smart grid gained tremendous attention among researchers and utility providers in recent years. With such a technology, advanced developments such as sensing, control, digital communications, and networking are integrated into the power systems to effectively and intelligently control and monitor the power grid. Generally speaking, the power grid consists of three major components: power generation, power transmission, and power distribution [1]. Typically, wired communications such as optical networking are adopted to support the power backbone consisting of the power generation and transmission systems [2]; but for the power distribution network, which provides power directly to customers, both wired and wireless communications are adopted.

Smart grid brings new features into the power grid: renewable-based generation, demand-response, wide area protection, and smart metering, just to name a few [3]. Within a smart grid, utility companies can send alerts to notify customers and may further ask them to reduce their power consumption by temporarily turning off some devices during the periods of peak energy consumption [4]. The certain critical control actions can be sent from the control center to smart meters, in which the actions are expected to be taken immediately for safe operations, and the wide area protection schemes can be deployed to prevent cascaded failures and provide better interconnections. However, despite the attractive features provided by smart grid technologies, challenges, especially those in cyber security and privacy [5], are still present. For example, it has been reported that the pervasively adopted integrated Supervisory Control and Data Acquisition (SCADA)/ Energy Management systems [6] are vulnerable to significant security threats [7–10].

As paper [11] pointed out, we need new technologies to protect the confidentiality of the customer's data. Also, customer's privacy should be preserved when data are collected for marketing purpose. It has been demonstrated by [12] that, even without a priori knowledge of household activities, it is still possible to extract customers' usage patterns from the data uploaded by smart meters once every 15 minutes.

Utility companies need customer's energy consumption data for billing purpose. Third-party service providers may need to collect electricity usage records of certain smart devices to monitor device's status and detect potential problems. Some other data analysis companies may need user's energy consumption data to do market research. From customer's perspective, customer should have control over their own data. It means that customer knows and controls the access to his own energy consumption data. If the data is needed for marketing purpose, customer should be informed and guaranteed that his own data are anonymized. Traditionally, smart meter needs to learn receiver's identity (e.g., smart meters should know the certificate of the utility company) and decides whether to send its data or not. For such a large communication network, it may not desirable for smart meters to learn all the identities. And the wide used public-key infrastructure based on X.509 protocol on Internet does not provide enough security guarantees since a fake or stolen certificate may cause tremendous damage and loss in smart grid communication network. On the other hand, all the data can be uploaded into a data repositories [13–15], which store customers' data and distribute them to the third-party service providers under the supervision of a fine-grained access control. It is the data repositories' responsibility to enforce the access control policies and distribute customers' data based on customer's choice and the related regulations and laws, which certainly put tremendous burden on the data repository servers since the compromise of a data repository server reveals all the data it maintains.

To tackle the challenges, we take a fundamentally different approach by employing attribute based crypto system: attribute based encryption (ABE) enables the smart meters to encrypt its data on a set of descriptive attributes, which determine the access privilege of the data. All the legitimate users that may have different identities but possess appropriate sets of attributes can decrypt the data independently. This successfully implements a secure multicast of the customer's data to multiple users, and the smart meters even do not need to know the receiver's detailed identity. Attribute based signature (ABS), in which a signature attests not to the identity of the individual who endorsed a message, but instead to a (possibly complex) claim regarding the attributes she possesses [16], provides a strong unforgeability guarantee for the verifier that the signature was produced by a single party whose attributes satisfy the claim being made. Also, the signature reveals nothing about the identity and even the attributes of the signer beyond what is explicitly revealed by the claim being made. This successfully solves the problem of data anonymity, so that the marketing companies only know that the data comes from the desired group of customer and customer's identity is fully preserved.

Attribute based encryption, more specifically, Ciphertext-Policy Attribute Based Encryption (CP_ABE) [17], provides a secure multicasting and role-based access control. Data stored on data repositories are encrypted and the compromise of data repositories only leaks the encrypted data. It does not need to use a software approach that checks an entities' privilege and decide whether access is granted or not. Attribute based signature is more preferable than other privacy preserving signature schemes such as group signatures [18, 19], ring signatures [20], and mesh signatures [21]; that is, ABS is more practical and provides a stronger guarantee on privacy. Group signature needs a predefined group of people and a group manager. Ring signature needs a predefined group of people too. And the group should be large enough to achieve anonymity. As for mesh signature, it explicitly allows collusion [16], which is not desirable in our case.

ABE and ABS need attribute authority to issue secret keys for attributes so the entity with proper set of secret keys can decrypt and sign a message. In a large scale communication network like smart grid, the attribute authority might become the bottleneck of the entire system. It is desirable to have attribute authority distributed. The decentralized attribute based encryption proposed by Lewko and Waters [22] makes multiauthority possible, and attribute authority does not need to trust each other in the system. Multiauthority ABS has been proposed by Maji et al. [16] that enables multiauthority settings too. In our paper, we mainly focus on implementing and analyzing the decentralized ABE and multiauthority ABS in smart grid communication network.

The contributions of this paper are summarized as follows:(1)We propose a secure and scalable communication architecture involving multiple authorities, smart meters, data consumers, and data repositories for smart grid systems. Our architecture emphasizes customers' control on their data and privacy.(2)We implemented decentralized attribute based encryption scheme [22] and multiauthority attribute based signature [16] scheme. We described the communication protocols to achieve customer controlled access control and data anonymity.(3)We measured the performance of the implemented schemes on different types of curves and groups. We analyzed the efficiency of the implemented schemes and provide future research directions.

The remainder of this paper is structured as follows. In Section 2, we discuss the related work. In Section 3, we introduce the required preliminaries and the system model. Section 4 proposes the secure communication mechanism and presents a scheme to ensure access control for the sensitive data. Section 5 gives performance analysis, followed by the conclusions in Section 6.

RELATED WORK

In smart grid communication network, security problems mainly lie in the subjects of sensor networks, wireless networks, and Internet. A significant amount of research has been carried out to protect the smart grid systems. Multicast authentication schemes such as TELSA, Biba, HORS, and OTS [3, 23] were proposed for authenticating entities such as utility companies and control centers when messages or control commands are sent to smart meters. To authenticate smart meters or other smart devices to the control center, batch verification schemes [24–27] were developed to improve the efficiency. Data aggregation based on homomorphic encryption, secret sharing, and other technologies [13, 25, 26] was designed to aggregate customers' data and to protect their privacy.

Recently, ABE has received significant amount of attention in securing smart grids because it does not require certificates and it can be used to construct a fine-grained access control mechanism. Actually, the original motivation for ABE scheme is to design an error-tolerant (or fuzzy) identity-based encryption scheme [28] that could be applied to biometric identities. However, the original threshold ABE scheme in [28] is not very impressive as it is limited from designing more general systems. A more general idea called key-policy attribute based encryption (KP_ABE) was proposed by Goyal et al. [29] to embed a general secret sharing scheme for a monotonic access tree instead of the Shamir secret sharing scheme used in [28].

Later, Bethencourt et al. proposed the Ciphertext-Policy Attribute Based Encryption (CP_ABE or BSW CP_ABE) scheme [17] that reverses the KP_ABE construction: the encrypted data (the ciphertext) carries an access structure over attributes; meanwhile, a user's private key is associated with a set of descriptive attributes. The owner or the encryptor now has more control over the data by constructing an access structure for every data to be encrypted.

Later, ABE has been utilized to fit practical problems. Pirretti et al. implemented the threshold ABE system [30] while Chase [31] provided a construction for a multiauthority attribute based encryption system. A decentralized Ciphertext-Policy Attribute Based Encryption scheme was proposed in [22], which deals with the fact that, in practice, there may be more than one attribute authority. And we implemented the decentralized ABE in prime order group in this paper and further analyze the computational cost in different curves and groups.

ABS was introduced by Maji et al. in [16] to achieve a strong unforgeability guarantee for the verifier, which means that the signature was produced by a single party whose attributes satisfy the claim being made. And the privacy of the singer is fully preserved since the signature reveals nothing about the identity or attributes of the signer beyond what is explicitly revealed by the claim being made. However, the security proof in [16] is in generic model group. Later Li et al. proposed an ABS scheme that is selective secure in standard model. But the scheme deals with only (t,l)-threshold, which means that it may not be as expressive as Maji et al.'s ABS scheme, which uses an monotone access structure. Moreover, since we prefer large universe construction in smart grid communication network, it is hard and unpractical to implement schemes that are secure in standard model (usually we need to have a polynomial P(x) with degree d and the size of public parameters grows with d). We implemented and analyzed Maji et al.'s multiauthority ABS [16] scheme in this paper. One has to notice that, in multiauthority attribute based crypto system, attribute authorities are completely independent from each other, which is a desirable feature for large scale, distributed smart grid communication network.

As a promising technique, identity/attribute based crypto system has been proposed to solve problems in smart grid communication network. A scheme that employs IBE to provide a zero-configuration encryption and authentication solution for end-to-end secure communications was proposed in [32]. The concept of IBE was utilized by [25] to construct a signature and later verify the signature. KP_ABE was adopted by [33] to broadcast a single encrypted message to a specific group of users. Reference [13] utilizes the Linear Secret Sharing to construct the access policy [22, 34] and then enforce access control. However, most of the works done before have no implementation and real life performance analysis. This paper serves as a step that brings the discussion to a more practical stage: implementation and performance analysis. Essentially, the decentralized ABE scheme and multiauthority ABS scheme have their own set of parameters. There are works which have been done to combine ABE with ABS [35], which can be a potential future research direction.

PRELIMINARIES

In this section, we mainly introduce the preliminaries related to our actual implementation. Theoretical preliminaries can be found in [16, 22, 36, 37].

Bilinear Maps

Let $\mathbb{G}, \mathbb{G}_T$ be cyclic groups of prime order r. Let g be a generator of $\mathbb{G}$. A symmetric bilinear map [38] e is an efficiently computable function:

$$e : \mathbb{G} \times \mathbb{G} \longrightarrow \mathbb{G}_T, \quad (1)$$

such that

Float 1:*(Nondegeneracy)* $e(g, g) \neq 1$;

Float 2:*(Bilinearity)* $e(g^a, g^b) = e(g, g)^{ab}$ for all $a, b \in \mathbb{Z}$.

A asymmetric bilinear map is that e is an efficiently computable function:

$$e : \mathbb{G}_1 \times \mathbb{G}_2 \longrightarrow \mathbb{G}_T, \quad (2)$$

and the property of Nondegeneracy and Bilinearity still hold. We run our implementation on both symmetric and asymmetric pairings and analysis the efficiency.

Access Structure

We mainly discuss monotone access structure (MAS) [39] here.

Let $\{P_1, \ldots, P_n\}$ be a set of parties. A collection $\mathbb{A} \subseteq 2^{\{P_1,\ldots,P_n\}}$ is monotone if $\forall B, C,\ B \in \mathbb{A}$ and $B \subseteq C$ imply $C \in \mathbb{A}$. An monotone access structure is a monotone collection $\mathbb{A}$ of nonempty subsets of $\{P_1, \ldots, P_n\}$. The sets in $\mathbb{A}$ are called authorized sets, and the sets not in $\mathbb{A}$ are called unauthorized sets. For example, let $P = \{A, B, C, D\}, \{\{A, B\}, \{B, C\}, \{C, D\}, \{A, B, C\}, \{A, B, D\}, \{A, C, D\}, \{A, B, C, D\}\}$ be a MAS. More importantly, we use a Boolean formula (with only AND and OR gates) to describe a MAS. For example, we are using (A AND B) OR (B AND C) OR (C AND D) to represent the MAS mentioned before.

We are more familiar with (t,l)-threshold gate and a threshold gate in [17] can be represented as Boolean formula. For example, an (2, 3)-threshold gate of $\{A, B, C\}$ can be expressed as (A AND B) OR (B AND C) OR (A AND C). In this paper, we are using Boolean formula to express an access structure.

Further, we are using the linear secret sharing schemes (LSSS) proposed in [39, 40], which means we will parse a Boolean formula into a access matrix A and a mapping $\rho()$, where A is called the share-generating matrix and $\rho()$ maps rows of the matrix to the elements in the Boolean formula. Formally, A has ℓ rows and n columns, and the xth row of A will be mapped to an elements in Boolean formula by the function $\rho(x)$. When we consider the column vector v=$(s, r_2, \ldots, r_n)$, where $s \in \mathbb{Z}_p$ is the secret to be shared and $r_2, \ldots, r_n \in \mathbb{Z}_p$ are randomly chosen, $A \cdot v$ is the vector of ℓ shares of the secret s. The share $(A \cdot v)_x$ belongs to the element $\rho(x)$.

We use the converting method in [22] and the detailed algorithm is described in Section 5.6. Here is an example: consider an access structure (A AND (D OR (B AND C))); the corresponding access matrix and $\rho()$ will be

$$\begin{pmatrix} 1 & 1 & 0 \\ 0 & -1 & 0 \\ 0 & -1 & 1 \\ 0 & 0 & -1 \end{pmatrix} \overset{\rho()}{\longmapsto} \begin{bmatrix} A \\ D \\ B \\ C \end{bmatrix}. \tag{3}$$

For an authorized set $\{A, B, C\}$, the corresponding matrix A_{matched}

$$\begin{pmatrix} 1 & 1 & 0 \\ 0 & -1 & 1 \\ 0 & 0 & -1 \end{pmatrix} \tag{4}$$

has a vector $(1, 0, \ldots, 0)$ in their span. In other words, there is a vector c_x, which in this case is $(1, 1, 1)$, and $c_x \times A_{\text{matched}} = (1, 0, \ldots, 0)$. In this case,

$$[1 \ \ 1 \ \ 1] \times \begin{pmatrix} 1 & 1 & 0 \\ 0 & -1 & 1 \\ 0 & 0 & -1 \end{pmatrix} = [1 \ \ 0 \ \ 0]; \tag{5}$$

$(c_x \times A_{\text{matched}}) \times v^T = s$ means that once we have A_{matched} and c_x, we can recover s. The processing described above is called *linear reconstruction.*

Note that we do not lose any efficiency by using the LSSS matrix as opposed to the previously used tree access structure descriptions in [17]. The reason is that the computational cost is directly related to the number of attributes involved in the encryption or sign, and the computational cost of linear reconstruction or polynomial interpolation is negligible. Section 5 will go through a detailed analysis of computational cost.

Security Notions and Models

There are two security notions in identity-based encryption: selective-ID secure and fully secure. Selective secure, introduced by Canetti et al. [41, 42], is weaker than fully secure, which was introduced by Boneh and Franklin in [43]. Generally speaking, fully secure means that the scheme is secure even if the adversary adaptively selects identity to attack based on previous secret keys. For selective secure, the adversary must commit ahead of time to the identity that he will attack. In other words, adversary in fully secure is more powerful since he can query even after he receives the identity to attack.

There are several security models for public-key crypto system. The *random oracle model* was first introduced by Bellare and Rogaway [44]. It assumes that the adversary has the access right to a public, truly random hash function H, which is based on SHA-1. Random oracle model is very useful in practice, but from a theoretical perspective, the *standard model* is more preferred. In the standard mode, security is proven using only standard complexity assumptions. For example, [45] is built on *Decisional Bilinear Diffie-Hellman Assumption* and *Computational Diffie-Hellman Assumption*.

Even if standard model is desirable from the perspective of theory, random oracle model is more practical especially when it comes to large universe construction. Paper [46] is fully secure under standard model. But we need to random a set of group elements for attributes in the system. It means that attributes are defined at the setup and published in the public parameters. We call this kind of construction as "small universe construction." In practice, especially in a communication network like smart grid, it is desirable to dynamically use any attribute as we want. The easy way to do this is to use a hash function that we model as a random oracle to map an attribute to a group element. However, we end up with a scheme that secure in random oracle model.

If we still adopt the standard model, we can use a polynomial P(x) with degree d [46] and map attributes in $\mathbb{Z}_q$ to elements in $\mathbb{G}$ by setting $H(\text{attribute}) := g^{P(\text{attribute})}$, where g is the generator of group $\mathbb{G}$. The public parameters would then include $\{g^{P(x)}\}$ for d+1 points x so that $H(\text{attribute})$ could be computed for any attribute by polynomial interpolation. One has to notice that, in practice, we not only need to map an attribute into a group element, but also need to map an identity (which we call it uid in this paper) into a group element. Since $P(x)$ is a $(d + 1)$-wise independent function modulo primes, the system is vulnerable to collusion attacks when a user has d+1 secret keys or more than d+1 users get together to collude. To prevent this from happening, we need to set d large enough so that no users will have more than d+1 secret keys and it is impossible for more than d+1 users to get together and collude. This will boost the size of public parameters and the assumption that no more than d+1 users will collude sounds less convincing than random oracle model and a SHA-1 hash function.

Generic Group Model, Composite, and Prime Order Groups

Besides random oracle model and standard model, there is a model called *generic group model*, proposed by Shoup [47]. The model relies on hardness of problems related to finding the discrete logarithm in a group with bilinear pairings. In the model, algorithms can only manipulate group elements via canonical group operations (including the bilinear pairing). We are using prime order groups here in our paper since prime order subgroups of general elliptic curve groups are good examples of groups where all known attacks against the discrete log problem are not significantly better than attacks in the generic group. The multiauthority ABS [16] is secure in the generic group model. The decentralized ABE [22] in prime order groups is secure in generic group model too.

Bilinear groups of composite order were introduced by Boneh et al. [48]. Since the elliptic curve group order n must be infeasible to factor in composite order group, it must be at least 1024 bits. On the other hand, the size of a prime order elliptic curve group that provides an equivalent level of security is 160 bits. It is not practical to implement the decentralized ABE scheme on composite order group since group operations and especially pairing computations are prohibitively slow on composite order curves [49]. A Tate pairing on a 1024-bit composite order elliptic curve is roughly 50 times slower than the same pairing on a comparable prime order curve [49]. The small universe construction of decentralized ABE is fully secure in standard model in composite order groups. However, we implemented the decentralized ABE scheme in prime order group and the security reduced to the generic group model.

In summary, we implemented the decentralized ABE and multiauthority ABS that are secure in generic group model. We test and analyze the performance of implementation under both symmetric groups and asymmetric groups. And we are using LSSS matrix and linear reconstruction in our implementation.

ARCHITECTURE AND PROTOCOL

In this section, we introduce our architecture and communication protocol. Generally speaking, we use decentralized ABE to achieve a fine-grained access control on data collected by smart meters. Also, multiauthority ABS has been used to achieve data anonymity when data consumers or marketing companies need data from certain area or subset of smart meters while user's privacy needs to be preserved.

System Model

We consider the architecture in Figure 1 as the basis of our following discussion. Figure 1 reproduced from Hu et al. (2017) [14]. There are different entities in the communication structure: attribute authorities (AAs), smart meters, data repositories, and data consumers. Data consumers mainly refer to the utility companies (UCs) and third-party service providers (TPDCs). The following sections are a brief introduction to all the entities.

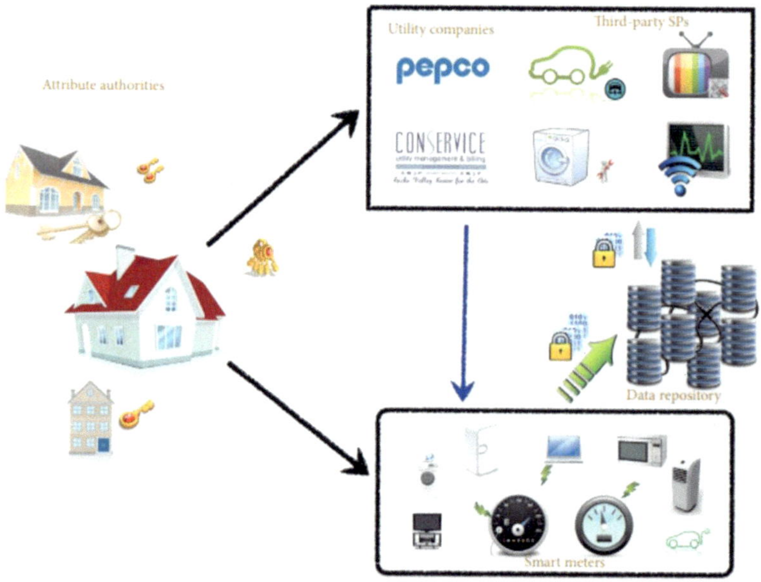

Figure 1: A communication architecture in smart grid systems (Hu et al. (2017) [14]).

(1) *Attribute Authorities (AAs).* AAs are responsible for generate and distribute secret keys for smart meters and data consumers. There are multiple AAs in the system and they may not know each other or trust each other. An AA is only responsible for generating secret keys for attributes. We assume that every entity in the system has a unique identifier (GID or uid), and any entity should prove its identity to AA if it needs secret key for its attributes. In this, we do not discuss how to obtain the GID or uid for an entity and how to prove its identity for AAs. Generally speaking, in a communication network like smart grid, every entity (e.g., smart meter) has a unique ID and registered in certain

government authorities. The distribution of secret keys can be done by preestablished channel.

Note that a signature trustee should be deployed besides AAs in a multiauthority ABS system. The signature trustee is responsible for issuing an "ID" to the entity. We model the signature trustee as an attribute authority in our architecture.

(2) *Smart Meters.* Smart meters are the key entities in a smart grid communication network. Smart meters collect user's energy consumption information and other pieces of information. In a home area network, smart meters are the center controller. Smart meters monitor the activities of every smart device in the home area. In our architecture, smart meters mainly collect user smart devices' energy consumption information. The total energy consumption can be used by UCs to charge the bill. Energy consumption by some smart devices (e.g., e-cars, TV, and PCs) can be used by third-party SPs to analysis device's working status and diagnose potential problems. Also, TPSPs can use those data to do market analysis and further guide the marketing. However, in this case, anonymity should be enforced to preserve user's privacy and we are proposing multiauthority ABS to achieve data's anonymity.

Each smart meter has a unique officially certified ID, which registers in the system. The communication between smart meters (and any other entities that need secret keys from AAs) and AAs is preestablished secured channel, which is out of our paper. Identity-based encryption/signature systems are an intriguing candidate to establish a secure channel between smart meters and AAs since every single entity is uniquely identified. We leave the integration of identity-base encryption/signature schemes as one of the future works. In the same time, smart meters use attribute based encryption to encrypt its data to achieve a user defined fine-grained access control. For example, the user can construct an access structure ("ARLINGTON.22202" and "ARLINGTON.UC") and encrypt data with it. Only the entity that has corresponding valid set of key can decrypt the data. The data consumers may need data for market purpose and want to protect users' privacy too. Smart meters can sign a data with the secret key for attributes and claim that the secret keys it process satisfy the predicate, which is the access structure or access matrix. One has to notice that we trust smart meters to honestly encrypt and sign a message. The compromise of a smart meter may cause some misbehavers. For example, the attacker controls some smart meters to encrypt and sign any data at any frequency. Further mechanisms should be adopted to secure smart meter and detect the attacks, which is also beyond our discussion here.

(3) *Data Repositories.* Data repositories are storage facility that stores the encrypted or signed data. In attribute based crypto system, the data needs to be encrypted or signed once and later any entities with appropriate set of secret keys can decrypt. Instead of store all the data themselves, smart meters can upload the data to the data repositories and data consumers can retrieve the data from the repository. Data repositories should have higher network throughout capacity. It is certainly more reasonable to have some data repositories with high network bandwidth than having all communication between smart meters and data consumers directly, which may require every smart meter to have higher network processing capacity.

The deployment of data repositories does not affect the confidentiality of the data encrypted under an access structure. The data uploaded by smart meters are encrypted with ABE and only the entities with appropriate set of secret keys can be decrypted. ABE reduces the trust we traditionally put on a data repositories, which has software to enforce the access policy based on the records to describe every entity's privilege. ABE's key feature is the fine-grained access control provided by underlying cryptography algorithms. The data repositories handle the request and deliver the data. Even if a data repository is compromised, the data are safe since they are encrypted.

Note that the data is already protected by ABE and we do not need to have a secured channel between data repositories and other entities. However, the assumption is that every entity in the system has a unique identifier and every entity has the ability to verify the sender's identity. This can be done with identity-based encryption/signature, of which we leave the integration as one of the feature works.

(4) *Data Consumers.* Data consumers refer to utility companies (UCs) and third-party service providers (TPSPs). Generally speaking, UCs need the data collected by smart meters to do the billing. TPSPs may need the data collected by smart meters regarding a specific device to understand their working status and detect potential problems. Also, TPSPs may need data to do market research while they protect user's privacy. Briefly, if data consumers need data, they can retrieve data from data

repositories and decrypt it if they have required secret keys to satisfy the access structure. Data consumers verify the signature on data during anonymity data collection, too.

Protect Smart Meter's Data with ABE

We implemented appendix D of paper [22]. Decentralized ABE scheme will enable user defined access control to the data. We will talk about how we use decentralized ABE scheme to protect the data collected by smart meters in this subsection. The scheme we implemented can be found at Appendix of this paper and the following subsections briefly describe the algorithm and the communication protocol.

(1) Global Setup. Global setup in DABE will output $\mathsf{ABE\_GP}$, which contains the generators, an hash function we model as a random oracle. Also, $e(g_1, g_2)$ is precalculated.

(2) Authority Setup. We describe the AAs as the issuer of secret keys for attributes. One has to notice that AAs are independent with each other and even if two AAs issue secret keys for the same attribute called "TV," they are essentially different and one should specify which AA the attribute belongs to during the encryption and decryption. We are using the format "Arlington. TV" to represent an attribute. The first part of the attribute name is the name of the AA and the second part is the description of the attribute. In this way, attribute "WashingtonDC.TV" is different from "Arlington.TV" and it becomes much more clear during the encryption and decryption regarding which AA an attribute belongs to. In the attribute authority setup, the AA will generate two random exponents for each attribute and publishes PK, which contains all the public keys for attributes and AA will save exponents as the secret key. For example, given an input,

{$\mathsf{ABE\_GP}$, attributes = {"22201", "22202", "*CAR*", "*TV*"}, *AA*name = "Arlington"}

Algorithm ABE_AuthoritySetup() will output:

$\mathsf{ABE\_APK}_{\text{Arlington}}$:=
{"Arlington.22201": $\{\mathbb{G}_T \times \mathbb{G}_1\}$, "Arlington.22202": $\{\mathbb{G}_T \times \mathbb{G}_1\}$, "Arlington.CAR": $\{\mathbb{G}_T \times \mathbb{G}_1\}$, "Arlington.TV": $\{\mathbb{G}_T \times \mathbb{G}_1\}$}

$\mathsf{ABE\_ASK}_{\text{Arlington}}$:=
{"Arlington.22201": $\{\mathbb{Z}_q \times \mathbb{Z}_q\}$, "Arlington.22202": $\{\mathbb{Z}_q \times \mathbb{Z}_q\}$, "Arlington.CAR": $\{\mathbb{Z}_q \times \mathbb{Z}_q\}$, "Arlington.TV": $\{\mathbb{Z}_q \times \mathbb{Z}_q\}$}

as the APK and ASK for AA "Arlington," which are python dictionaries indexed by the name of the attribute (the concatenation of AA's name and attribute's name) and $\{\mathbb{G}_T \times \mathbb{G}_1\}$, mean the that they contain an element in $\mathbb{G}_T$ group and an element in $\mathbb{G}_1$ group.

(3) *Attribute Generation.* In order to decrypt a data block encrypted by smart meters with an access structure, data consumers need to process a proper set of secret keys. Data consumers obtain secret keys from AAs first. We assume that data consumers and AAs can establish a secured communication by other ways via identity-based encryption/signature or traditional PKI. Figure 2 illustrates the protocol between data consumer and AAs. Data consumer "UC_Pentagon" needs secret keys for attribute "Arlington.22202," "Arlington.22201," and "Arlington.TV." The AA "Arlington" will first check if the attributes belong to it or not and it will only generate secret keys for attributes it has.

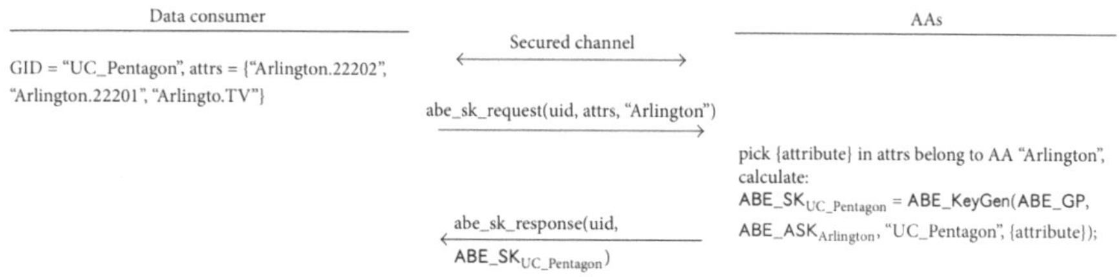

Figure 2: DABE: secret key generation.

(4) *Encryption.* Smart meters can upload encrypted data to data repository. Data will be encrypted by ABE with an access structure (AS). The AS will be converted into a access matrix A in the encryption algorithm. Figure 3 illustrates the protocol between smart meter and data repository. There is no need to establish a secured channel forehead since the data transmitted are already encrypted. The MAC in the protocol serves as a proof of sender's identity and protects the integrity of the payload and so do all the MAC described in the following section. If we have identity-based signature in our system, we can use the identity-based signature to sign a digest of the payload. We leave the integration of identity-based encryption/signature as one of the future works.

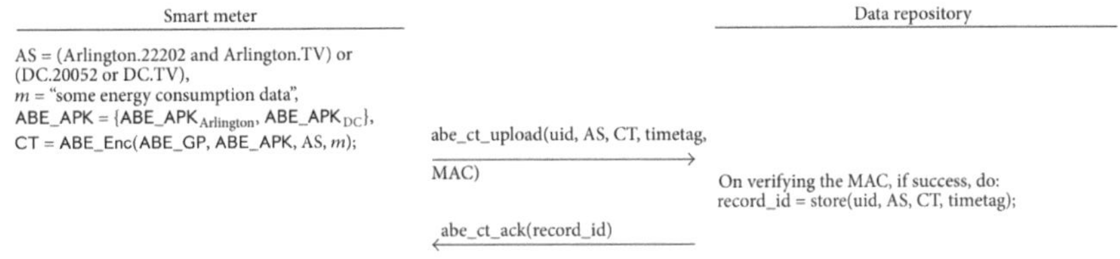

Figure 3: DABE: encryption.

(5) *Decryption.* Data consumer can retrieve the data from data repositories by using the record_id or (uid and (startTime, endTime)). The data repository will return the ciphertext. On receiving the ciphertext, data consumer will decrypt the data with the secret keys it has. Figure 4 illustrated the communication between data consumer and data repository. One has to notice that data consumer will have secret keys from different AAs. And the decryption should distinguish keys from different AAs.

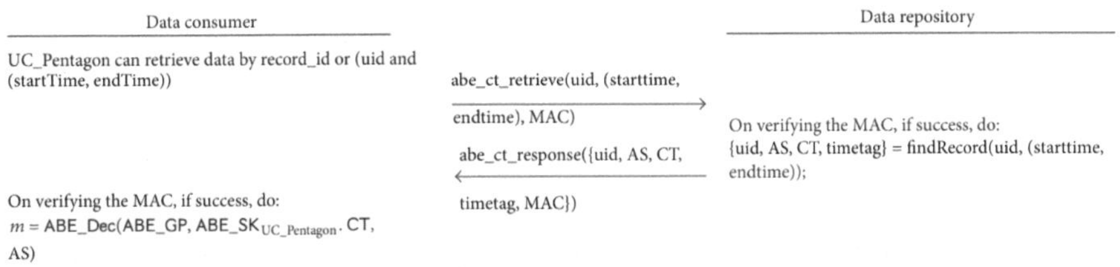

Figure 4: DABE: decryption.

Protect Data Anonymity by ABS

We use the ABS to provide data anonymity and achieve sender's verification. On verifying the signature, the receiver knows that the secret keys the sender have satisfy the access structure and nothing more. ABS provides a strong privacy guarantee. The following subsection describes the communication between entities. The code can be found in Appendix of the paper and we will only highlight the communication protocols in the following subsections. There are researchers working on ABE and ABS that share the same set of parameters [35], but for now, we treat ABE and ABS as separate systems, which means that the global parameters, keys, and other parameters are different.

(1) *Global Setup.* The difference between decentralized ABE and multiauthority ABS is that ABS has one more entities,

which is called "signature trustee." Signature trustee will issue a token to a user based on its "uid" and the token must be provided when a user requests secret keys from AAs. In our implementation, we model the signature trustee as one of the AAs (AA "signature trustee") too. And AA "signature trustee" will run the $\mathsf{ABS\_GlobalSetup}(t_{\max}) \rightarrow \mathsf{ABS\_GP}, \mathsf{ABS\_TSK}$. AA "signature trustee" will save $\mathsf{ABS\_TSK}$ and publish $\mathsf{ABS\_GP}$. $t_{\max}$ is the max number of columns in an access structure, which is related to the numbers of AND gate in the access structure. In our implementation, we first give t_{max} a value and the value can be changed to a larger value in the future if needed.

(2) *Authority Setup.* The authority setup of multiauthority ABS is similar to the authority setup of decentralized ABE except that there is no need to explicitly specify the set of attributes at the setup. One has to notice that the decentralized ABE and multiauthority ABS scheme we implemented are both in large universe construction, which means that we can have as much attributes as we want. AA in multiauthority can issue keys for any attributes. However, in $\mathsf{ABS\_Sign}()$ and $\mathsf{ABS\_Verify}()$, one must explicitly specify the source of the attributes, which means that, for every attribute, one needs to specify which AA it belongs to.

(3) *Token Register and Attribute Generation.* Before entities request secret keys for attributes, the entity needs to register itself at AA "signature trustee." The signature trustee will produce a token for an entity. With the token, an entity can request secret keys from any AAs in the system. One has to notice that secret key for attribute "Arlington.22201" in multiauthority ABS is different from the secret key for "Arlington.22201" in decentralized ABE system even for the same entity. They belong to different scheme and we donate them separately as $\mathsf{ABS\_SK}_{\mathrm{uid}}$ and $\mathsf{ABE\_SK}_{\mathrm{uid}}$. Also, the communication happens in a secured channel. Figure 5 illustrates that the smart meter "SM_RiverhouseApt" requests its token and secret keys from AAs.

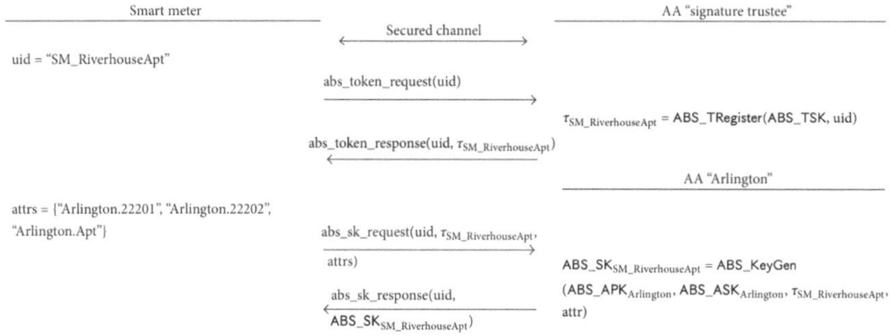

Figure 5: Multiauthority ABS: token and secret key generation.

(4) *Sign.* To sign a message m, the smart meter must have proper set of secret keys. If it does not have, ABS_Sign() will abort at the first stage. Also, AS will be parsed into an access matrix A with a mapping function $\rho()$. As what we did in decentralized ABE, the AS here also explicitly tells the AA of an attribute by using an attribute like "Arlington.22201." Signed data will be uploaded to the data repository too. Figure 6 illustrates the communication between smart meters and data repository.

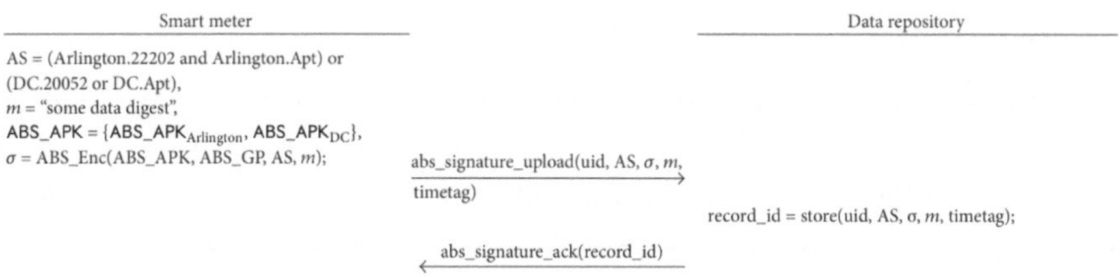

Figure 6: Multiauthority ABS: sign.

(5) *Verify.* In verify, if the ABS_Verify() returns reject, the verification failed. The verification is successful if it passes all the "checkpoint." Figure 7 illustrates the communication between data consumer and data repository.

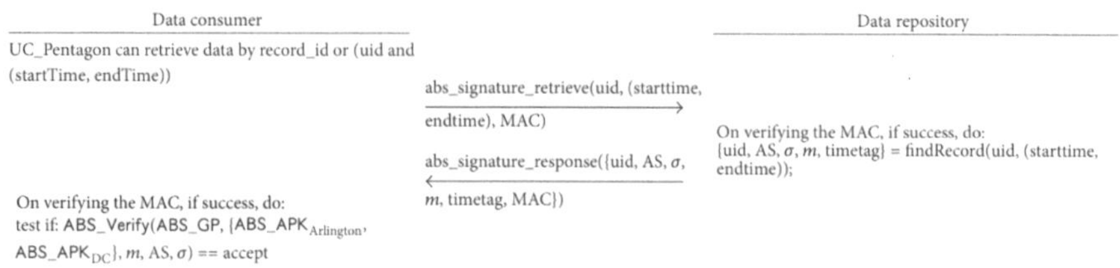

Figure 7: Multiauthority ABS: verify.

Combine ABE and ABS

If a customer wants his smart meter to anonymously sign a data and, in the meanwhile, control the access by an access structure, the smart meter can combine ABE and ABS. It can either sign first and then encrypt or encrypt first and then sign. Since every entity in the system can verify a signature but only the entities with proper set of secret keys can decrypt, our recommendation is to sign and then encrypt. The reason is simple: for those entities that cannot decrypt, we do not want them to know that the signature ever existed. From the perspective of data analysis companies, it can only collect data that intended to been sent to them.

Eliminate the MAC in the Protocol

We use a MAC in the communication protocol. Actually, if we have identity-based signature (IBS) in our system, we can use IBS to sign the digest of the payload. To integrate an IBS into our current architecture, we may need a trustee that certifies every identity. We leave it as a future research direction.

In our current architecture, one can remove the MAC by using our ABS and set the access structure to be the sender's identity. The sender (in this case, a smart meter) can obtain a secret key for its identity and sign the message with an access structure that involved only its identity. However, the computational cost of doing this is larger than using IBS and we discourage this particular method.

Security Analysis

Both schemes we implemented are secure in generic group model. In actual large university construction of attribute based crypto system, security in standard is hard to achieve since we need to introduce a polynomial and assumptions that no more than certain amount of user will get together and collude. Generic group model and random oracle model are practical in real-left applications.

PERFORMANCE ANALYSIS

We implemented the decentralized attribute based encryption scheme in prime order group, the scheme in appendix D in [22]. Also, we implemented the multiauthority ABS in Section 4.2 of paper [16]. This section discusses the implementation details and performance analysis.

Implementation Details

The implementation is based on a python library, Charm crypto [50], which is framework that is prototyping advanced cryptosystems such as IBE and IBS. The core mathematical functions behind Charm are from the Stanford Pairing-Based Cryptography (PBC) library [38], which is a free C library that performs the mathematical operations underlying pairing-based cryptosystems. At the same time, there is a project called TinyPBC [51] that has a better performance in terms of elements multiplication in groups. The efficiency of multiplication was improved by a factor of 4-5 and so does the Exponential operation. However, the current release of Charm does not have TinyPBC integrated. We are still using the PBC library for underlying mathematical operations.

The implementation of the decentralized ABE scheme is a little bit different from the original scheme due to the fact that the original paper describes the scheme in symmetric groups. We implemented the decentralized ABE scheme in asymmetric groups and add some precalculated values into the public parameters to reduce the computational cost in Enc() and Dec(). The detailed implementation can be found at Appendix. Since we are using the prime order group instead of the composite order group, the scheme implemented is secure in generic group model. As mentioned before, using composite order groups will largely increase the element size in groups. The computation cost will be boosted especially when we want higher security level; for example, A Tate pairing on a 1024-bit composite order elliptic curve is roughly 50 times slower than the same pairing on a comparable prime order curve [49]. As argued above, generic group model and random oracle model are practical in real life applications. The implementation of multiauthority ABS can also be found at Appendix. Some notations have been changed to avoid confusion.

We are running the code on 32-bit Ubuntu 12.04, which is a virtual machine running in VMWare fusion on a MACbook Air with 1.8 GHz Intel i5 and 4 GB memory. The virtual machine has access to one core of CPU and maximum 1 GB of memory. The PBC library provides a preprocessing mode for Exponential and Pairing. However, we did not use any preprocessing here since Charm did not integrate it. One has to notice that the preprocessing improves the performance by precalculating some value, which means that the preparation itself takes a long time. Preprocessing is recommend when there are a lot of Exponential and Pairing operations to compensate the cost of preparation itself.

Groups and Curves

We implement based on both symmetric groups and asymmetric groups. We will use "SS512" to denote the symmetric group that has a 160-bit order and 512-bit long of base field. A group with order of 160 bits equals 80 bits of NIST symmetric encryption security. For asymmetric groups, we use MNT curve [52] with degree 6 and BN curve [53] with degree 12. To have 80 bits of symmetric security, we use "MNT159," which has a 159-bit base field size in $\mathbb{G}_1$. The field size of $\mathbb{G}_2$ should be 6 times longer than the field size of $\mathbb{G}_1$. However, the PBC library actually implemented $\mathbb{G}_2$ to be 3 times longer. One has to know that the shorter an element in a group is, the faster the multiplication will be and so does the Exponential. As we will see in the following subsections, choosing groups and curves has a great influence to efficiency. The BN curve has a field size of 160 bits in $\mathbb{G}_1$ and the NIST symmetric security is 80 bits too. The degree of BN curve is 12, which means that the operation in $\mathbb{G}_T$ group is more expensive than operations in $\mathbb{G}_T$ in MNT curve, which has a degree of 6. Table 1 is the real world benchmark in Charm of different operations in different groups and curves.

Table 1: Benchmark on curves and groups, time unit is ms. Run 1000 trials and the average is recorded

	$\mathbb{G}_1$ Exp.	$\mathbb{G}_2$ Exp.	$\mathbb{G}_T$ Exp.	Pairing
SS512	3.7289 ms	3.6653 ms	0.4652 ms	3.9136 ms
MNT159	1.1093 ms	9.8533 ms	2.6308 ms	8.5181 ms
BN	1.1875 ms	2.3718 ms	10.5363 ms	46.1763 ms

People care about the number of Pairing in an identity/attribute based scheme. Table 1 shows that the Exponential operation consumes equal computational cost. Usually the Pairing operation takes longer than Exponential, but the underlying mathematical function of Charm, which is the PBC library, has no optimizations to the multiplication operation, and the Exponential takes longer than we expect. More discussions of optimization can be found at Section 5.5. For now, the python based Charm crypto is our choice to do the implementation.

From Table 1, the $\mathbb{G}_1$ Exponential is expensive in SS512 since the field size of SS512 is 512 bits while MNT159 and BN have 160 bits of field size. Also, the $\mathbb{G}_2$ Exponential in MNT159 is expensive compared to SS512 even if the element in $\mathbb{G}_2$ is only 3 times longer than $\mathbb{G}_1$, which is about 480 bits. In terms of $\mathbb{G}_2$ Exponential, BN curve is better. BN curve is better in both $\mathbb{G}_1$ Exponential and $\mathbb{G}_2$ Exponential. That is why BN curve is a good candidate when the top priority is to minimize bandwidth (e.g., shorter signature) and faster the schemes that have most of the operation in $\mathbb{G}_1$ and $\mathbb{G}_2$. Another advantage of BN curve is that if finite field discrete log algorithms improve further, MNT curves need to use larger fields, but BN can still remain short [38]. However, $\mathbb{G}_T$ Exponential and Pairing in BN curve take much more longer time than in SS512 and MNT159. If a scheme has heavier operations in $\mathbb{G}_T$ and a large amount of Pairing, we should avoid using BN curve. Different identity/attribute based crypto schemes have different amount of Exponential and Pairing operations in key generation (sometimes called key extraction), encryption, decryption, signature, and verification. We are going to analyze the performance of the decentralized ABE scheme and multiauthority ABS scheme in the following subsection.

Performance of Decentralized ABE

Different curves have different computational costs for Exponential operation in groups. The chosen curves will affect the performance of decentralized ABE scheme. Since Table 1 lists the Exponential operations in $\mathbb{G}_1$, $\mathbb{G}_2$, and $\mathbb{G}_T$, we start with the number of Exponential operations in KeyGen(), Enc(), and Dec() of decentralized ABE scheme. Table 2 lists the number of operations for KeyGen(), Enc(), and Dec() of the scheme we implemented, which can be found at Appendix.

Table 2: Number of operations of decentralized ABE scheme

KeyGen()	Enc()	Dec()
$2\mathbb{G}_2$/attribute	$1\mathbb{G}_T + (3\mathbb{G}_1 + 2\mathbb{G}_T) \cdot \ell$	$(1\mathbb{G}_T + 2\text{Pairing}) \cdot \ell_r$

ℓ is the number of attributes involved in the access structure, and it is also the number of rows in an access matrix. ℓ_r is the number of required attributes to decrypt a message. The receiver may not need all the attributes in the access structure to decrypt the message since the minimal set that satisfies the access structure will work.

The key generation needs $2\mathbb{G}_2$ per attribute. The Exponential in $\mathbb{G}_2$ in asymmetric groups is slower than Exponential in $\mathbb{G}_1$. To make the key generation faster, one can play a trick and swap $\mathbb{G}_1$ with $\mathbb{G}_2$. After we swapped $\mathbb{G}_1$ with $\mathbb{G}_2$, the key generation is operations in $\mathbb{G}_1$. However, Enc() will have $3\mathbb{G}_2$ per attribute instead of $3\mathbb{G}_1$. Table 3 is the running time of key generation under SS512, MNT159, and MNT159.S. MNT159.S means that we swapped $\mathbb{G}_1$ with $\mathbb{G}_2$ in the scheme. The swap will not affect the security of the scheme. It will affect only the efficiency and the length of parameters. Note that there are some inconsistency between Tables 1 and 3. The reason that the key generation in MNT159 is about 25 ms longer than we expect is that we need to map an identity to an element in $\mathbb{G}_2$ using a random oracle. And the time of mapping depends on the target group $(\mathbb{G}_1, \mathbb{G}_2, \mathbb{G}_T, \text{or } \mathbb{Z}_q)$ and the curve (SS512, MNT, or BN) been used. And the mapping is the reason to the variance in Figure 9 too.

Table 3: Key generation per attribute of decentralized ABE scheme

SS512	MNT159	MNT159.S
15.61 ms	53.08 ms	2.31 ms

In Figures 8 and 9, the error bar means the standard deviation of the Enc() and Dec(). As we expected, the running time of Enc() and Dec() grows with the number of attributes involved and the number of attributes required, respectively. One can see that MNT159 has the best performance in Enc(), but the worst in KeyGen(). As for Dec(), SS512 is better. MNT159 and MNT159.S should have the same performance in Dec() according to Table 3 since Dec() involves no Exponential in both $\mathbb{G}_1$ and $\mathbb{G}_2$. However, in Dec(), we do need to map an identity to an element in the target group (its $\mathbb{G}_2$ in MNT159 and $\mathbb{G}_1$ in MNT159.S), and as mentioned before, the mapping takes 25 ms when mapping the identity to an element in $\mathbb{G}_2$ in MNT159. This explains why the red line is about 25 ms above the brown line in Figure 9.

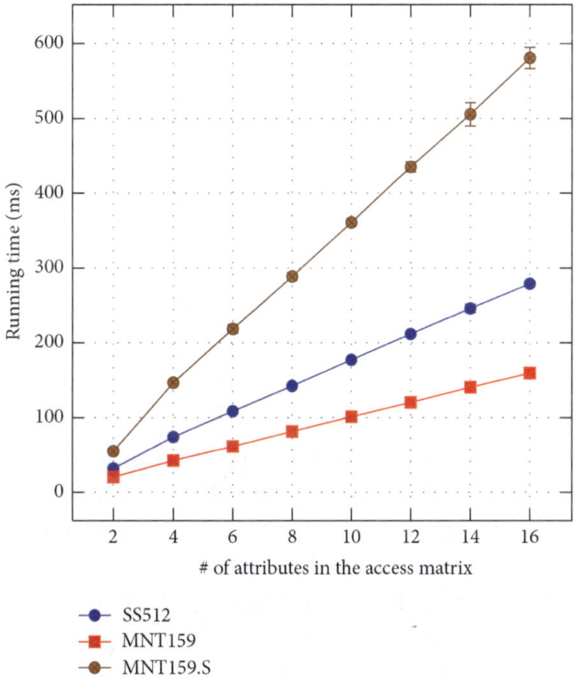

Figure 8: Decentralized ABE: Enc().

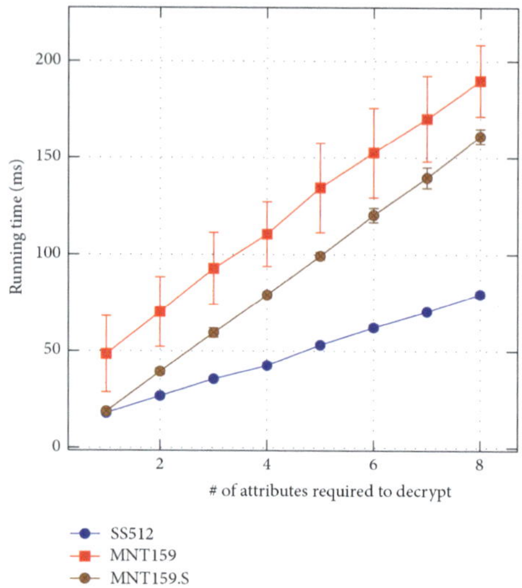

Figure 9: Decentralized ABE: Dec().

In the architecture we proposed, the KeyGen() is performed by the attribute authorities (AAs), and the Enc() and Dec() are performed by smart meters or data consumers. The intuition is that Enc() and Dec() are performed distributively, and the AAs might have bottleneck issues with the fact that there are a large amount of users need secret keys from the AAs. Situation becomes worse if we take key and user revocation into consideration. For example, if the secret keys issued by the AA have a time tag attached (e.g., Arlington.TV.Jan 2013), which means that this attribute will expire in certain amount of time and users should obtain the secret key for the next time period of this attribute by contacting the AA or we integrate some real-time user (or key) revocation scheme just as paper [54] did, the KeyGen() will certainly cause a lot of pains to the AAs. Our recommendation here is to use MNT159 curve and swap $\mathbb{G}_1$ with $\mathbb{G}_2$ to achieve the best efficiency in KeyGen(). Even the performance of MNT159.S in Enc() is the worst, it will be acceptable due to the fact that the encryption will only need to be performed once and the computation is totally distributed.

Performance of Multiauthority ABS

We also implemented the multiauthority ABS [16] and ran the performance test on our implementation. The difference between the decentralized ABE scheme and multiauthority ABS scheme is that the multiauthority ABS scheme has a signature trustee, which handles the user registration part.

Given the token from signature trustee, AAs can generate secret keys for attributes the user requested. Since the TSetup() and ASetup() only happen once, we mainly focus on the TRegister(), AttrGen(), Sign(), and Verify(). One has to notice that the verification we implemented is the probabilistic verification mentioned in Section 3.3.1 of paper [16], which has at most 1/p probability to make a false positive. The computational cost of verification reduced by one degree: from $\ell \cdot t + 2$ to $\ell + 4$, where ℓ is the number of rows in the access matrix and t is the number of columns.

Table 4 summarizes the number of operations for TRegister(), AttrGen(), Sign(), and Verification() of the scheme we implemented, which can also be found at the Appendix.

Table 4: Number of operations of multiauthority ABS scheme

TRegister()	$1\mathbb{G}_1$/user
AttrGen()	$1\mathbb{G}_1$/attribute
Sign()	$2\mathbb{G}_1 + 3(\ell_r)\mathbb{G}_1 + 2(\ell - \ell_r)\mathbb{G}_1 + 2(\ell \cdot t)\mathbb{G}_2$
Verify()	$1\mathbb{G}_1 + 2(\ell \cdot t + t)\mathbb{G}_2 + (\ell + 4)$Pairing

ℓ is the number of attributes involved in the access structure, and it is also the number of rows in an access matrix. t is the number of columns of the access matrix. t increases by one when the algorithm meets an "AND" gate in an access structure. ℓ_r is the number of required attributes to sign a message.

Also, we start with the AttrGen(), which may be the bottleneck of our system. The TRegister() has the same amount of computational cost to the AttrGen() according to Table 4. However, we need to use a random oracle to map the identity into an element in groups and the discussion in the previous subsection. However, this mapping may take a long time. Meanwhile, a user needs to contact the signature trustee to get this token, then the user needs more than secret key for the attributes. The computational cost of TRegister() should be less than the computational cost of AttrGen(). We focus on AttrGen() now and we can generalize the performance of TRegister() from Table 5.

Table 5: Key generation per attribute of multiauthority ABS scheme

SS512	MNT159	MNT159.S	BN.S
3.67 ms	9.72 ms	1.13 ms	2.30 ms

As what has been discussed in the previous subsection, MNT159.S swaps $\mathbb{G}_1$ with $\mathbb{G}_2$. and BN curves are brought into discussion as it has its advantages in $\mathbb{G}_1$ and $\mathbb{G}_2$ Exponential operations.

From Figures 10 and 11, error bar means standard deviation. Computational cost in Sign() and Verify() is higher than the Dec() and Enc() in the decentralized ABE scheme just as we expected. The multiauthority ABS signature has a lot more Exponential operations in $\mathbb{G}_1$ and $\mathbb{G}_2$. Particularly in Sign(), it grows with $\ell \cdot t$. It also grows with the number of attributes required to sign. In the access structure we using, the number of required attributes to sign is $\ell_r = \ell/2$. The verification is the probabilistic verification which has a reasonable and negligent probability to produce a false positive. Both MNT159.S and BN.S have a better performance in AttrGen(). As for Sign(), BN.S has the lowest cost since the Exponential operations in $\mathbb{G}_1$ and $\mathbb{G}_2$ are less expensive than other schemes. MNT159.S has better performance in the verification. If considering the performance of Sign() as the priority, BN curve should be used and $\mathbb{G}_1$ should be swapped with $\mathbb{G}_2$. However, the sender needs only to generate one signature for a message and verification might happen more than one time. One can also consider the verification as the priority; MNT159.S would be a better choice.

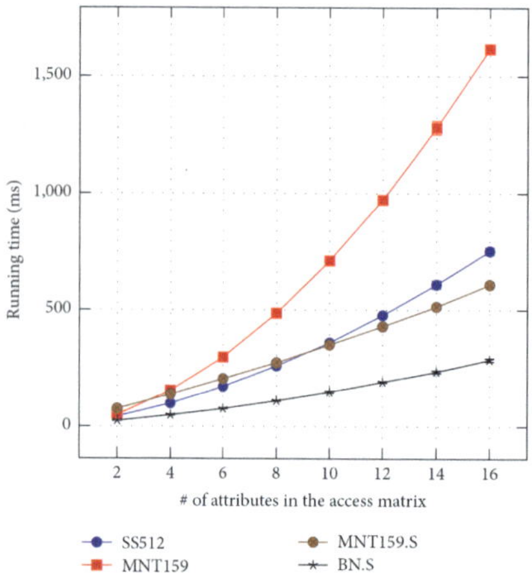

Figure 10: Multiauthority ABS: Sign().

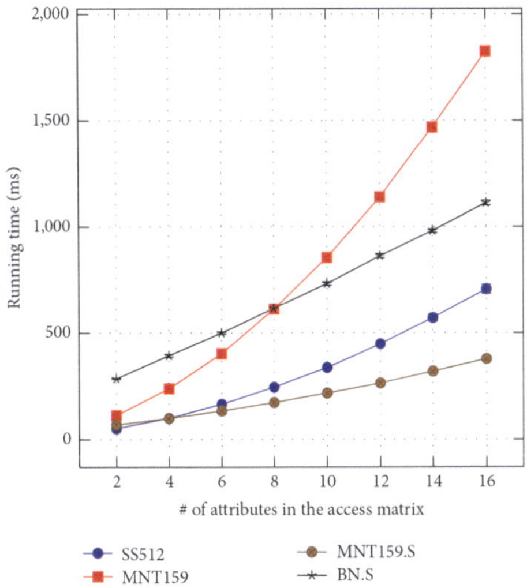

Figure 11: Multiauthority ABS: Verify().

In the smart grid communication network, the AttrGen() is centralized and may be the bottleneck. Both MNT159.S and BN.S can fit the task. If the efficiency of Sign() matters, one should use BN curve and swap $\mathbb{G}_1$ with $\mathbb{G}_2$. Also, the signature size can be reduced to compare with MNT curve: in $\mathbb{G}_1$, elements have the same length. However, the elements in $\mathbb{G}_2$ are 2 times longer than elements in $\mathbb{G}_1$ in BN curve instead of 3 times longer in MNT curve. If the resource on smart meters is very limited, BN curve will be a good choice. However, if the efficiency of verify() is the priority, MNT159.S should be used. In the scenario that data consumers need to collect anonymity data from a group of users that satisfy an access structure, the verification is performed per user and MNT159.S will save a lot of computational cost.

More about Efficiency

Efficiency can be further improved by using the preprocessing provided by PBC library or using the Lopez-Dahab algorithm [55], which is TinyPBC's optimization on multiplication. Both of them are not in Charm's current release. In PBC library, we can prepare an element for Exponential operation or Pairing operation. For example, if we preprocess the generator g_1, the exponential operation based on g_1 will be roughly 6-8 time faster, which is shown in Table 6. See Table 6 for details.

Table 6: Preprocessing in PBC library

Average, ms	MNT159	SS512
G1 Exponential	1.18	4.25
After preprocessing	0.16	0.59
Preprocessing itself	5.47	20.37
G2 Exponential	10.07	4.08
After preprocessing	1.45	0.55
Preprocessing itself	46.74	19.01
Pairing	8.48	4.33
After preprocessing	6.92	1.83
Preprocessing itself	1.68	3.86

According to TinyPBC's implementation on multiplication, the speed of multiplication will be 4–6 times faster. If we combine the preprocessing and Lopez-Dahab algorithm, we expect the implemented scheme to be 20 times faster. Further work needs to be done to optimize the underlying mathematical functions to make the multiplication and pairing faster.

Converting an AS to an Matrix More Efficiently

Algorithm 1 is converting an access structure to an access matrix.

Algorithm 1: Boolean formula 2LSSS $(node, v, c, A, \rho())$.

```
Let v = (1), which represents the root of T.
Let c = 1, which is a counter for the converting process.
A is the LSSS matrix, initialized to null; ρ() is the function that maps the rows of A to attributes.
BooleanFormula2LSSS(root(T),v, c, A, ρ());
function (BooleanFormula2LSSS (node, v, c, A, ρ()))
    if node is an AND gate then
        pad v with 0's (if necessary) to make its length = c;
        label the left child with the vector v ‖ 1, where ‖ means concatenation;
        label the right child with the vector (0,...,0) ‖ −1, where (0,...,0) denotes the zero vector of length c;
        c = c + 1;
    else if node is an OR gate then
        label the left child and the right child with the vector v;
    end if
    for node's child node as n_c do
        if n_c is an attribute then
            A.push(v); // add v to the end of matrix A
            add the mapping ρ(v) = n_c;
        else
            BooleanFormula2LSSS(n_c,v, c, A, ρ())
        end if
    end for
end function
```

From decentralized ABE and multiauthority ABS scheme, the size of matrix influences the efficiency. The project [56] reduces the size of access matrix and the computational cost of our implemented schemes will reduce too. We leave the implementation of a more efficient transformation of access matrix as one of the potential future research directions too.

In summary, we run the implementation on different curves and groups and analyze the performance in this section. Note that the decentralized ABE and multiauthority ABS do not share the common parameters such as group generator and so on. Reference [35] has some initial work in combining ABE with ABS, and ABE shares public parameters, even the secret keys with ABS. Combined ABE and ABS can be a potential next step in our future work. However, once we combine the ABE and ABS, even the storage for secret keys reduced, computational cost would be higher than using ABE and ABS schemes, separately. Different schemes have different performance under different curves and groups.

CONCLUSION

In this paper, we describe a smart grid communication architecture and then present a secure and scalable data communication scheme in smart grids, which is employed decentralized attribute based encryption. The security analysis demonstrated that the scheme ensures security and privacy. The performance analysis shows that the scheme is efficient in terms of computational cost.

Our future research lies in the following directions: design a decentralized CP_ABE scheme with constant size of ciphertext length to reduce the storage and communication cost. Examine more attacks on the architecture we proposed and defend those attacks. Cooperate our current scheme with other broadcast authentication schemes and signature schemes to make a more comprehensive and applicable architecture. The communication architecture for smart grids proposed in this paper serves at the basis of our future research and we shall further propose new approaches to enhance and extend this architecture.

APPENDIX

Here are some detailed implementation.

A. Duplicated Attributes in an AS

For the duplicated attributes in an AS, we will extend the attribute and make them different. For example, if we have two "Arlington.22202" in AS, we will encode them into "Arlington.22202_1" and "Arlington.22202_2." However, in the encryption, we will treat "Arlington.22202_1" as "Arlignton.22202" and later in the decryption or sign, if the entities have the secret key for "Arlington.22202," it can decrypt or sign both "Arlington.22202_1" and "Arlington.22202_2."

B. The Decentralized ABE Scheme Implemented

The differences between the scheme we implemented and the original paper are as follows:

(i) The original paper was described under symmetric group settings. We implemented it under asymmetric group settings.

ii. Hash function H maps an identity into an element in $\mathbb{G}_2$.

iii. Secret keys for attributes are elements in $\mathbb{G}_2$.

iv. $e(g_1, g_2)$ is precalculated in our implementation.

Here is the scheme we implemented:

(1) ABE_GlobalSetup(λ):

$g_1 \leftarrow_R \mathbb{G}_1$, $g_2 \leftarrow_R \mathbb{G}_2$, $H: \{0,1\}^* \rightarrow \mathbb{G}_2$;
return: ABE_GP := $\{g_1, g_2, H, e(g_1, g_2)\}$;

(2) ABE_AuthoritySetup(ABE_GP, {attribute},*AAname*):

ABE_APK$_{AAname}$ = {}; # init a python dictionary
2. ABE_ASK$_{AAname}$ = {}; # init a python dictionary
for *i* **in**{attribute}:

$\alpha_i, y_i \leftarrow_R \mathbb{Z}_q$;

ABE_APK[*AAname.i*] = $\{e(g_1, g_2)^{\alpha_i}, g_1^{y_i}\}$;
ABE_ASK[*AAname.i*] = $\{\alpha_i, y_i\}$;

return: ABE_APK$_{AAname}$, ABE_ASK$_{AAname}$;

(3) ABE_KeyGen(ABE_GP, ABE_ASK, uid, attribute):

ABE_SK$_{\text{uid}}$ = {}; # init a python dictionary
for *i* **in**{attribute}:

ABE_SK$_{\text{uid}}[i] = g_1^{\alpha_i} H(\text{uid})^{y_i}$;

return: ABE_SK$_{\text{uid}}$;

(4) ABE_Enc(ABE_GP, {ABE_APK}, AS, m):

parse(AS) $\rightarrow$ $(A, \rho())$; # we implemented the converting in Appendix C of the paper.

$\vec{v} \leftarrow_R \mathbb{Z}_q^{\ell}$; # A is a $n \times \ell$ matrix.

$\vec{w} \leftarrow_R \mathbb{Z}_q^{\ell}$, $w[0] = 0$;

$\vec{r} \leftarrow_R \mathbb{Z}_q^{\ell}$;

for A_x as row of A:

$\lambda_x = A_x \cdot \vec{v}$;

$w_x = A_x \cdot \vec{w}$;

$C_1, C_2, C_3 = \{\}, \{\}, \{\}$;

$C_0 = \mathrm{M}e(g_1, g_2)^{v[0]}$;

for x as row of A:

$C_1[\rho(x)] = e(g_1, g_2)^{\lambda_x} e(g_1, g_2)^{\alpha_{\rho(x)} r_x}$;

$C_2[\rho(x)] = g_1^{r_x}$;

$C_3[\rho(x)] = g_1^{y_{\rho(x)} r_x} g_1^{w_x}$;

return: (CT = $\{C_0, C_1, C_2, C_3\}$, AS);

(5) ABE_Dec(ABE_GP, ABE_SK$_{\mathrm{uid}}$, CT, AS):

parse(AS) $\rightarrow$ $(A, \rho())$;

find subset A' of A has $(1, 0, \ldots, 0)$ as the span.

$\vec{c} \cdot A' = (1, 0, \ldots, 0)$;

for x as row of A':

numerator $* =$ $((C_{1,x} \cdot e(C_{3,x}, H(GID)) / e(C_{2,x}, K_{\rho(x),GID}))^{c_x}$

return: ($m = C_0$/numerator).

C. The Implement of Multiauthority ABS Scheme

The differences between the scheme we implemented and the original paper are as follows:

1. in original paper is $\mathbb{G}_1$ in our implementation. H is $\mathbb{G}_2$.
2. We have two hash functions. H_1: $\{0, 1\}^* \rightarrow \mathbb{Z}_q$ will be used to map attributes into elements in $\mathbb{Z}_1$; H_2: $\{0, 1\}^* \rightarrow \mathbb{G}_1$ will be used to map identity into elements in $\mathbb{G}_1$.

3. Secret keys for attributes are elements in $\mathbb{G}_2$.
4. In Sign(), the computing of S_i will not compute $(K_u[H_1(\rho(i))])^{v[i]})^{r[0]}$ if $v[i]$ is 0, which means that the signer does not have the corresponding secret key for the attribute. We save 1 $\mathbb{G}_1$ Exponential by doing so.

Here is the scheme we implemented:

(1) ABS_GlobalSetup($t_{\max}$):

$g_1, C \leftarrow_R \mathbb{G}_1,\ h_0, \ldots, h_{t_{\max}} \leftarrow_R \mathbb{G}_2$;
H_1: $\{0,1\}^* \rightarrow \mathbb{Z}_q$, H_2: $\{0,1\}^* \rightarrow \mathbb{G}_1$;
$\alpha_0 \leftarrow_R \mathbb{Z}_q,\ A_0 = h_0^{\alpha_0}$;
ABS_GP := $((\mathbb{G}_1, \mathbb{G}_2), H_1, H_2, g_1, A_0, h_0, \ldots,$
$h_{t_{\max}}, C)$;
ABS_TSK := (α_0);
return: ABS_GP, ABS_TSK;

(2) ABS_AuthoritySetup(ABS_GP, *AAname*):

$a, b \leftarrow_R \mathbb{Z}_q$;
A, B = list(), list(); # A, B are two list.
for j in $[1, t_{\max}]$:
  $A_j = h_j^a,\ B_j = h_j^b$;
ABS_ASK$_{AAname}$:= (a, b);
ABS_APK$_{AAname}$:= $\{A, B\}$;
return: ABS_ASK$_{AAname}$, ABS_APK$_{AAname}$;

(3) ABS_TRegister(ABS_TSK, uid):

$K_{\text{base}} = H_2(\text{uid}),\ K_0 = K_{\text{base}}^{1/\alpha_0}$;
return: $\tau_{\text{uid}} = (K_{\text{base}}, K_0)$

(4) ABS_KeyGen(ABS_APK, ABS_ASK, τ_{uid}, {attr}):

K_u = {}; # init a python dictionary
for i **in** {attr}:
  $K_u[i] == K_{\text{base}}^{1/(a+bH_1(i))}$;
return: ABS_SK$_{\text{uid}}$:= $\{K_{\text{base}}, K_0, K_u\}$;

(5) ABS_Sign(ABS_GP, {ABS_APK}, ABS_SK$_{\text{uid}}$, AS, m):

if ABS_SK$_{\text{uid}}$ does not satisfies AS:

Abort;

parse(AS) $\rightarrow$ $(A, \rho())$; # A is an $\ell \times t$ matrix

$\mu = H_1(m \parallel AS)$

$\overrightarrow{v} \cdot M = (1, 0, \ldots, 0)$;

$\overrightarrow{r} \leftarrow_R \mathbb{Z}_q^{\ell+1}$;

$Y := K_{\text{base}}^{r[0]},\ W := K_0^{r[0]}$;

S, P = list(), list(); # init two list.

for i in$[1, \ell]$:

$S_i := ((K_u[H_1(\rho(i))])^{v[i]})^{r[0]} \cdot (Cg_1^{\mu})^{r[i]}$;

for i in$[1, t]$:

$P_j = \prod_{i=1}^{l} (A_j B_j^{H_1(\rho(i))})^{M_{ij} \cdot r[i]}$;

return: $\sigma = (Y, W, S, P)$;

(6) ABS_Verify(ABS_GP, {ABS_APK}, AS, m, σ):

parse(AS) $\rightarrow$ $(A, \rho())$; # A is an $\ell \times t$ matrix

$\mu = H_1(m \parallel AS)$

if $Y == 1$:

return: reject;

$r_1, r_2, \ldots, r_l \leftarrow_R \mathbb{Z}_q$;

if $e(W, A_0)! = e(Y, h_0)$:

return: reject;

if $\prod_{i=1}^{\ell} e(S_i, \prod_{j=1}^{t} (A_j B_j^{H_1(u(i))})^{M_{ij} \cdot r_j})! \quad =$
$e(Y, h_1^{r_1}) e(Cg_1^{\mu}, \prod_{j=1}^{t} P_j^{r_i})$:

return: reject;

return: accept.

ACKNOWLEDGMENTS

This research was partially supported by the National Natural Science Foundation of China under Grants 61702062, 61471028, 61672119, and 61771077, National Program on Key Basic Research Project of China (973 Program), the State Key Program of National Natural Science of China (no. U1766215), the Major Science and Technology Program of Guangxi (Grant no. GKAA17129002), and the Key Research Program of Chongqing Science and Technology Commission (Grant no. CSTC2017jcyjBX0025).

REFERENCES

1. S. M. Kaplan, "Electrical power transmission," in *Background and policy issues , TheCapital.Net*, vol. Government, p. 42, Electrical power transmission, Background and policy issues, 2009.
2. X. D. Wang and P. Yi, "Security framework for wireless communications in smart distribution grid," *IEEE Transactions on Smart Grid*, vol. 2, no. 4, pp. 809–818, 2011.
3. Q. Li and G. Cao, "Multicast authentication in the smart grid with one-time signature," *IEEE Transactions on Smart Grid*, vol. 2, no. 4, pp. 686–696, 2011.
4. A. Ipakchi and F. Albuyeh, "Grid of the future," *IEEE Power & Energy Magazine*, vol. 7, no. 2, pp. 52–62, 2009.
5. "Guidelines for smart grid cyber security (vol. 1 to 3)," *NIST, Tech. Rep. NIST IR-7628*, Aug 2010, http://csrc.nist.gov/publications/PubsNISTIRs.html.
6. G. Baker and A. Berg, "Supervisory control and data acquisition (scada) systems," *The Critical Infrastructure Protection Report*, vol. 1, no. 6, pp. 5-6, 2002.
7. C. Zimmer and F. Mueller, "Fault Tolerant Network Routing through Software Overlays for Intelligent Power Grids," in *Proceedings of the 2010 IEEE 16th International Conference on Parallel and Distributed Systems (ICPADS)*, pp. 542–549, Shanghai, China, December 2010.
8. Y. Liu, P. Ning, and M. K. Reiter, "False data injection attacks against state estimation in electric power grids," *ACM Transactions on Information and System Security*, vol. 14, no. 1, article 13, 2011.
9. X. Zheng, Z. Cai, J. Yu, C. Wang, and Y. Li, "Follow but no track: privacy preserved profile publishing in cyber-physical social systems," *IEEE Internet of Things Journal*, 2017.
10. Y. Deng, C. Hu, R. Deng, and D. Liang, "A secure communication architecture in the smart grid," in *Proceedings of the 2017 4th International Conference on Information, Cybernetics and Computational Social Systems (ICCSS)*, pp. 668–672, Dalian, July 2017.
11. J. Liu, Y. Xiao, S. Li, W. Liang, and C. L. P. Chen, "Cyber security and privacy issues in smart grids," *IEEE Communications Surveys & Tutorials*, vol. 14, no. 4, pp. 981–997, 2012.

12. A. Molina-Markham, P. Shenoy, K. Fu, E. Cecchet, and D. Irwin, "Private memoirs of a smart meter," in *Proceedings of the the 2nd ACM Workshop*, p. 61, Zurich, Switzerland, November 2010.
13. S. Ruj, A. Nayak, and I. Stojmenovic, *A security architecture for data aggregation and access control in smart grids*, 1111.2619, ArxivpreprintarXiv, 2011.
14. C. Hu, Y. Huo, L. Ma, H. Liu, S. Deng, and L. Feng, "An Attribute-Based Secure and Scalable Scheme for Data Communications in Smart Grids," in *Wireless Algorithms, Systems, and Applications*, vol. 10251 of *Lecture Notes in Computer Science*, pp. 469–482, Springer International Publishing, Cham, 2017.
15. C. Hu, X. Cheng, Z. Tian, J. Yu, K. Akkaya, and L. Sun, "An Attribute-Based Signcryption Scheme to Secure Attribute-Defined Multicast Communications," in *Security and Privacy in Communication Networks*, vol. 164 of *Lecture Notes of the Institute for Computer Sciences, Social Informatics and Telecommunications Engineering*, pp. 418–437, Springer International Publishing, Cham, 2015.
16. H. K. Maji, M. Prabhakaran, and M. Rosulek, "Attribute-based signatures," in *Topics in cryptology-CT-RSA 2011*, vol. 6558 of *Lecture Notes in Comput. Sci.*, pp. 376–392, Springer, Heidelberg, Germany, 2011.
17. J. Bethencourt, A. Sahai, and B. Waters, "Ciphertext-policy attribute-based encryption," in *Proceedings of the IEEE Symposium on Security and Privacy (SP '07)*, pp. 321–334, May 2007.
18. D. Chaum and E. van Heyst, "Group Signatures," in *Advances in Cryptology — EUROCRYPT '91*, vol. 547 of *Lecture Notes in Computer Science*, pp. 257–265, Springer Berlin Heidelberg, Berlin, Heidelberg, 1991.
19. C. Hu, X. Cheng, J. Yu, Z. Tian, and R. Bie, Achieving privacy preservation and billing via delayed information release , submittedtoIEEETransactionsonServiceComputing,.
20. R. L. Rivest, A. Shamir, and Y. Tauman, "How to leak a secret," in *Advances in Cryptology---ASIACRYPT*, vol. 2248 of *Lecture Notes in Comput. Sci.*, pp. 552–565, Springer, 2001.
21. X. Boyen, "Mesh signatures: how to leak a secret with unwitting and unwilling participants," in *Advances in cryptology---{EUROCRYPT} 2007*, vol. 4515 of *Lecture Notes in Comput. Sci.*, pp. 210–227, Springer, Berlin, 2007.

22. A. Lewko and B. Waters, "Decentralizing attribute-based encryption," in *Advances in Cryptology—EUROCRYPT 2011*, vol. 6632 of *Lecture Notes in Computer Science*, pp. 568–588, Springer, Heidelberg, Germany, 2011.
23. M. Kgwadi and T. Kunz, "Securing RDS broadcast messages for smart grid applications," *International Journal of Autonomous and Adaptive Communications Systems*, vol. 4, no. 4, pp. 412–426, 2011.
24. H. Guo, Y. Wu, H. Chen, and M. Ma, "A batch authentication protocol for v2?g communications," in *in NewTechnologies, Mobility and Security(NTMS, 2011 4th IFIP International Conference*, pp. 1–5, IEEE, Paris, France, 2011, 1em plus 0.5em minus 0.4em.
25. R. Lu, X. Liang, X. Li, X. Lin, and X. Shen, "EPPA: an efficient and privacy-preserving aggregation scheme for secure smart grid communications," *IEEE Transactions on Parallel and Distributed Systems*, vol. 23, no. 9, pp. 1621–1632, 2012.
26. D. Li, Z. Aung, J. R. Williams, and A. Sanchez, "Efficient authentication scheme for data aggregation in smart grid with fault tolerance and fault diagnosis," in *Proceedings of the IEEE PES Innovative Smart Grid Technologies (ISGT '12)*, pp. 1–8, IEEE, January 2012.
27. X. Zheng, Z. Cai, J. Li, and H. Gao, "Location-privacy-aware review publication mechanism for local business service systems," in *Proceedings of the IEEE INFOCOM 2017 - IEEE Conference on Computer Communications*, pp. 1–9, Atlanta, GA, USA, May 2017.
28. A. Sahai and B. Waters, "Fuzzy identity-based encryption," in *Advances in Cryptology – EUROCRYPT 2005*, vol. 3494 of *Lecture Notes in Computer Science*, pp. 457–473, Springer, Berlin, Germany, 2005.
29. V. Goyal, O. Pandey, A. Sahai, and B. Waters, "Attribute-based encryption for fine-grained access control of encrypted data," in *Proceedings of the 13th ACM Conference on Computer and Communications Security (CCS '06)*, pp. 89–98, November 2006.
30. M. Pirretti, P. Traynor, P. McDaniel, and B. Waters, "Secure attribute-based systems," *Journal of Computer Security*, vol. 18, no. 5, pp. 799–837, 2010.
31. M. Chase, "Multi-authority attribute based encryption," in *Theory of Cryptography*, vol. 4392 of *Lecture Notes in Computer Science*, pp. 515–534, Springer, Berlin, Germany, 2007.

32. H. K. So, S. H. Kwok, E. Y. Lam, and K. Lui, "Zero-Configuration Identity-Based Signcryption Scheme for Smart Grid," in *Proceedings of the 2010 1st IEEE International Conference on Smart Grid Communications (SmartGridComm)*, pp. 321–326, Gaithersburg, MD, USA, October 2010.
33. Z. M. Fadlullah, N. Kato, R. Lu, X. Shen, and Y. Nozaki, "Toward secure targeted broadcast in smart grid," *IEEE Communications Magazine*, vol. 50, no. 5, pp. 150–156, 2012.
34. C. Hu, H. Li, Y. Huo, T. Xiang, and X. Liao, "Secure and Efficient Data Communication Protocol for Wireless Body Area Networks," *IEEE Transactions on Multi-Scale Computing Systems*, vol. 2, no. 2, pp. 94–107, 2016.
35. C. Chen, J. Chen, H. W. Lim, Z. Zhang, and D. Feng, "Combined public-key schemes: the case of abe and abs," in *Provable security*, vol. 7496 of *Lecture Notes in Comput. Sci.*, pp. 53–69, Springer, Heidelberg, 2012.
36. C. Hu, X. Liao, and X. Cheng, "Verifiable multi-secret sharing based on {LFSR} sequences," *Theoretical Computer Science*, vol. 445, pp. 52–62, 2012.
37. C. Hu, A. Alhothaily, A. Alrawais, X. Cheng, C. Sturtivant, and H. Liu, "A secure and verifiable outsourcing scheme for matrix inverse computation," in *Proceedings of the IEEE INFOCOM 2017 - IEEE Conference on Computer Communications*, pp. 1–9, Atlanta, GA, USA, May 2017.
38. B. Lynn, *On the implementation of pairing-based cryptosystems [Ph.D. thesis]*, Stanford University, Calif, USA, 2007, Ph.D. dissertation.
39. A. Beimel, *Secure schemes for secret sharing and key distribution [Ph.D. thesis]*, Israel Institute of Technology, Technion, Haifa, Israel, 1996, Ph.D. dissertation.
40. C. Hu, N. Zhang, H. Li, X. Cheng, and X. Liao, "Body area network security: a fuzzy attribute-based signcryption scheme," *IEEE Journal on Selected Areas in Communications*, vol. 31, no. 9, pp. 37–46, 2013.
41. R. Canetti, S. Halevi, and J. Katz, "A forward-secure public-key encryption scheme," in *Advances in cryptology---{EUROCRYPT} 2003*, vol. 2656 of *Lecture Notes in Comput. Sci.*, pp. 255–271, Springer, Berlin, 2003.

42. C. H. Tan, "Chosen ciphertext security from identity-based encryption without strong condition," in *Advances in information and computer security*, vol. 4266 of *Lecture Notes in Comput. Sci.*, pp. 292–307, Springer, Berlin, 2006.
43. D. Boneh and M. Franklin, "Identity-based encryption from the Weil pairing," in *Advances in Cryptology—CRYPTO 2001*, vol. 2139 of *Lecture Notes in Computer Science*, pp. 213–229, 2001.
44. M. Bellare and P. Rogaway, "Random oracles are practical," in *Proceedings of the the 1st ACM conference*, pp. 62–73, Fairfax, Virginia, United States, November 1993.
45. B. Waters, "Efficient identity-based encryption without random oracles," in *Advances in cryptology---{EUROCRYPT} 2005*, vol. 3494 of *Lecture Notes in Comput. Sci.*, pp. 114–127, Springer, Berlin, 2005.
46. B. Waters, "Ciphertext-policy attribute-based encryption: an expressive, efficient, and provably secure realization," in *Public Key Cryptography—PKC 2011*, D. Catalano, N. Fazio, R. Gennaro, and A. Nicolosi, Eds., vol. 6571 of *Lecture Notes in Computer Science*, pp. 53–70, Springer, Berlin, Germany, 2011.
47. V. Shoup, "Lower bounds for discrete logarithms and related problems," in *Advances in cryptology---{EUROCRYPT} '97 (Konstanz)*, vol. 1233 of *Lecture Notes in Comput. Sci.*, pp. 256–266, Springer, Berlin, 1997.
48. D. Boneh, E.-J. Goh, and K. Nissim, "Evaluating 2-DNF formulas on ciphertexts," in *Theory of Cryptography*, vol. 3378 of *Lecture Notes in Computer Science*, pp. 325–341, Springer, Berlin, Germany, 2005.
49. D. M. Freeman, "Converting pairing-based cryptosystems from composite-order groups to prime-order groups," in *Advances in Cryptology—EUROCRYPT 2010*, vol. 6110 of *Lecture Notes in Computer Science*, pp. 44–61, Springer, Berlin, Heidelberg, Germany, 2010.
50. J. A. Akinyele, C. Garman, I. Miers et al., "Charm: a framework for rapidly prototyping cryptosystems," *Journal of Cryptographic Engineering*, vol. 3, no. 2, pp. 111–128, 2013.
51. L. B. Oliveira, D. F. Aranha, C. P. L. Gouvêa et al., "TinyPBC: pairings for authenticated identity-based non-interactive key distribution in sensor networks," *Computer Communications*, vol. 34, no. 3, pp. 485–493, 2011.

52. A. Miyaji, M. Nakabayashi, and S. Takano, "Characterization of Elliptic Curve Traces Under FR-Reduction," in *Information Security and Cryptology — ICISC 2000*, vol. 2015 of *Lecture Notes in Computer Science*, pp. 90–108, Springer Berlin Heidelberg, Berlin, Heidelberg, 2001.
53. P. S. Barreto and M. Naehrig, "Pairing-friendly elliptic curves of prime order," in *Selected areas in cryptography*, vol. 3897 of *Lecture Notes in Comput. Sci.*, pp. 319–331, Springer, Berlin, 2006.
54. X. Liang, R. Lu, X. Lin, and X. S. Shen, "Ciphertext policy attribute based encryption with efficient revocation," Tech. Rep., University of Waterloo, 2010.
55. J. López and R. Dahab, "High-speed software multiplication in ," in *Progress in cryptology---{INDOCRYPT} 2000 (Calcutta)*, vol. 1977 of *Lecture Notes in Comput. Sci.*, pp. 203–212, Springer, Berlin, Germany, 2000.
56. Z. Liu and Z. Cao, "On efficiently transferring the linear secret-sharing scheme matrix in ciphertext-policy attribute-based encryption , Cryptology ePrint Archive," Tech. Rep. 2010/374, Tech., 2010.

Chapter 12

S-DPS: An SDN-Based DDoS Protection System for Smart Grids

Hassan Mahmood[1,2], Danish Mahmood[1,2], Qaisar Shaheen[3], Rizwan Akhtar[4], and Wang Changda[1]

[1]School of Computer Science and Communication Engineering, Jiangsu University, Zhenjiang, China

[2]Department of Computer Science, SZABIST, Islamabad, Pakistan

[3]Department of Computer Science, Superior College, Lahore, Pakistan

[4]Department of IT and Computer Science, Pak-Austria Fachhochschule Institute of Applied Sciences and Technology, Haripur, Pakistan

ABSTRACT

Information Communication Technology (ICT) environment in traditional power grids makes detection and mitigation of DDoS attacks more challenging. Existing security technologies, besides their efficiency, are not adequate to cater to DDoS security in Smart Grids (SGs) due to

Citation: Hassan Mahmood, Danish Mahmood, Qaisar Shaheen, Rizwan Akhtar, Wang Changda, "S-DPS: An SDN-Based DDoS Protection System for Smart Grids", Security and Communication Networks, vol. 2021, Article ID 6629098, 19 pages, 2021. https://doi.org/10.1155/2021/6629098.

highly distributed and dynamic network environments. Recently, emerging Software Defined Networking- (SDN-) based approaches are proposed by researchers for SG's DDoS protection; however, they are only able to protect against flooding attacks and are dependent on static thresholds. The proposed approach, i.e., Software Defined Networking-based DDoS Protection System (S-DPS), is efficiently addressing these issues by employing light-weight Tsallis entropy-based defense mechanisms using SDN environment. It provides early detection mechanism with mitigation of anomaly in real time. The approach offers the best deployment location of defense mechanism due to the centralized control of network. Moreover, the employment of a dynamic threshold mechanism is making detection process adaptive to the changing network conditions. S-DPS has demonstrated its effectiveness and efficiency in terms of Detection Rate (DR) and minimal CPU/RAM utilization, considering DDoS protection focusing smurf attacks, socket stress attacks, and SYN flood attacks.

INTRODUCTION

There is a drastic increase in energy dependence from very minute to huge activity especially the cloud-based data centers and Internet of Things (IoT), which have a dire need of availability, reliability, and efficiency of power systems. This requirement paved path towards the Smart Grid (SG) paradigm that ensures two-way communication within power systems. It holds the capability to remove the constraints of a traditional grid infrastructure and provide power systems that are scalable, dynamic, situation-aware, and flexible. On the other hand, such facilities give birth to complexity, heterogeneity, and interconnectivity of diverse ICT requirements due to which the existing network paradigms and security strategies are marked as ineffective [1, 2]. Moreover, the IP-enabled communication infrastructure in SGs raises the likelihood of malicious activities and attacks. Such attacks may result in wrong smart meter readings or incorrect demands or responses to or from electricity company or they can be severe for power generation systems [3].

Millions of consumers are serviced by SG. The service provided by SG is crucial and the availability of such a service is extremely important. The SG makes up a cyber-physical system (CPS) and a single error in any part of the system can lead to a direct or indirect catastrophic effect on human life [4]. Distributed Denial of Service (DDoS) attacks are also making more frequent appearances and are becoming more sophisticated and severe because of the fact that the existing protection mechanisms are not capable to deal with such threats. Hence, detection, mitigation, and prevention of DDoS attacks are now on the top most priority of engineering industries and research arena. Researches have come up with SDN-based approaches to handle DDoS attacks in SG [5]. However, still experimental validation of the proposed approaches is lagging. Also, SGs are safeguarded against High-Rate (HR)-DDoS attacks only during their detection and mitigation approach [1] [6]. This level of safety is not enough for a sensitive ICT infrastructure such as SG that carries mission critical information. Because of these reasons, there exists a desperate need for further research regarding SDN-based security protocols in SG to ensure a safe and light-weight mechanism against DDoS, having a capability to detect in the early stages and mitigation of varied level of DDoS attacks.

The remainder structure of the paper is organized as follows: Section 2 discusses related work with critical evaluation of literature, motivation with problem statement, and approach with contribution. Section 3 discusses system model with implementation constraints. Experimental setup with evaluation criteria is discussed in Section 4. Results and discussion are presented in Section 5. Section 6 discusses performance evaluation of the approach. Finally, Section 7 presents conclusion with future work.

LITERATURE REVIEW

Considering the wide spread of ICT and upcoming IoT devices, applications, and scenarios in almost every field of life, the authors in [7] showcased the vulnerabilities that may attract negative attentions. Moreover, the authors discussed state-of-the-art work regarding mitigation of such malicious activities. Researchers from academia and industry have shown interest and utilize new network paradigm, i.e., SDN to deal with underlying security risks in SG communication network [5]. The authors in [8] presented a comprehensive survey focusing SG communication security measures and privacy breaches. Major emphasis is given to privacy handling within SG communication networks. In [9], the authors presented a taxonomy of network attacks focusing fog-based smart grid SCADA systems. The authors in this study classify intrusion detection systems (IDSs) as major solution for the attacks; however, they focused mainly on machine learning approaches which at times are more time consuming and compel in comparison to entropy-based IDSs.

The authors in [10] used blockchain for securing the data and SDN to deal with control issues and scalability. Furthermore, the authors in [11] also used blockchain to secure energy sector, mainly authentication and privacy issues.

SDN is recently used in SG to provide a resilient SDN-based security framework/simulators and communication architecture. Researchers have utilized either single or multicontroller architecture to establish the underlying network infrastructure [12] which is considered to be the first SG-enabled simulator which is resilient and secured. The proposed security module is able to detect and resolve DoS attack within 60 seconds with no impact on bus system. However, maximum power capacity allowed on each bus/branch is not mentioned to address how much additional load other branches can bear in case of failure. Static threshold of 40% for number of packets/sec is used in detection mechanism. Similarly, in [1], a novel SDN-based communication architecture for resiliency and security of microgrid operations is proposed. They have used three applications, i.e., self-healing mechanism, network verification, and intrusion detection. Self-healing mechanism uses rapid network configuration changes to mitigate further penetration by doing traffic isolation. Network verification is implemented using consistent updates feature, to avoid network instability by ensuring consistency of packets. Specification-based intrusion detection system is proposed; however, experimental validation is missing in the work.

Entropy-based approaches have been widely in traditional networks to provide DDoS protection. These approaches have proved to be useful in SDN environment as well, providing better detection efficiency [13]. The authors in [14] used open flow SDN controller to detect DDoS attacks on SG. As this is basic feature of SDN framework, the proposed methodology is also situation aware; however, for anomaly detection, an entropy-based mechanism is proposed which not only detects but also mitigates the attacks. However, the authors have not enhanced their proposed model to adoptively change according to situation and environment.

The authors in [15] presented a DDoS traceback mechanism under the umbrella of SDN architecture. The authors established an anomaly tree by analysing the communication flow changes via base station nodes. Once the anomaly tree is formulated, traceback scheme calls out any of different DDoS protection algorithms depending upon the nature and severeness of the attack. The authors claim that proposed scheme is better than the state-of-the-art frameworks regarding detection and trace back time with minimal usage of resources. In the future, the authors intend to optimize this approach by making it adoptive such that it can detect most types of the DDoS attacks.

A scheduling algorithm based on two levels is proposed in [16] to make sure better QoS regarding power services and communication network. For this purpose, an SDN controller is utilized and in first level, a scheduling mechanism is devised focusing priority in terms of delay constrained power services. Once priority-based services are schedules, then congestion and queueing control mechanism follows which ensures minimal delay with respect to the priority assigned. The authors used Mininet and Ryu controllers for simulation purposes. The proposed approach reduced delay and packet loss ratio with respect to state-of-the-art work. An elliptic curve cryptography (ECC) is presented in [17] and the proposed scheme which is based on mutual authentication by using biometric system. The authors claim to eliminate many authentication attacks.

Moreover, ECC technique supersedes other state-of-the-art protocols considering the performance metrics of communication and computational time considering SG environment. The authors in [18] proposed multilevel autoencoders-based IDS for DDoS attacks in SGs. The authors claim to have better accuracy in predictive analysis with other state-of-the-art methods. In [19], the authors presented a novel SDN-based IDS for SG. Basic feature of SDN, i.e., centralized controller in control plane, is made distributed by using blockchain approach. The proposed model is simulated using AnyLogic and results declare it as more effective in terms of DDoS detection with state-of-the-art frameworks. Moreover, this approach also reduces the controller overhead; however, the delay in decision making is the trade-off that is not required in demand responsiveness feature of SG.

The authors in [20] present a novel entropy-based statistical approach in multicontroller SDN environment approach which is proposed for early detection and mitigation of DDoS attack. Apart from early detection, it is able to identify the attack path as well to apply the mitigation strategy instantly after detection. Shannon entropy with experimental static threshold against "DA" is used as detection mechanism, whereas Drop/block ports mechanism is used for mitigation purpose. Experimental validation for backup controller functionality, in case of primary controller failure, is missing. Threshold mechanism should have been adaptive rather than static, considering dynamic nature of modern networks. The authors have not addressed the efficacy of approach in protecting against LR-DDoS attacks.

Further, approach should have been validated using performance metrics like DR and FPR.

Apart from SDN-based approaches for DDoS protection, it is important to discuss existing entropy-based approaches that have been successful in detecting DDoS attacks in traditional networks. Different variants of entropy are available for detecting DoS/DDoS attacks, i.e., generalized entropy, Tsallis entropy, and normalized entropy. Each of them can achieve varying DR and FPR for HR and LR-DDoS attacks. Some are able to detect both types of DDoS attacks with better DR and FPR, where some are best suited for a single type only. These solutions depend on the traffic features and perform statistical procedures on normal and attack traffic to do the comparison in order to find the anomaly. The authors in [21] present a generalized entropy-based feature selection technique which is used to detect network anomalies from real-life WAN traffic data with a high DR and low FPR. An outlier score function is used to detect the anomalies. The algorithm was evaluated against other techniques like LOF and ORCA using dataset Zoo. They achieved DR of 94.11% and FPR of 2.38%, higher than the other two approaches. However, user-defined parameters for threshold values are used. Although these values are set after conducting training on datasets, but still, it poses a limitation with respect to dynamic nature of networks and underlying attacks. The approach did not directly work on categorical and mixed types data.

A variant of Renyi entropy is proposed in [22], as a light-weight detection system utilizing extended entropy-based metric to detect HR-DDoS flooding attack and IP traceback. The proposed approach is evaluated against other entropy metrics like Shannon entropy and Kullback–Leibler divergence using both simulated and real-time DDoS datasets. Another important variant of parameterized entropy, i.e., Tsallis entropy, is utilized by researchers for anomaly detection. A feature-based Anomaly Detection System (ADS) using Tsallis entropy at device level is proposed in [23] and is capable of detecting and classifying known and unknown anomalies with additional information regarding network usage. Primitive properties of flows like SA, DA, SP, and DP and derived flow properties at device and network level, i.e., out-degree, in-degree, per flow, per packet, per byte, packet per sample (pps), etc., are used in flow extraction process. Based on the discussion above, it can be observed that the authors have not highlighted the capability of their approach to detect both HR and LR-DDoS attacks. Moreover, static thresholds based on experiments are utilized in their approaches which cannot prevail in dynamic and complex environments like SG. Real dataset for DDoS attacks based on SG networks are not publically available easily [24]; hence, researchers have used simulated datasets for validation of their work.

Tsallis entropy metric has performed well, as per validation metrics, compared to other entropic metrics in detecting varying number of DDoS attacks, i.e., both LR-DDoS and HR-DDoS attacks. For effective DDoS defense mechanism, mitigation strategies should also be incorporated with intrusion detection system. Placement of detection mechanism is way important for efficient detection and in-time capitalization of DDoS attack. Such fact has not been addressed by many of the researchers. Finally, Tsallis entropy metric, besides its efficiency with respect to DR and FPR, has not been tested in an SDN environment. Utilizing SDN for securing SGs is in focus for energy engineering industries and research arena as well. The authors in [25] orchestrate a strategic connection, monitoring SDN controllers and sources of new flow requests that are threatening for DDoS attack. Compromised switches are identified and a noncooperative game is orchestrated using dynamic Bayesian network. The authors in [26] proposed a DDoS detection mechanism in SGs using Convolutional Neural Network. Variance fractal dimension trajectory is used as a preprocessing tool, whereas postprocessing of data is conducted by employing support vector machine. The authors claimed to achieve 87.35% accuracy in DDoS detection. Critical evaluation of literature is given in Tables 1 and 2.

Table 1: SDN-based security approaches for DDoS protection

Approach	Security parameter	Network/dataset	Experimental setup	Tools/simulators	Parameters/approach for intrusion detection	Limitations
DSSnet: microgrid simulator [1]	DoS and resilience	IEEE-13 bus power distribution system with two subsystems, i.e., wind turbine and energy storage system	Developed a simulator for evaluation of microgrid operation. Applications: (i) Self-healing network management (ii) Communication network verification (iii) Specification-based intrusion detection	OpenDSS, Mininet, virtual time system (Linux-based kernel)	(i) Network slicing (ii) Traffic isolation	A little literature on specification-based intrusion detection provided experimental validation of intrusion detection is not provided.
PYGRID: SG simulator [12]	DoS protection and resilience	Simulated IEEE-14 bus power system	Scenarios: normal operation, bus failure, and bus attack Result: successfully mitigated DDoS attack	Mininet, PYPOWER	(i) Number of packets/second=40% threshold (ii) Flows count	(i) Maximum power capacity allowed on each bus/branch is not mentioned; rationale for using fixed threshold limit for number of packets/sec is missing. (ii) All traffic flows are being monitored for rapid detection. Computation overhead cost is associated with the approach since all flows will go to application layer.

Multicontroller-based SDN [20]	UDP/ TCP/ ICMP flood attacks	Simulated	Design components entropy-based DDoS detection algorithm (i) Virtualized network environment of 3 switches and 32 hosts (ii) Set of mitigation actions (block traffic/ ports) (iii) UDP flood attack simulated	POX controller, Mininet 2.0, and Scapy tool for traffic generation	Analysis metric (i) Destination IP address entropy	(i) Experimental validation for backup controller functionality, in case of primary controller failure, was missing. (ii) Threshold value should have been changed dynamically as per the changing network environment. (iii) The authors did not address the efficacy of approach in protecting against LR-DDoS attacks (iv) Flash crowds may be detected by the algorithm as an attack, resulting in extra FPR. (v) The proposed approach should have been validated against performance metrics like DR, FPR, etc.

Table 2: Entropy-based approaches

Year	Technique	Anomalies addressed	Dataset	Data source	Source tool	Flow properties for anomaly detection	Comparison	Validation metrics (%)	Conclusion
2016 [21]	Generalized entropy	DDoS, probe	Real: KDDcup99, NSL-KDD, UCI machine learning repository datasets Simulated: Testbed dataset (TUIDS) for DDoS and probe attacks	IP packet/IP flow	Netflow data	Dynamic selection of features through mutual information and GE	LOF for τ=0.58 at dataset Zoo	DR=82.35 FPR=19.04	Proposed approach achieved better DR and FPR metrics compared to other outlier approaches
							ORCA	DR=88.23 FPR=13.09	
							Proposed approach	DR=94.11 FPR=2.38	
							Shannon entropy	DR=55 FPR=15	
							Kullback–Leblier divergence	DR=70 FPR=15	
2015 [22]	Extended entropy	DDoS, port scan, network scan, DoS, worm, and spam	Legitimate traffic from tsinghua University Campus network	IP flow	Netflow	Source IP address, source port, destination IP, address, destination port, flow byte, flow direction, protocol number, and TCP control bit	—	DR=93.46 FPR=5	2015
2017 [23]	Tsallis entropy	Real and simulated versions: DDoS, alpha flow, port scan, network scan	**Real** Campus network data, i.e., UTFPR/Toleda Campus and FISTSC/GW campus	IP flow	Netflow v9	Source address, destination address, source port, destination port, number of packets, number of flows, number of bytes, in-degree	Tsallis entropy	DR=100 FPR=1	Achieved better DR and FPR compared to Shannon entropy validation metrics dropped a little with sampling effects
							Shannon entropy	DR=25 FPR=2.2806	
							After incorporating sampling effects in technique	DR=99.45 FPR=0.12	

Motivation and Problem Statement

In light of the above discussion, it can be concluded that SDN significantly addresses the deployment locality requirement to its centralized controller architecture. Further, entropy-based techniques used by the researchers rely on experimental-based thresholds and do not adapt to changing network conditions. Therefore, it necessitates developing an adaptive light-weight entropy-based defence mechanism using SDN environment for SG, providing early detection and mitigation of anomaly in real time. Real-time reconfiguration based on network conditions is required to change static thresholds and also to make it appropriate for high Detection Rate (DR) and low False Positive Rate (FPR). Here DR measures portion of the attacks that are detected correctly by the system and FPR provides the percentage of events that are reported as negative events where actually they are positive events. This factor makes it highly inappropriate for SG due to dynamic nature and heterogeneity.

With the advent of IoT, security concerns related to user and network resources have become even more critical and prone to attacks. SG being one important application of IoT also shares the same security threats that exist in traditional IoT environment. However, protection of DDoS attacks in SG has grabbed more intention of researchers. The reason is the occurrence of massive DDoS attack on Ukraine power grid in 2015. Existing security protocols/techniques provide network protection at Internet edge only and are not sufficient enough to prevent dynamic attacks, considering borderless architecture of IoT. Additionally, current approaches of security, i.e., firewall zoning and intrusion detection and prevention system (IDPS) are too constrained by traditional network architecture. They are computationally heavy when considering the increase in network devices [27]. If appropriate security actions are not taken, then attacks like DDOS, service unavailability, and most importantly threat to human life might happen. Moreover, early detection and mitigation are deemed necessary for infrastructure like SG since deep penetration to SG network can lead to devastating consequences.

Entropy-based techniques used by the researchers rely on experimental-based thresholds and do not adapt to changing network conditions. Moreover, utilization of static experimental thresholds and Shannon entropy do not provide adequate security against both HR and LR-DDoS attacks for an ICT infrastructure like SG. Static thresholds need to be reconfigured on changing network conditions to adjust for high DR and low FPR and that makes it unsuitable for SG. Moreover, Shannon entropy provides low DR and FPR as compared to Tsallis entropy [23] and on detection of DDoS attack, it is important to mitigate it as well to prevent its penetration further in the network; that is missing in DDoS protection approaches. In order to improve the security and reliability of SG in reference to DDoS attacks, researchers have suggested an SDN-based approach to handle the glitches in the conventional network paradigms. However, these approaches [6, 12, 20] are still only capable of handling HR-DDoS attacks, i.e., TCP/UDP/ICMP-based flooding attacks only, not catering stealthy and low-rate DDoS attacks, and also rely heavily on static thresholds. Hence, it makes it necessary to develop a light-weight DDoS defence mechanism for SG that is fueled by SDN environment and using Tsallis entropy for better DR and FPR. Additionally, the SDN environment should be adaptive and capable of providing early detection and mitigation of both HR and LR-DDoS attacks.

Solution and Contributions

Considering our proposed solution, DDoS detection application uses Tsallis entropy metric with traffic features, i.e., Source Address (SA), Destination Address (DA), Source Port (SP), and Destination Port (DP) for efficient detection of varying DDoS attacks. For mitigation approach, IP address and port blocking mechanism is available in SDN controller software; i.e., OpenFlow is utilized. Blocking data is provided by the local list maintained in the SDN controller. Since SDN controller provides a global view of the whole network and is centrally located, the proposed approach significantly addresses the locality problem of DDoS defense mechanism that is missing in literature. A novelty in approach is added by using dynamic thresholds for traffic features using Exponential Weighted Moving Average (EWMA) instead of static threshold values for detection purposes. Moreover, to the best of our knowledge, Tsallis entropy in SDN environment has not been used previously. The proposed approach provides a near real-time detection within 250 packets with mitigation of anomaly in real time. In the following section, the proposed system model is discussed in detail. The following contributions are made in this work:(i)A light-weight entropy-based detection approach is developed underlying SDN environment(ii)Adaptive threshold mechanism is proposed to achieve better DR and False Positive Rate (FPR) using Exponentially Weighted Moving Average (EWMA) and Tsallis entropy(iii)Low-rate (LR)- and HR-DDoS attacks are successfully detected(iv)In addition to the real-time protection mechanism, a DDoS mitigation mechanism is also explained in terms of proposed model(v)Resource utilization (CPU and RAM utilization is optimized without compromising LR- or HR-DDoS protection)

SYSTEM MODEL

In existing SDN-based solutions, a limited level of DDoS protection, i.e., against flooding-based DDoS attacks only, is being provided. Further, deployment locality of DDoS defence mechanism is critical in efficient and in-time detection. Most of the researchers did not fully address the issue of locality. SDN controller has a global view of the network and is responsible for all routing and filtering features of a network. In other words, it is a brain and central point of network. Therefore, SDN network utilization can provide optimal deployment locality for DDoS defence mechanism. Lastly, software-based control of SDN provides IP address/port blocking mechanisms as a built-in feature.

So, these mechanisms can be optimally utilized as a DDoS mitigation approach. The conceptual framework is divided into two parts, namely, SDN-based environment for consumer-utility provider network and intrusion detection and prevention system (IDPS) as depicted in Figure 1. In this section, a detailed description of the system model is presented, following the implementation constraints and limitations.

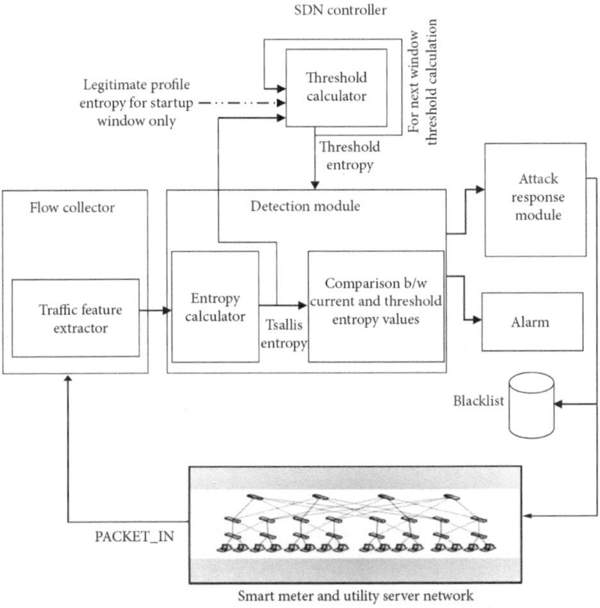

Figure 1: System model S-DPS.

IDPS is divided into three modules, namely, flow collector ("FC"), anomaly detector ("AD"), and anomaly mitigation ("AM"), as depicted in Figure 1. "FC" module is located in controller and collects network flows/packets and statistics from each connected switch through Netflow standard protocol, utilized by the controller. These flows are stored in the local database of the controller and relevant features, i.e., Source Address ("SA") and Destination Address ("DA"), are extracted for further processing by "AD." "AD" calculates Tsallis entropy value per traffic feature in current window of 50 packets and compares it with corresponding feature threshold value for that window. In case of a mismatch as per conditions discussed in subsequent section, an alarm is generated and further action is taken by the "AM" module. "AM" module performs drop/deny action on the flows and pass it on to v-switch performing forwarding decisions. It also stores the blacklisted IPs in blacklist database maintained locally in the controller for scrutiny of incoming packets. Threshold calculator calculates threshold values per feature for next window by applying Exponentially Weighted Moving Average (EWMA) on current window entropy value and previous window threshold value and passes it on to "AD" module for comparison purposes. Table 3 describes the primitive flow properties being used in the analysis. After extraction of required details, data is parsed to "AD" module which follows the mechanism as discussed in upcoming sections.

Table 3: Primitive flow properties for AD

No. (m)	Primitive flow property (xm)	Detail
1	SA	Source IP address of packet
2	DA	Destination IP address of packet

Anomaly Detector Module (AD Module)

After extraction of traffic features ("SA," "SP," "DA," and "DP") by "FC" from new packets destined to the controller from OF switches through Netflow protocol supported by POX controller, data is fed to the "AD" module. Data in anomaly detector is processed based upon window size that can be based upon either time stamp of packet received or number of packets. For this work, it is based upon number of packets and set to 50 packets per window for efficient detection and memory foot-print [20]. Moreover, the experimental setup constitutes not more than 50 hosts (smart meters and utility server), so 50 packets per window is an appropriate window size. Therefore, consider W as the set of data with n elements in which each data element x_{mi} signifies the event pertaining to specific traffic feature as can be seen in (1). Probability of x_{mi} happening in window W can be calculated using (2). Further, Tsallis entropy is denoted by "H_q" which can be calculated by (3) [23]. For $q>1$, higher probabilities have more impact on the final entropy value compared to lower probabilities and vice versa. Here value of q is set as −1.3 or −0.8 for high DR and low FPR.

$$w = x_{mi}, \quad m = 1, 2, 3, 4, i = 1, 2, \ldots n, \tag{1}$$

$$P_{mi} = \frac{x_{mi}}{n}, \tag{2}$$

$$H_q = \frac{1}{q-1} 1 - \sum_{i=1}^{n} p_{mi}^{q}. \quad (3)$$

In each window, entropy values of four traffic features are calculated and compared with respective normal entropy value, using (4). Here $H_{q-(n)}^{x_m}$ is the entropy value of specific traffic feature taken in normal traffic conditions, i.e., without any abnormal traffic, and λ signifies the difference of entropies. In case the value of λ is positive, it means the entropy value of feature for current window has decreased; i.e., data distribution is concentrated. However, in case value of λ is negative, it means the entropy value of feature for current window has increased; i.e., data distribution is dispersed.

$$H_{q-(n)}^{x_m} - H_q = \lambda. \quad (4)$$

Application of (4) is different for each traffic feature as depicted in Table 4. In case of DDoS attack, value of λ is positive for "DA" and negative for "SA," whereas for different types of DDoS attacks, values of "SP" and "DP" are variable. Equations (1)–(4) are calculated for subsequent windows (50 packets per window) and in case value of λ is positive for "DA" and negative for "SA" for five consecutive windows, an alert for DDoS attack is generated. A counter for subject purpose is utilized, which is incremented on meeting the set conditions in each window. In case set conditions for "SA" and "DA" are not met in any 5 consecutive windows, counter is set to zero and process starts again with counter=0.

Table 4: Interpretation of value of λ

No. (m)	Flow property	Value of	Impact	Result
1	SA	Negative real number	Data dispersion	Attack from multiple IPs
2	DA	Positive real number	Data concentration	Attack towards specific IP

Anomaly Mitigation Module

A specific action is associated by the controller with each flow in flow tables of the controller as discussed in background section related to OF protocol. In case an alarm is generated by the "AD" module, then the "SA" with maximum number of occurrences in the 5 windows is extracted and "drop/

deny" action is set by the controller against the matched flows associated with that "SA" in run time.

Dynamic Threshold

Initially, threshold values for each traffic feature are set by simulating the network environment in normal conditions, i.e., without any attack traffic. These threshold values are used to detect DDoS attack in progress as per the conditions discussed previously. Value of threshold dictates the performance of entropy-based detection approach in terms of DR and FPR. So, choosing optimal thresholds is most significant and important to achieve desired results. One approach is to conduct multiple experiments using attack traffic (tool or datasets) with normal traffic to tune these thresholds, while another approach is to utilize current network conditions in real time and system automatically updates these thresholds. The latter is more convenient and effective, considering the dynamic nature of SG network. So, in order to make the anomaly detection adaptive, consider a mean entropy value for each traffic feature as and for each subsequent window, mean entropy value for each traffic feature as a threshold, is calculated using (5). EWMA filter is used for calculating the average mean, and β value of 0.1 is used for catering current network conditions and is more reactive in nature, considering highly critical networks such as SG. Value of constant c depends upon the network characteristics.

$$H_{q(i)}^{x_m} = \left(\beta \times H_{q(i-1)}^{x_m} + (1-\beta) \times H_{q(i)}^{x_m}\right) + c. \tag{5}$$

Threshold values, calculated as per (5), are based upon current network conditions with β value set to 0.1 (very reactive) and can result in high FPR for burst channel. Similarly, in case of stealthy attack pattern such as increasing and decreasing DDoS attacks, detection will be difficult. So, there is a need to tune the value of threshold in real time. In order to achieve optimum DR and FPR and keep the threshold in acceptable bounds, a maximum change/difference of current threshold from the normal entropy threshold (calculated during normal conditions) should not exceed by 1.5, considering 90% confidence interval for normal distribution. In case it exceeds more than 1.5 times, value of current threshold will become 1.5 times to normal entropy threshold; otherwise, it will remain the same as per the calculated mean threshold value. However, for decreasing entropy with respect to normal entropy threshold (calculated during normal conditions) for more than 1.5 times, value of current threshold will be normal entropy threshold value divided by 1.5 to normalize the threshold; otherwise, it will remain the same as per the calculated mean threshold value. The multiplication and division factor of 1.5 is used to keep the thresholds within reasonable bounds with respect to normal threshold value. The increasing entropy check is applicable for SA entropy, whereas decreasing entropy check is applicable for DA entropy. The reason is that DDoS attack tends to decrease DA entropy while increasing SA entropy values. Flow chart for the algorithm is presented in Figure 2. In OF-based v-Switch, a packet for which no flow entry exists is passed on to controller for decision making. So, in the algorithm packet in flow step signifies entry of new packet in the controller. Traffic features of the packet as per Table 3 are checked for existence of entries already in the system. In case entries exist in the lists; then occurrence counters for each feature are incremented. Otherwise, new entries in the corresponding lists are made. If the number of packets count has reached 50 as per the set window, entropy value for each traffic feature using the corresponding traffic feature list is calculated. It is then compared with the threshold value. In case current DA entropy value is less than mean DA threshold value and current SA entropy value is greater than mean SA threshold value; then the consecutive window counter is incremented and the same cycle starts again for next window with number of packets count set to zero. Moreover, in case the current DA, entropy value is greater than mean DA threshold value and SA entropy is less than mean SA threshold value; then the cycle starts again with number of packets count set to zero.

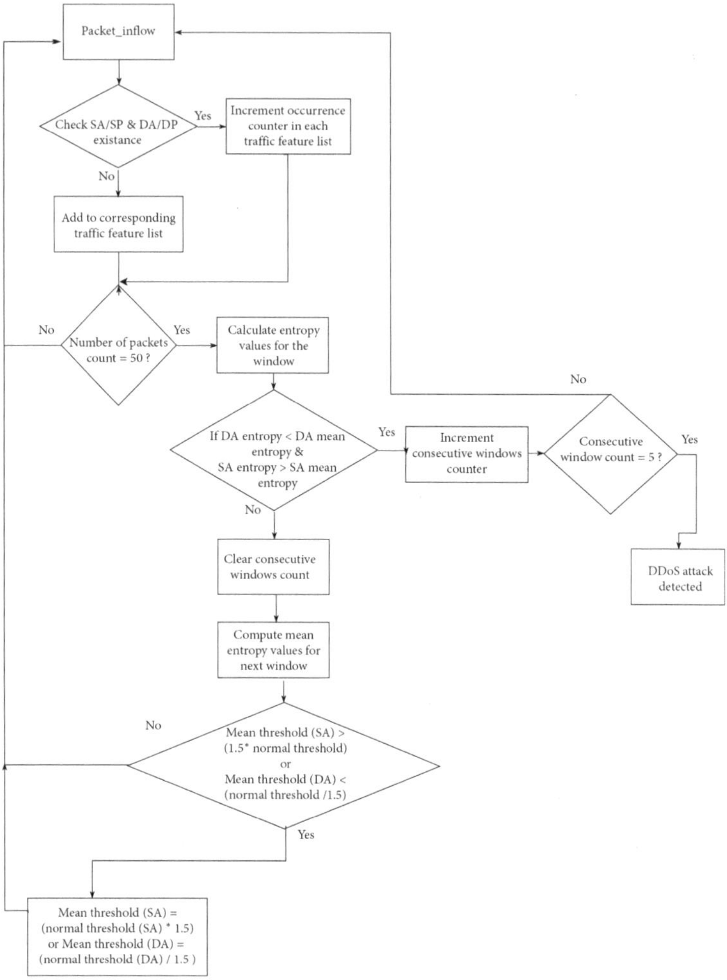

Figure 2: Proposed IDS: flow chart.

Implementation Constraints and Limitations

As previously mentioned in related work, DDoS related datasets for SG are not publicly available and datasets like MIT Lincoln, FIFA, DDoSTB, and CAIDA datasets are not SG related [12]. Therefore, [1, 12] relied on simulated traffic to test the viability of their proposed approach. Similarly, in the paper, simulated normal and attack traffic is being generated using Scapy tool to test the proposed model because it is python-based and can be integrated with Mininet. Single topology is tested for different types of DDoS attacks and the traffic is simulated one. Results obtained may vary in real-time traffic. Moreover, model presented is independent of any protocol (tested for TCP/UDP/ICMP-based packets) and threshold for DDoS detection is being adjusted automatically with varying network conditions. So, solution is viable for dynamic network conditions as in modern networks. Apart from it, the solution is tested using a software-based simulator. Its capability will further be improved with powerful hardware-based SDN controller available in the markets.

Furthermore, the approach is based on single controller architecture, wherein it can present a bottleneck and security constraint when dealing with large-scale network like SG. Difference between using single-controller and multicontroller architecture is linked to load balancing, high availability, and security of controller. However, for this research it is outside the scope of work and the approach can be integrated and tested with multicontroller architecture for future research.

EXPERIMENTAL SETUP

In this section, an experimental setup for validation of S-DPS against Utility-Consumer Communication Network implementation is discussed. For this purpose, a series of steps are followed in order to establish simulation for performing the experiments using test cases as discussed in the following section.

Simulation Steps

Controller

POX is used as SDN controller for the experiments. It is an open-source and python-based controller that is widely used in experiments. It is lightweight and developed as a platform to be customizable, meeting desired needs of a controller. It supports famous operating system like Windows, Linux, and MAC OS and has a network discovery feature installed. Apart from this, another two famous controllers like Floodlight and Beacon are also available. However, in most SDN-based papers highlighted in literature review, NOX controller, a predecessor of POX, is used. So, for the research POX controller is selected.

Network Emulator

Mininet 2.2.2 is used as a network emulator for the experiment. It is an open-source platform with support for SDN environment and OF protocol. It treats each network component as a kernel process and can be installed easily on a laptop or Personal Computer (PC) using kernel namespace feature. Each network namespace has its own Network Interface Card (NIC), Address Resolution Protocol (ARP) table, ping service, scripts, and routing table. Both Graphical User Interface (GUI) and command line interfaces are available to create network topologies. As a default, NOX controller is embedded in Mininet.

Traffic Generation

Scapy is used as a traffic generator tool, both for normal and for attack traffic. It has features of scanning, packet spoofing, packet forging, sniffing, etc. Here, TCP packets are generated using the tool. It supports python programming language and POX controller also uses python. So, both controller and traffic generation tool can be integrated. Spoofed source IP addresses and Host IP addresses are generated using python function "random."

This function returns a uniform random float in the range of 0.0 to 1.0. These random floats are joined together to form a spoofed IP address. Other options, i.e., type of packets and packets interval available in Scapy, are used to create normal and attack traffics. TCP/UDP/ICMP is set for type of packets and 0.4 seconds as interval for normal traffic between smart meter and utility server. Moreover, TCP/UDP/ICMP-based DDoS attacks with attack rates ranging between 200 and 4000 packets/sec are simulated in existing researches, i.e., [6, 20, 27, 28]. Such variations of traffic generation are catered for in existing experiments, covering both LR- and HR-DDoS attacks.

Network Setup

Network is set up on Laptop Dell Inspiron with Core i3 2.41 Ghz processor, 4GB RAM, and 100/1000Gbps NIC. Operating System is Windows 8.1 with VirtualBox 6.0.4 installed. Mininet 2.2.2 on Linux Ubuntu 14.04.4 is installed in the VirtualBox for setting up the environment. Mininet 2.2.2 supports OF version 1.3. Moreover, "mn" command is used in Mininet to set up the network. As a default, two hosts with one switch are configured. However, custom network is set up using different filters available in "mn" command, i.e., related to controller either local or remote, type of switch, number of hosts/switches, etc.

Network Topology

A tree-type network constituting smart-meter-utility server communication network is depicted in Figure 3. It has a depth of 2 with 10 switches and 54 hosts (smart meters and utility server). Here "*smart meters*" are the core of SGs because they are smart and possess the ability to sense, measure, and examine the usage of electricity, continuously transmit the data and information collected to the central location, and perform two-way communications with all other components of the SG and the consumer. Meanwhile, "*utility server*" has a dual-role to play; i.e., it has a two-way communication with smart meter as well as with power generation facility. It is located at control center and provides live consumption data to both users and to power generation facility. Finally, "*controller*" is the brain of the overall network managing all OF-enabled switches/routers by installing forwarding rules and performs centralized network and configuration management for better performance and security in the network.

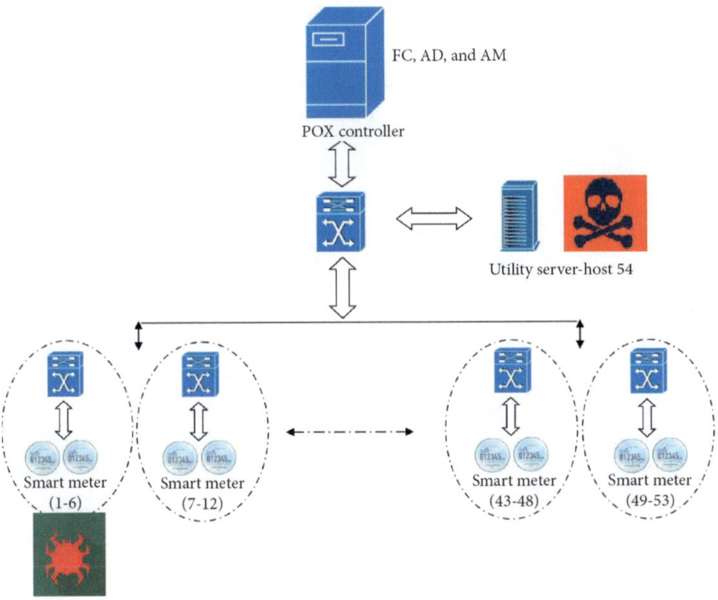

Figure 3: Network topology.

Utility server is connected to Switch-1, whereas Switches 2–10 are used to connect 53 smart meters, evenly divided, i.e., 6 smart meters each. However, last switch consists of 5 smart meters. OF-enabled v-Switch available in Mininet is used to connect hosts. L3-learning module of POX controller with addition of two functions, i.e., traffic feature collection and entropy calculation, is used for the controller function of the network.

Evaluation Criteria

The S-DPS is evaluated using DR and FPR metrics where DR measures portion of the attacks that are detected correctly by the system and represented by (6) and FPR provides the percentage of events that are reported as negative events where actually they are positive events and represented by (7). In the equations, True Positive (TP) event means that the system has detected a correct anomalous event, whereas False Positive (FP) means system has detected an incorrect anomalous event; i.e., actually the event is legitimate but detected otherwise. Similarly, True Negative (TN) event means that the system has detected a correct legitimate event, whereas False Negative (FN) means system has detected an incorrect legitimate event, i.e., actually the event is anomalous but detected otherwise. Varied levels of both LR- and HR-DDoS attacks, i.e., smurf, socket stress, and SYN flood attacks, are launched against the utility server for early detection and real-time mitigation of attack.

$$DR = \frac{TP}{TP + FN}, \quad (6)$$

$$FPR = \frac{FP}{TN + FP}. \quad (7)$$

RESULTS AND DISCUSSION

The experiment covers topology highlighted in Figure 3, which contains 10 switches and 54 hosts. Each host from h1-h53 represents a smart meter, where host h54 is a utility server with which each smart meter sends its observed values. Each switch in the topology is OF-enabled v-Switch centrally connected to POX controller c0. In order to simulate the traffic between connecting entities, certain parameters like frequency of communication between smart meter and utility server, type of protocol, etc., need to be defined. AMI infrastructure does not have any standardized architecture and varying implementations exist defining the network and dynamics of communication. Frequency of communication between smart meter and utility server is also set to different intervals, i.e., 1 second, 4 seconds, 60 seconds, 5 minutes, and 15 minutes, depending upon the scheduling criteria set by utility service provider [29]. Considering the periodic traffic profile in most common architectures, smart meters are scheduled to transmit and receive at interval of 0.4 seconds [30]. Further, both UDP and TCP protocols are used in two-way communication between smart meter and utility server. Seven sets of traffic profiles are generated in the experiment, i.e., normal traffic, smurf attack, socket stress attack, and SYN flood attack. Traffic profile for the experiments is shown in Table 5. These traffic profiles are simulated using UDP/TCP/ICMP-based packets at destination port 80/21 using random spoofed source IP addresses and source ports.

Table 5: Traffic profiles

Type of traffic	Protocol	DP	SP	Payload: number of packets	Source IP address	Destination IP address	Traffic interval	Attack type
Normal	UDP	80/21	2/3	None	10.0.0.54 or random (10.0.0.0/24)	10.0.0.54 or random (10.0.0.0/24)	0.4 sec	—
Smurf attack	ICMP	—	—	6000 bytes	10.0.0.54	10.0.0.255	—	DDoS
Socket stress	TCP	21	Random (0–65535)	None	10.0.0.4	10.0.0.54	—	DDoS
SYN flood	TCP	80	Random (1000–9000)	None	Random	10.0.0.54	—	DDoS

Normal Traffic Profile

In normal traffic profile, a total of 5 runs of experiment are performed, each containing 1250 packets with window size of 50 packets. Packet interval between smart meter(s) and utility server and reverse is set to 0.4 seconds. In normal circumstances, at any given point in time, a utility server is sending probe to any smart meter and any smart meter is sending its readings to utility server.

Therefore, two-way packets are generated randomly using one of the IP addresses of smart meter and of utility server with interval of 0.4 seconds to obtain average normal entropy value. The whole experiment is covering observation of 6,250 packets. The results of normal traffic separately for source IP (SrcIP) and destination IP (DestIP) are presented in Figure 4. Here, average entropy values per window for both SrcIP and DestIP are used to plot the graphs. As can be seen from Figure 4, entropy value for

DestIP ranges from 1011.923 to 1372.990 and average normal entropy value is being utilized as a base entropy for DestIP in attack scenarios. Similarly, entropy value for SrcIP ranges from 1066.081 to 1722.328 and average normal entropy value is 1320.678, being utilized as a base entropy for SrcIP in attack scenarios.

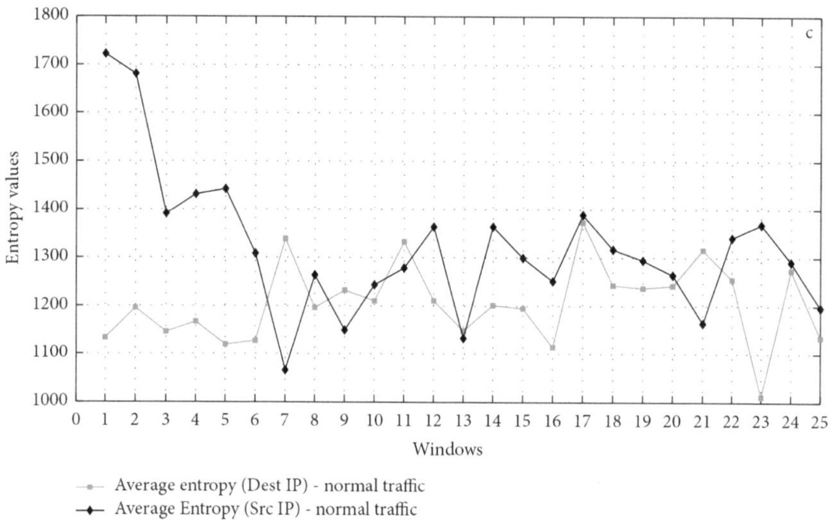

Figure 4: Normal traffic profile.

Smurf Attack

A smurf attack is a type of DDoS attack in which vulnerabilities in Internal Protocol (IP) or Internal Control Message Protocols (ICMP) are exploited as such that it makes the overall computer network inoperable. For smurf attack to work, a false IP packet with spoofed IP is created. IP packet is basically an ICMP ping message that tells the network nodes to receive and send back echo reply. These echoes are then sent back to all network devices creating an infinite loop in the network. To further amplify the attack, IP broadcasting technique can be used.

In the experiment, an ICMP echo request is generated towards the broadcast address of all switches/routers, i.e., 10.0.0.255 using the spoofed IP address, i.e., of target address (10.0.0.54), *which is a utility server*. In this case, all the smart meters lying under these switches/routers will send their ICMP echo replies towards the target, i.e., *utility server*. In order to further amplify the attack, each smart meter is relaying 6000 bytes of junk IPv4 packets towards the target. Two separate scripts are being run manually using two random hosts, e.g., h1 and h4. At h1, normal traffic generation is carried out, whereas at h4 (*attacker machine*), smurf attack towards target address (*utility server*) is launched. Traffic profiles both for source and destination IP for the scenario are depicted in Figure 5. It can be seen from Figure 5(b) that destination IP current entropy is far below the threshold value between windows 6 and 25, meaning the number of packets with same DestIP/window, i.e., towards the target host, has increased exceptionally resulting in decrease of overall DestIP address entropy. So, the attack is detected in these windows. Further, to verify whether it is a DoS or DDoS attack it can be seen from Figure 5(a) that source IP current entropy is above the threshold value between windows 11–22, meaning the number of packets with multiple SrcIPs/windows for the target host exists, resulting in increase of overall SrcIP address entropy from the threshold. Therefore, the attack detected is DDoS. In case it is below the threshold values, the attack is considered as DDoS attack.

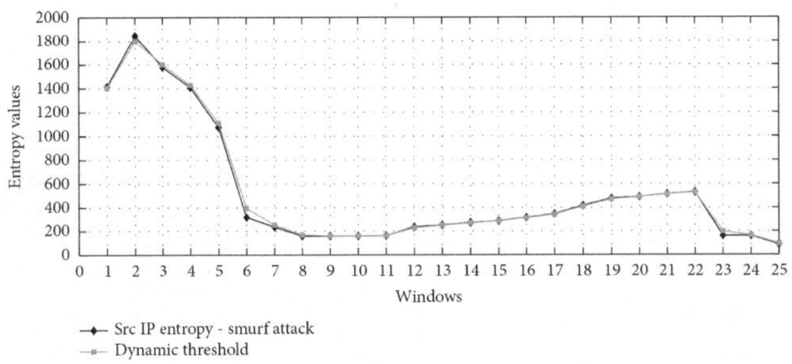

(a)

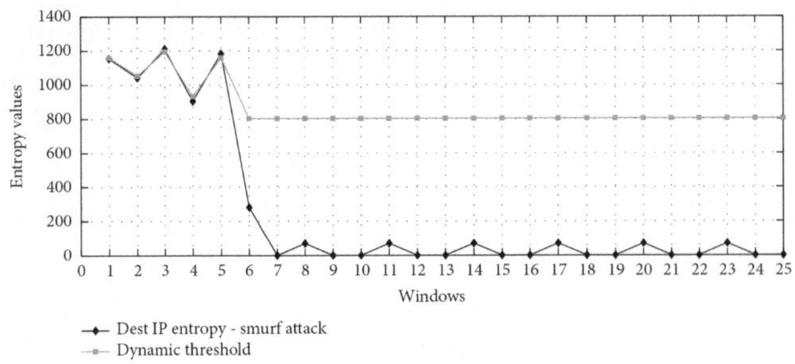

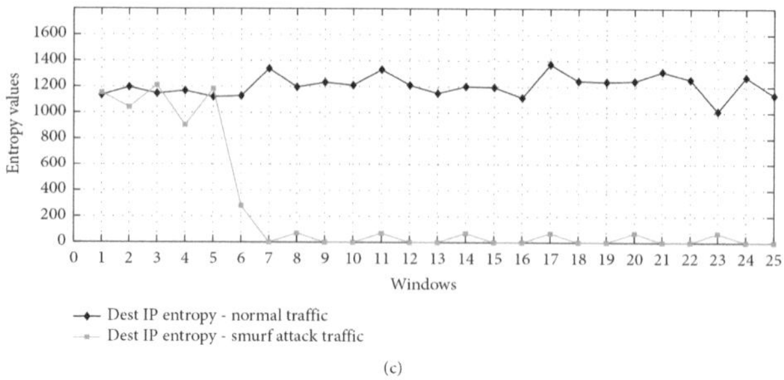

Figure 5: Smurf attack detection- static vs. dynamic thresholds. (a) Source IP address traffic profile; (b) destination IP address traffic profile; (c) destination IP-normal vs. smurf attack traffic.

Moreover, comparison between the normal and attack traffic for destination IP is depicted in Figure 5(c). It can be been seen that entropy values/window for attack traffic has declined significantly compared to normal traffic since most of the traffic/window is directed towards a single DestIP, resulting in decline of entropy. Moreover, it can be observed from Figures 5(a) and 5(b) that S-DPS-based threshold is changing as per the current network conditions compared to experimental static threshold that remains fixed no matter how much the network environment varies. So, S-DPS-based threshold is able to provide true picture of the network while achieving DR of 100% with 0% FPR for simulated traffic.

Socket Stress Attack

Considering socket stress attack, raw sockets are used to establish a connection with the target machine. It is an asymmetric resource consumption attack, where asymmetric refers to less requirement of resources at attacker end verses a great deal of resource consumption on target machine. For such attack to work, it should be targeted to an open port in victim's machine. In the attack, attacker advertises a zero window at the end of three-way handshake; meaning it has not received the data so the victim will tend to open the connection and probe the client periodically to check whether data is received or not. Similarly, multiple connections at the victim machine are opened, consuming many resources on the victims' machine. Socket stress attack script is executed randomly on h4 (*attacker machine*) targeting utility server *(victim machine)* at IP address 10.0.0.54 and port 80. In the attack, 20 random connections using random source ports are created with a timeout value of 1 minute. Timeout value defines the time before which new connection is established to the target. So, at any given point in time, a minimum of 20 connections remain active on the target machine. Two separate scripts are being run manually using two random hosts, say h1 and h4. At h1, normal traffic generation is carried out, whereas at h4, socket stress attack towards target address is launched. Traffic profiles both for source and destination IP for the scenario are depicted in Figure 6. It can be seen from the figure that destination IP current entropy is far below the threshold value between windows 5 and 25, meaning the number of packets with same DestIP/window, i.e., towards the target host, has increased exceptionally resulting in decrease of overall DestIP address entropy. So, the attack is detected in these windows. Further, to verify whether it is a DoS or DDoS attack, it can be seen from Figure 6(b) that source IP current entropy is not above the threshold value for consecutive windows from windows 1 to 25. It means that the number of packets with single SrcIP/window for the target host exists, resulting in decrease of overall SrcIP address entropy from the threshold. Therefore, the attack detected is DoS. Moreover, comparison between the normal and attack traffic for destination IP is depicted in Figure 6(c). It can be seen that destination IP entropy values/window for attack traffic has decreased significantly after the attack compared to normal traffic using the proposed S-DPS mechanism since most of the traffic/window is directed towards a single DestIP, resulting in decline of entropy.

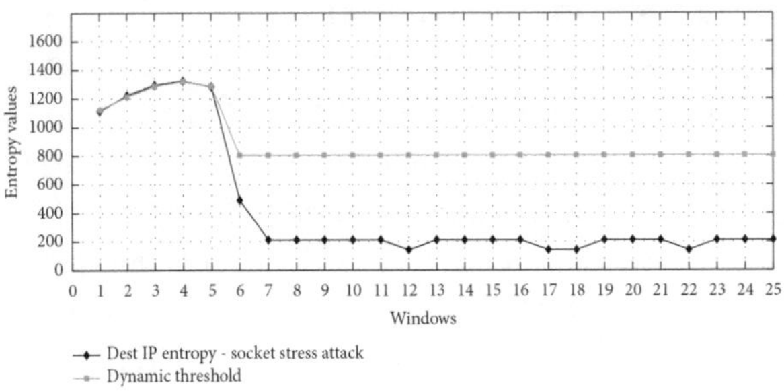

(a)

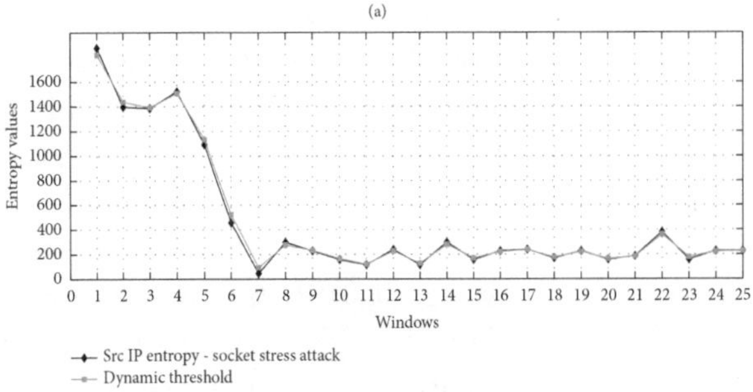

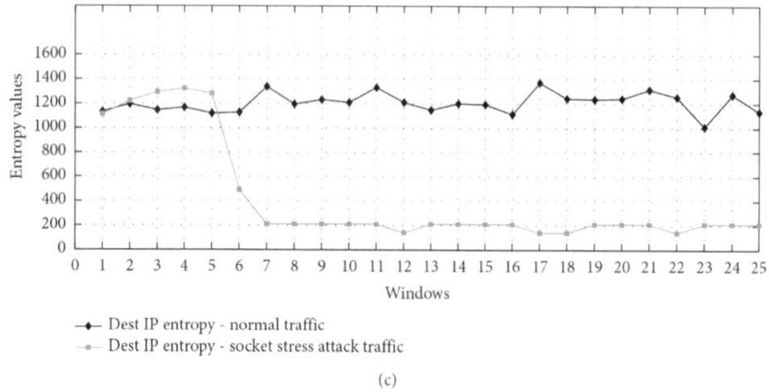

Figure 6: Socket stress attack detection-static vs. dynamic thresholds. (a) Destination IP address traffic profile. (b) Source IP address traffic profile. (c) Destination IP-normal vs. socket stress attack traffic.

SYN Flood Attack

In case of SYN flood attack, the attacker exploits part of normal TCP three-way handshake process by sending repeated SYN packets to the target machine with a frequency above its capacity to process. It can target all open ports or a specific port to block the service(s) of the target machine. The target machine responds to all received requests with SYN-ACK packets for that open port(s) and wait for ACK packets for some time. In most scenarios, source IP address and ports are malicious, i.e., spoofed, so ACK packets are never sent back or, in another case, ACK packets are not sent by the attacker deliberately to shut down the service of target machine. SYN flood attack script is executed randomly on h4 (*attacker machine*) targeting utility server (*victim machine*) at IP address 10.0.0.54 and port 80. In the attack, 10,000 packets with random source IP address and ports (ranging between 1000 and 9000) are sent to the utility server. These packets have random "seq" numbers and "window" size between 1000 and 9000. Two separate scripts are being run manually using two random hosts, say h1 and h4. At h1, normal traffic generation is carried out, whereas at h4 (*attacker machine*), SYN flood attack towards target address (*utility server*) is launched. Traffic profiles both for destination and source IPs for the scenario are depicted in Figure 7. It can be seen from Figure 7(a) that destination IP current entropy is far below the threshold value between windows 5 and 25; meaning the number of packets with same DestIP/window, i.e., towards the target host, has increased exceptionally resulting in decrease of overall DestIP address entropy. So, the attack is detected in these windows. Further, to verify whether it is a DoS or DDoS attack, it can be seen from Figure 7(b) that source IP current entropy is above the threshold value for consecutive windows from windows 6–25, meaning the number of packets with multiple SrcIPs/windows for the target host exists, resulting in increase of overall SrcIP address entropy from the threshold. Therefore, the attack detected is DDoS. Moreover, comparison between the normal and attack traffic for destination IP is depicted in Figure 7(c). It can be seen that destination IP entropy values/window for attack traffic has decreased significantly after the attack compared to normal traffic using the proposed S-DPS mechanism since most of the traffic/window is directed towards a single DestIP, resulting in decline of DestIP entropy.

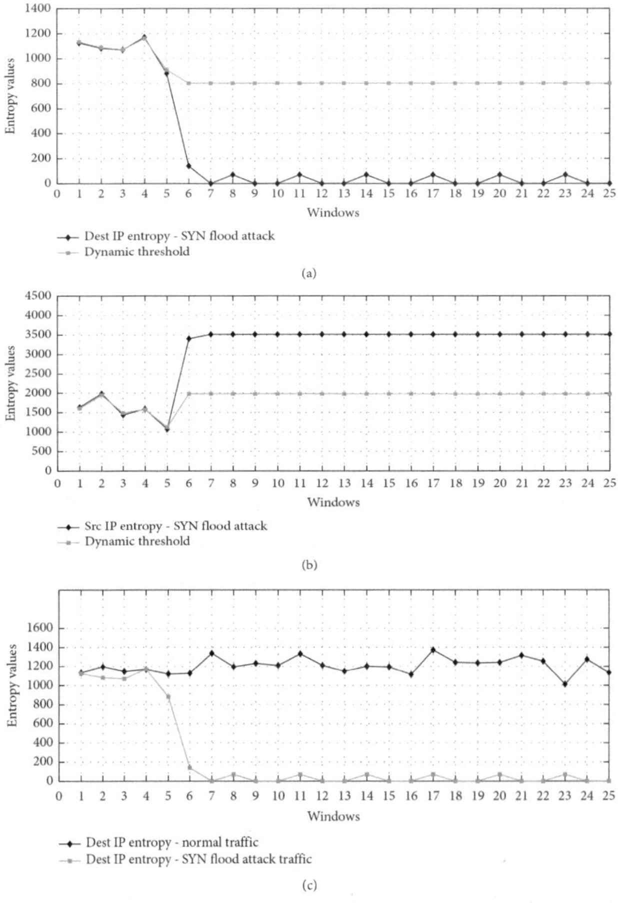

Figure 7: SYN flood attack detection-static vs. dynamic thresholds. (a) Destination IP address traffic profile; (b) source IP address traffic profile; (c) destination IP-normal vs. SYN flood attack traffic.

For all the attacks discussed above, although the target is *utility server (h54)*, controller being the brain of SDN network is processing all the normal and attack packets. So, in addition to utility server (*target machine*) controller is also being targeted in all attack scenarios discussed, but the detection and mitigation approach is implemented at the controller so attack is mitigated within near real time.

Mitigation of DDoS Attack

On detection of DDoS attack, it is important to mitigate it as well to prevent its penetration further in the network. OF protocol, due to its real-time reconfiguration feature, enables us to define flow rules that can block the switch ports in real time. The authors in [6, 27, 28] utilized OF port blocking or deletion of flows as a mitigation strategy, achieving time and space complexity of *O* (*n*), where "*n*" may be number of packets processed for port blocking or number of flows deleted. For that matter, port blocking strategy is utilized, achieving the same complexity of *O* (*n*). One timer variable of Boolean type, i.e., "timerSet," and two functions, i.e., Preventing() and _timerfunc(), are incorporated in the default L3_learningmodule of POX controller. By default, timerSet is set to "False" so that controller continues to operate without active DDoS defense mechanism till entropy of the window does not fall under threshold value of that window. In case entropy values of the windows from the entropy dictionary are less than the threshold values, Preventing() function is invoked with global Set_Timer set to "True"; otherwise, timerSet value is set to "False" to enable/allow normal operation of the controller again, i.e., without active DDoS defense mechanism. Eventually, _timer_ func is used to detect the happening of DDoS attack using the dictionary maintained by Preventing() function and block the switch ports with count greater than and equal to 5, occurring in five consecutive windows. Preventing() function is incorporated in POX controller using "*_handle_openflow_packetIn*" instance. Each time a new packet enters the controller, packet is accounted for in the dictionary being maintained. Dictionary constitutes a switch ID and port number with its counter. It has a form like switch ID (port number, count). Switch ID is recognized by OF parameter "*event.connection.dpid*" and port number by "*event.port*." It is used to detect whether DDoS attack has occurred or not. After creating the dictionary for 25 windows, _timer_func() is used to detect and mitigate DDoS attack. It iterates through all the items in the dictionary and if specific ports of a specific switch have its count greater than and equal to 5 and for five consecutive windows, DDoS attack is detected and these switch ports are blocked by sending message to controller using OF procedure calls, i.e., "*of.of p_packet_out*" and "*core.openflow.sendToDPID().*" The mitigation strategy is performed successfully on 25% rate attack on single host. As per results, dictionary maintained by the controller contains count of 69 for Switch-1 and Port-1, 12 for Switch-2 and Port-1, 60 for Switch-2 and Port-4, and 61 for Switch-8 and Port-9. All such switch ports are blocked by the controller as part of prevention strategy.

PERFORMANCE EVALUATION

The performance of S-DPS is evaluated using metrics like CPU/RAM utilization. CPU/RAM utilization is measured and compared with and without the approach using 25% attack rate on single host scenario. As mentioned in previous sections of conceptual framework, 4 additional functions are added to the L3-learning module of POX controller, i.e., traffic feature collection, entropy calculation, timer function, and preventing function, for DDoS detection and mitigation purpose. In order to see the effect of these functions on the overall CPU/RAM utilization of Mininet and on the controller, two simulations are run again. One simulation constitutes 25% rate attack on single host without the solution and other simulation with same setting with proposed solution. The elapsed time for both simulations is 25 seconds and normal traffic ran for 200 seconds. "Top" and "Htop" commands have been used to capture the CPU/RAM utilizations. Results are depicted in Table 6. It can be seen from Table 6 that overall CPU/RAM utilization is 55.5%/171 MB in case of simulation without the solution and controller instance has consumed 12.3%/1.4% of total memory. In case of simulation with the solution, overall CPU/RAM utilization is 55.2%/205 MB and controller instance has consumed 29.6%/1.7% of total memory. There is a slight increase in controller instance CPU/RAM utilization but it is still in acceptable limits.

Table 6: Resource utilization

	Controller instance		Mininet instance	
Resource utilization	CPU%	RAM%	CPU%	RAM (Mbs)
Without S-DPS	12.3	1.4	55.5	171
With S-DPS	29.76	1.7	55.2	2.5

DDoS detection and mitigation functions are incorporated in SDN controller, considering the low computational complexity of approach used, i.e., O (n) for both time and space complexity. It is verified by CPU/RAM utilization with and without the approach. At controller end, CPU utilization rises to only 29.6% from 12.3% with S-DPS. Similarly, there is a minimal increase in RAM utilization from 1.4% to 1.7%. Considering the facts, S-DPS can be both efficient and effective approach to provide DDoS protection in dynamic networks like SG communication network. The reason is its nondependency on any training requirements and due to adaptive nature of threshold calculations.

Several approaches to DDoS detection exist in literature. For example, Self-Organizing Maps (SOM), a machine learning approach, has been used by [31] to learn the behavior of network and decide whether network is attacked or not. Several hours of learning is required for better DR and FPR. In case of network or topology change, SOM is required to be trained again. With expansion of network, neurons used in SOM are also required to be increased, making the solution more expansive towards the network. The S-DPS is built in inside the controller and is easily adaptable to the changing network. No training is required upfront and computational complexity is lower than machine learning approach—SOM. Similarly, the authors in [32, 33] have utilized SNORT alongside SDN for DDoS detection. As highlighted previously, S-DPS has achieved better CPU/RAM utilization compared to SNORT. Moreover, DDoS protection mechanism is embedded in S-DPS, where in [33] separate SNORT detection system is integrated with SDN environment making it less transparent towards computational overhead, sampling requirements, and bandwidth limitations, if any. Both SOM and SNORT apply complex operations to learn the behavior of the network, e.g., processing large matrices or pattern matching schemes. In S-DPS, entropy-based mechanism is providing the same functionality without any of the complexities available in SOM and SNORT.

Benefits that are achieved through S-DPS are highlighted as follows:(i) High DR with no FPR(ii)DoS, LR-DDoS, and HR-DDoS attacks that have been successfully detected(iii)Threshold mechanism that is adaptive rather than static and without any experimental adjustment for better DR/FPR, thus making it more suitable for modern/dynamic networks(iv)DDoS mitigation mechanism also provided as an addition for real-time protection

CONCLUSION

Given the nature of current dynamic networks, DDoS attacks are constantly becoming more sophisticated and are rapidly growing. These attacks can prove to be devastating for the underlying networks, with special emphasis to communication networks existing in SGs. The conventional approaches to counter these attacks are not enough to provide sufficient safety. Many researchers have claimed that the evolving SDN-based approaches are successful in dealing with the DDoS attacks in such networks. But, it is reported that these approaches employ the static threshold mechanisms to detect the attacks which is not suitable, given the dynamic and heterogeneous networks of SGs. S-DPS has claimed to efficiently address and manage these issues by employing an SDN-based environment and using a light-weight entropy-based defense mechanism. The DDoS protection strategy is made more efficient and effective through the reconfiguration of network in real time and by providing the global view of SDN networks. It is capable of detecting the threat along with the mitigation of anomaly at the same time as early as the first 250 packets by blocking the ports. Additionally, the existing SDN-based approaches are unable to detect different level of DDoS attacks but with the use of Tsallis entropy and its sensitivity factor, detection becomes possible. DR of 100% with FPR of 0% is achieved through simulation of HR-DDoS attacks. The S-DPS is able to show its capability and productiveness in both protection against DDoS and computational costs through minimum usage of CPU and RAM.

Future Works

Single controller architecture is utilized in S-DPS, making it vulnerable to computational/bandwidth bottlenecks for very large networks. In order to add resiliency in S-DPS, a multicontroller architecture is recommended. Intercontroller communication mechanism is necessary to provide synchronized operations of the protection system, with necessary recovery and failsafe mechanism.

ACKNOWLEDGMENTS

This work was partially supported by the Natural Science Foundation of China (NSFC) under grant no. 62072217.

REFERENCES

1. D. Jin, Z. Li, C. Hannon et al., "Toward a cyber resilient and secure microgrid using software-defined networking," *IEEE Transactions on Smart Grid*, vol. 8, no. 5, pp. 2494–2504, 2017.
2. P. I. Radoglou-Grammatikis and P. G. Sarigiannidis, "Securing the smart grid: a comprehensive compilation of intrusion detection and prevention systems," *IEEE Access*, vol. 7, pp. 46595–46620, 2019.
3. H. Maziku, S. Shetty, and D. M. Nicol, "Security risk assessment for SDN-enabled smart grids," *Computer Communications*, vol. 133, pp. 1–11, 2019.
4. M. Z. Gunduz and R. Das, "Cyber-security on smart grid: threats and potential solutions," *Computer Networks*, vol. 169, Article ID 107094, 2020.
5. M. Abujubbeh, F. Al-Turjman, and M. Fahrioglu, "Software-defined wireless sensor networks in smart grids: an overview," *Sustainable Cities and Society*, vol. 51, p. 101754, 2019.
6. K. Giotis, C. Argyropoulos, G. Androulidakis, D. Kalogeras, and V. Maglaris, "Combining OpenFlow and 600s Flow for an effective and scalable anomaly detection and mitigation mechanism on SDN environments," *Computer Networks*, vol. 601, pp. 122–136, 2014.
7. R. Ande, B. Adebisi, M. Hammoudeh, and J. Saleem, "Internet of Things: evolution and technologies from a security perspective," *Sustainable Cities and Society*, vol. 54, Article ID 101728, 2019.
8. M. A. Ferrag, L. A. Maglaras, H. Janicke, J. Jiang, and L. Shu, "A systematic review of data protection and privacy preservation schemes for smart grid communications," *Sustainable Cities and Society*, vol. 38, pp. 806–835, 2018.
9. M. A. Ferrag, M. Babaghayou, and M. A. Yazici, "Cyber security for fog-based smart grid SCADA systems: solutions and challenges," *Journal of Information Security and Applications*, vol. 52, p. 102500, 2020.
10. S. K. Singh, Y. S. Jeong, and J. H. Park, "A deep learning-based IoT-oriented infrastructure for secure smart city," *Sustainable Cities and Society*, vol. 60, Article ID 102252, 2020.
11. F. A. Khan, M. Asif, A. Ahmad, M. Alharbi, and H. Aljuaid, "Blockchain technology, improvement suggestions, security challenges on smart

grid and its application in healthcare for sustainable development," *Sustainable Cities and Society*, vol. 55, Article ID 102018, 2020.

12. M. Samir, M. Azab, M. R. Rizk, and N. P. Y. G. R. I. D. Sadek, "A software development and assessment framework for grid-aware software defined networking," *International Journal of Network Management*, vol. 28, p. e2033, 2018.
13. N. Z. Bawany, J. A. Shamsi, and K. Salah, "DDoS attack detection and mitigation using SDN: methods, practices, and solutions," *Arabian Journal for Science and Engineering*, vol. 42, no. 2, pp. 425–441, 2017.
14. O. Jung, P. Smith, J. Magin, and L. Reuter, "Anomaly detection in smart grids based on software defined networks," in *Proceedings of the 8th International Conference on Smart Cities and Green ICT Systems*, vol. 1, pp. 157–164, SMARTGREENS, Heraklion, Crete, May 2019.
15. W. Chen, S. Xiao, L. Liu, X. Jiang, and Z. Tang, "A DDoS attacks traceback scheme for SDN-based smart city," *Computers & Electrical Engineering*, vol. 81, p. 106503, 2020.
16. X. Min, Z. J. Jin, R. H. Yang, M. R. Heng, and C. M. Xin, "Traffic scheduling strategy of power communication network based on SDN," *Computer Engineering and Applications Journal*, vol. 9, pp. 49–60, 2020.
17. K. Mahmood, S. A. Chaudhry, H. Naqvi, S. Kumari, X. Li, and A. K. Sangaiah, "An elliptic curve cryptography based lightweight authentication scheme for smart grid communication," *Future Generation Computer Systems*, vol. 81, pp. 557–565, 2018.
18. S. Ali and Y. Li, "Learning multilevel auto-encoders for DDoS attack detection in smart grid network," *IEEE Access*, vol. 7, pp. 108647–108659, 2019.
19. Z. Ahmed, N. Afaqui, and O. Humayan, "Detection and prevention of DDoS attacks on software defined networks controllers for smart grid," *International Journal of Computer Applications*, vol. 975, p. 8887, 2019.
20. T. Pandikumar, F. Atkilt, and C. A. Hassen, "Early detection of DDoS attacks in a multi-controller based SDN," *International Journal of Engineering Science*, p. 13422, 2017.
21. M. H. Bhuyan, D. K. Bhattacharyya, and J. K. Kalita, "A multi-step outlier-based anomaly detection approach to network-wide traffic," *Information Sciences*, vol. 348, pp. 243–271, 2016.

22. M. H. Bhuyan, D. K. Bhattacharyya, and J. K. Kalita, "E-LDAT: a lightweight system for DDoS flooding attack detection and IP traceback using extended entropy metric," *Security and Communication Networks*, vol. 9, no. 16, pp. 3251–3270, 2016.
23. A. A. Amaral, L. d. S. Mendes, B. B. Zarpelão, and M. L. P. Junior, "Deep IP flow inspection to detect beyond network anomalies," *Computer Communications*, vol. 98, pp. 80–96, 2017.
24. I. Aouini and L. B. Azzouz, "Smart grids cyber security issues and challenges," *International Journal of Electrical, Computer, Energetic, Electronic and Communication Engineering*, vol. 9, pp. 1255–1261, 2015.
25. R. A. Niazi and Y. Faheem, "A bayesian game-theoretic intrusion detection system for hypervisor-based software defined networks in smart grids," *IEEE Access*, vol. 7, pp. 88656–88672, 2019.
26. M. Ghanbari and W. Kinsner, "Detecting DDoS attacks using polyscale Analysis and deep learning," *International Journal of Cognitive Informatics and Natural Intelligence*, vol. 14, no. 1, pp. 17–34, 2020.
27. P. Bull, R. Austin, E. Popov, M. Sharma, and R. Watson, "Flow based security for IoT devices using an SDN gateway," in *Proceedings of the 2016 IEEE 4th International Conference on Future Internet of Things and Cloud (FiCloud)*, pp. 157–163, IEEE, Vienna, Austria, August 2016.
28. R. Wang, Z. Jia, and L. Ju, "An entropy-based distributed DDoS detection mechanism in software-defined networking," in *Proceedings of the 2015 IEEE Trustcom/BigDataSE/ISPA*, vol. 1, pp. 310–317, IEEE, Helsinki, Finland, 2015.
29. K. Balachandran, R. L. Olsen, and J. M. Pedersen, "Bandwidth analysis of smart meter network infrastructure," in *Proceedings of the 16th International Conference on Advanced Communication Technology*, pp. 928–933, IEEE, PyeongChang, South Korea, February 2014.
30. P. Cheng, L. Wang, B. Zhen, and S. Wang, "Feasibility study of applying LTE to smart grid," in *Proceedings of the 2011 IEEE First International Workshop on Smart Grid Modeling and Simulation (SGMS)*, pp. 108–113, IEEE, Brussels, Belgium, October 2011.
31. E. Mota, A. Passito, and R. Braga, "Lightwight DDoS flooding attack detection using NOX/Openflow," in *Proceedings of the IEEE 35th*

Conference on Local Computer Networks, IEEE, Denver, CO, USA, April 2010.

32. A. Mahajan, A. Gupta, and L. Sen Sharma, "Performance evaluation of different pattern matching algorithms of snort," *International Journal of Advanced Networking Applications*, vol. 10, no. 02, pp. 3776–3781, 2018.
33. D. Huang, L. Xu, C. Chung, and T. Xing, "SnortFlow: a openflow-based intrusion prevention system in cloud environment," in *Proceedings of the Second GENI Research and Educational Experiment Workshop*, Salt Lake, UT, USA, March 2013.

Chapter 13

Distribution System Reliability Analysis for Smart Grid Applications

Tawfiq M. Aljohani, Mohammed J. Beshir

Ming Hsieh Department of Electrical Engineering, University of Southern California, Los Angeles, USA

ABSTRACT

Reliability of power systems is a key aspect in modern power system planning, design, and operation. The ascendance of the smart grid concept has provided high hopes of developing an intelligent network that is capable of being a self-healing grid, offering the ability to overcome the interruption problems that face the utility and cost it tens of millions in repair and loss. In this work, we will examine the effect of the smart grid applications in improving the reliability of the power distribution networks. The test system used in this paper is the IEEE 34 node test feeder, released in 2003 by the

Citation: Aljohani, T. and Beshir, M., (2017), "Distribution System Reliability Analysis for Smart Grid Applications". *Smart Grid and Renewable Energy*, **8**, 240-251. doi: 10.4236/sgre.2017.87016.

Distribution System Analysis Subcommittee of the IEEE Power Engineering Society. The objective is to analyze the feeder for the optimal placement of the automatic switching devices and quantify their proper installation based on the performance of the distribution system. The measures will be the changes in the reliability system indices including SAIDI, SAIFI, and EUE. In addition, the goal is to design and simulate the effect of the installation of the Distributed Generators (DGs) on the utility's distribution system and measure the potential improvement of its reliability.

Keywords: Power System Reliability, Smart Grid Applications, Distribution System Reliability, Automatic Reclosers, Distributed Generation, Self-Healing Power Grids

INTRODUCTION

Power system reliability is a key aspect in power distribution system planning, design, and operation. Electric power utilities are required to provide uninterrupted electrical services to their customers at the lowest possible cost while maintaining an acceptable level of service quality. The importance of reliability arises as it can express the cost of service outages. A distribution system's quality of service can be judged by its reliability indices, which can be increased by automation of its feeder and associated parts, which eventually will lead to a desired reduction in the power interruptions. Reliable power distribution networks are those managing a high level of reliability.

The traditional power distribution grid is radial in nature, the power flows in one direction from the distribution substation to the load point. The radial system has low reliability, and those customers who are located at the end of the circuit, tend to be more prone to power outages than any other customers. Since there are no backup or alternative sources to back up the traditional distribution systems, there is a high chance that a major fault on the feeder would affect a substantial number of customers in the radial configuration [1] [2] [3] [4] .

The concept of reliability can be simply expressed as two states or conditions: up and down. The first, up, would mean that the system is available (functioning) while the latter, down, means the system is unavailable (failing). Electrically, when a device is interrupted by a fault, the state or condition of this specific equipment would be adjusted from the up state to the down state. The down state lasts until the equipment is fully repaired.

Once the equipment goes alive with the grid, the state condition returns to the up state again. The reliability problem could become even worse if the failure of one component influences the failure of others, increasing the possibility of cascading outages [5] [6] . The development of smart grid has raised questions recently about the opportunities this new technology offers to enhance the reliability of electric service. The term "smart grid" refers to the modernization of the electric power grid via the application of information and communication systems to incorporate alternative sources of energy into the power grid [7] .

LITERATURE REVIEW

There have been considerable volumes of research to quantify the benefits that can arise from the integration of the smart grid applications into the enhancement of the reliability of the power distribution grids. The U.S. Department of Energy states [2] :

Think of the smart grid as the internet brought to our electric system. Devices such as wind turbines, plug-in hybrid electric vehicles, and solar arrays are not part of the smart grid. Rather, the smart grid encompasses the technology that enables us to integrate, interface, with intelligently control these innovations and others.

Many efforts have been made to quantify the losses of the utilities due to the faults and outages in the distribution grid. Furthermore, there have been suggestions that many inconsistencies are found in the reported collected data that measure the utilities interruption events and its reliability indices' performance. Reference [8] suggested organization of the data used to make a comparison of distribution system reliability performances.

CASE STUDIES

Case Study 1: The Reliability Impact of the Optimal Placement of Automatic Reclosers.

In this paper, we analyze the potential effect of optimal installation of automatic reclosers on the distribution feeders, using the analytical technique. The application on this analysis is done using DISREL, an intelligent software that uses the concept of brute force in analyzing test systems, quantifies the effects of modifications, and provides recommendations for optimal installation of switches. Failure rates and MTTR values that have been used in modelling the original IEEE 34-node feeder in this work are based on the average values obtained from reference [1] . The automatic recloser (AR) is a protection device that has the ability to detect a fault and open for a pre-programmed time before closing again automatically and without the interference of the human factor. If optimally installed, ARs can help in achieving the concept of a self-healing power grid, which is one of the smart grid's main concepts according to the U.S. department of energy. It should be noted that several studies have been conducted to examine the effect of both optimal configuration and automatic switches on the distribution grid in the smart era and measure virtues of several scenarios [11] [12] [13] . In this work, we first model the original IEEE 34-node feeder using DISREL. Table 1 shows the results of the software recommendations for the optimal installations of automatic reclosers. The base case in this study is the original IEEE 34-node test system. Any contingency event will result in the loss of all the customers in service. SAIDI, SAIFI and other indices are simulated and provided in Table 1 for the original feeder following any outage event and before install any sectionalizing devices. Reference [14] discusses in detail an algorithm, developed by the authors of this paper, which is based on the analytical technique. The algorithm illustrates the main concepts of the isolation and restoration process in the distribution grid when installing automatic reclosers or switches. This is done by coordinating the MTTS of the automatic reclosers to be the time reclosers would take to locate the nearest sectionalizing devices. The provided results by the software suggest that the optimal location to install the automatic recloser is between nodes 832-858. Furthermore, it is shown that SAIDI has been reduced from 927.25 to 840.83 minutes per year, indicating an average of 9.32% reduction, or in other words improvement, to the duration of interruptions that the average customer will experience over the course of a year.

Table 1: The results for the optimal placements of the reclosers and switches

Case Description	SAIFI	SAIDI	CAIDI	ASAI	EUE (kW/yr)	Outage Cost ($)
Base Case	5.35187	927.25385	173.25789	0.998235822	25,245	251,769.00
Add AR [832-858]	4.91459	840.828	171.088	0.998400271	22,888	228,293.00
Add AR [858-834]	4.87825	848.89636	174.01645	0.998384893	23,109	230,501.00
Add AR [834-860]	4.84965	865.86755	178.54211	0.998352587	23,572	235,100.00

Add AR [860-836]	4.90973	873.88715	177.99095	0.998337328	23,791	237,276.00
Add AR [834-842]	5.08451	886.42841	174.33902	0.998313487	24,134	240,694.00

This will allow the utility to improve its service to the customer, in case of a fault occurring downstream of the automatic recloser installed on this specific location. The recloser would automatically isolate the affected area of the feeder; thus, restoring the service to the upstream customers, which improves SAIDI in this case. However, if the fault is located upstream of the automatic recloser, then there is no way to restore the power to the downstream customer unless there is another source of power that can feed these customers while isolating this fault when possible. This will give virtues to the utilities in reducing the repair hours by detecting the location of the fault quicker than the case of the original feeder, where there is no sectionalizing switch at all.

Yet, this does not necessarily mean that this option will yield the best outcomes in regard to SAIFI. SAIFI measures the sustained interruptions an average customer will experience. For the best option provided by the software, which is installing the automatic recloser in between 832-858, SAIFI also witnessed a reduction from 5.35 to 4.92 interruptions per customer over the course of a year, which is equal to 8.16% improvements, while in some other options results in greater reduction (9.34% when install the auto-recloser in between 858-834 instead, as shown in Figure 1). However, it is expected that installing another automatic sectionalizing device would yield better improvements and savings to the system.

To investigate such option, wemodify the test system to include the first recloser as DISREL suggests in between 832-858 and then model it to examine the effect of adding other automatic sectionalizing devices on the grid. Table 2 shows the results of the modified IEEE test system shown in Figure 2. Based on the results, the installation of two automatic reclosers projects more improvement to the system indices, with higher revenues. The simulated results suggest that installing the recloser at 834-860 will raise a total savings of $34,112. This can be justified by the fact that the utility would be able to isolate faults that are probably located near the densest area in the feeder, where around 40% of the total load is located in the distance between 858 to the end of the lateral at 848. In addition, this would reduce the repair hours for the utilities to fix the issues as fast as possible and restore the power more quickly in the areas where the customers are still experiencing service interruption.

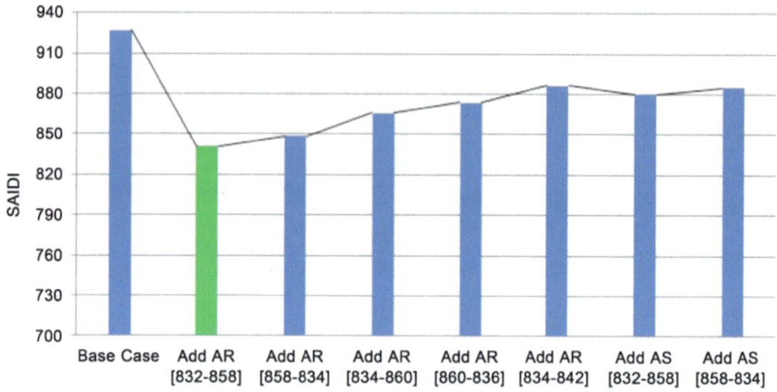

Figure 1: Around 10% improvements in SAIDI considering one AR.

Table 2: The results of the optimal placements for the modified test feeder shown in Figure 2

Case Description	SAIFI	SAIDI	CAIDI	ASAI	EUE (kW)	Outage Cost ($)
Base Case	5.35187	927.25385	173.25789	0.998235822	25,245	251,769.00
Base Case + one AR	4.94058	833.42004	168.68883	0.998414338	22,686	226,282.00
Add AR [834-860]	4.66379	801.66541	171.89131	0.998474777	21,822	217,657.00
Add AR [860-836]	4.69352	804.74646	171.45895	0.998468876	21,906	218,496.00
Add AR [834-842]	4.78457	817.78149	170.92068	0.99844408	22,262	222,051.00
Add AR [832-888]	4.79635	816.11261	170.15291	0.998447299	22,213	221,571.00
Add AR [842-844]	4.85976	825.29413	169.82184	0.998429835	22,465	224,082.00
Add AS [834-860]	5.32914	812.24915	152.41666	0.99845463	22,110	220,465.00
Add AS [860-836]	5.39499	816.2149	151.29141	0.998447061	22,218	221,545.00

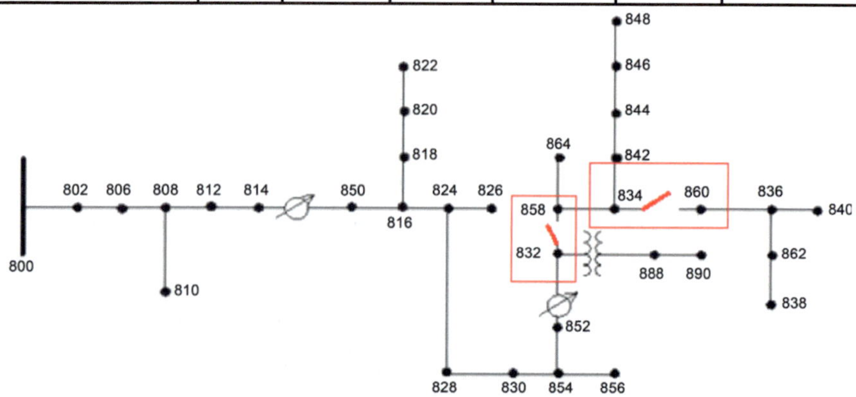

Figure 2: The optimal locations of ARs based on the results.

An outage would make the automatic recloser opens to isolate and clear the fault. If the customers who are upstream of the automatic recloser (located between 832-858) are still out of service after isolating the fault, then mostly the fault will be located in the area upstream this recloser to the distribution substation.

This concept can be applied to all the automatic reclosers (or switches) until the location of the fault is detected, which would reduce the repair hours, thus improving SAIDI index. The installation of these two automatic devices is considered the optimal solution in this case, since it both reduces SAIFI and SAIFI by 12.9% and 13.54% respectively. Figure 3 shows the projected savings for considering two sectionalizing devices, while Figure 4 shows a multi-scale graph comparing the savings to the outage costs for this case. It can be shown that when we install a second AR at 834-860, the system achieves the highest savings by closing the gap with the outage costs.

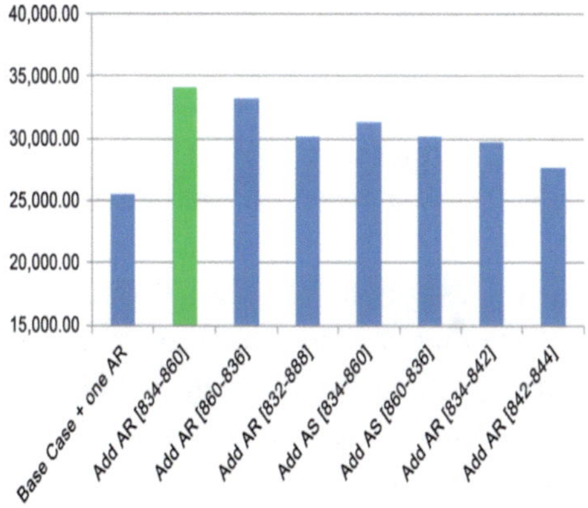

Figure 3: Projected savings for case study 1 considering two ARs.

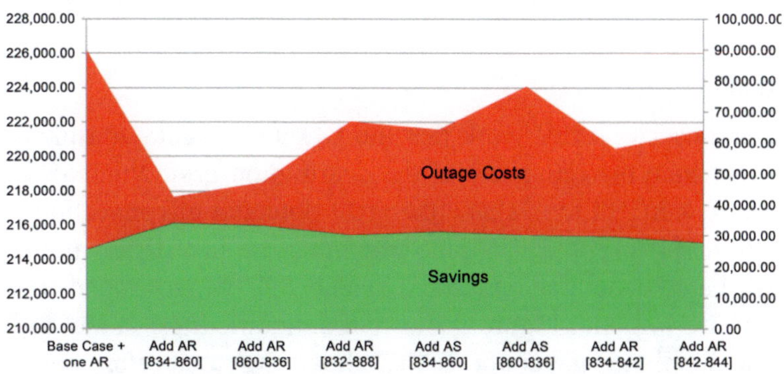

Figure 4: The savings to the outage costs for case study 1.

It is also noted that there are no significant changes applied to CAIDI. This is due to the fact that CAIDI mathematically equals to SAIDI divided by SAIFI, which means that when we have improvements on both SAIDI and SAIFI, these improvements are not going to be reflected in CAIDI unless there are outstanding improvement in SAIDI only. Regard ASAI, it is noted that this index maintains higher reliability values (over 0.99 in all the cases) since the measure here is how much energy an average customer receives to the amount this customer demanded from the utilities. Thus, we can conclude that both CAIDI and ASAI are not good reliability measures as SAIFI and SAIDI.

One of DISREL's virtues is that it can quantify the outage costs for each option based on, EUE, which represents the expected unserved kW per year due to interruptions. According to [15] , the U.S. utilities' losses due to energy not being served to the customer is estimated to be in between \$80 to \$188 billion a year, which does not include the damages that might happen to the equipment. The outage cost in this study is assumed to be \$10 for each kWh lost, which is reasonable when compared with the real life outage costs that were estimated in reference [16] . The installation of an effective, high-quality automatic recloser costs \$20,000-\$30,000 in total [17] [18] . Thus, it is more likely that this kind of investment will ensure a payback to the electric utility in less than 1.8 years from the installation taking place. However, it is worth mentioning that the return might be sooner than suggested, since the savings in our study are marked for only one distribution feeder, which is only considered as generalized saving values, whereas the local utilities, in most cases, have hundreds of distribution feeders in their electrical infrastructure. Thus, this study should be considered effective in evaluating the impact of this smart-grid concept on the overall distribution system that consists of hundreds of similar radial distribution feeders.

Case Study 2: The Reliability Impact of the Distributed Generators on the Radial Feeder

We emphasize in this work that the DG units would considered great tools to enhance the reliability of the distribution grid. Numerous studies have been conducted to evaluate the great potential of connecting DGs on the reliability [19] [20] [21] . At the beginning, we investigate modeling a 1 MW distributed generator, connected to node 890, where around 30% of the customers were found. Figure 5 shows the modified test model to include the 1 MW distributed generator. Table 3 shows the results after the modeling of the test system. The base case would be here the modified system shown in Figure 5 above. The results of the base case illustrate the need for the automatic reclosers/CBs when we install a DG to the distribution system; otherwise, there would be no benefit since the fault will certainly block the connection of the DG units during outages.

The DG unit could be sized based on the need, whereas in this feeder, a 1 MW DG unit provides approximately the same benefits that could be added by the installation of a 6 MW for example, since the amount of load connected to the feeder is around 1.7 MW, which would make no sense to connect a DG unit that provides power more than the customers demand.

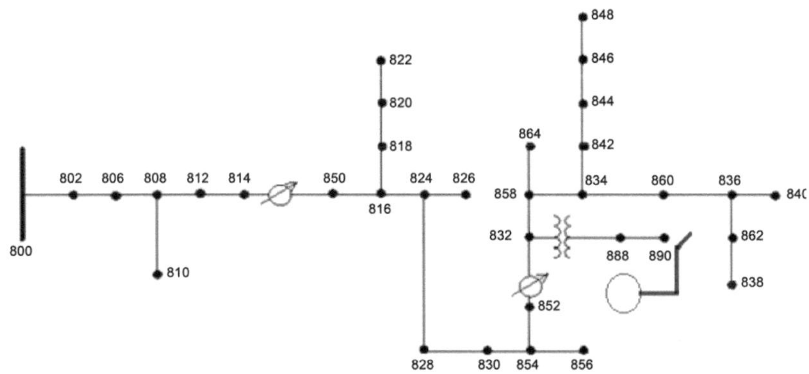

Figure 5: The test feeder with one DG unit connected.

Table 3: Results of the installation of one 1 MW DG unit at node 890

Case Description	SAIFI	SAIDI	CAIDI	ASAI	EUE (kW)	Outage Costs ($)
Original Case	5.35187	927.25385	173.25789	0.998235822	25,245	251,769.00
Base Case + one DG	5.35187	764.83942	142.91068	0.998544812	20,840	207,732.00
Add AR [852-832]	4.24777	516.23724	121.53123	0.999017835	14,082	140,250.00
Add AR [854-852]	4.30945	520.5567	120.79435	0.999009609	14,198	141,414.00
Add AR [830-854]	4.49053	546.98199	121.80791	0.998959303	14,913	148,562.00
Add AR [828-830]	4.6694	567.53998	121.54451	0.998920202	15,474	154,167.00
Add AR [888-890]	4.94313	593.51715	120.06919	0.99887079	16,203	161,381.00
Add AS [830-854]	5.35186	644.85278	120.49136	0.998773098	17,576	174,718.00

Figure 6 shows the customer minutes per year for the feeder considering 1 MW DG unit at node 890. For the original IEEE 34 node feeder, the estimated interruption minutes per year for the whole feeder are simulated to be 753,000 minutes year. When we considered connecting a DG unit to the system, as shown in Figure 5, the total interruption minutes per year were greatly reduced when considering several scenarios for adding automatic reclosers/or switches. For instance, the installation of one automatic recloser between nodes 852-832 to the modified system will reduce the interruption minutes to almost 419,000 minutes per year, which is approximately 44.32% reduction than the original test system (with no DG or AR connected at all). In the case of any contingency event, the DG unit will provide the system the ability to operate as a small microgrid, providing service to the unaffected parts of the feeder and improving the system indices, specifically SAIDI, which will provide huge savings to the involved parties (both the utility and the IPPs) as shown in Table 4. To better understand the potential effects of adding more DG units in the system, we model the test feeder after installing three 1-MW DG units at nodes 890, 844, and 820 which constitute the most populated nodes in the feeder.

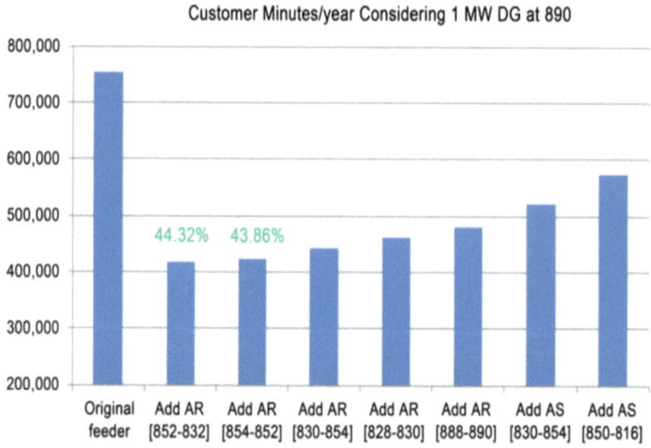

Figure 6: Customer interruption improvements per year for the feeder.

Table 4: The savings in case of connecting 1MW DG unit to the grid

Case Description	Savings ($)
Base Case with DG	44,037.00
Add AR [852-832]	111,519.00
Add AR [854-852]	110,355.00
Add AR [830-854]	103,207.00
Add AR [828-830]	97,602.00
Add AR [888-890]	90,388.00
Add AS [830-854]	77,051.00
Add AS [850-816]	59,996.00

Based on the results, when we install an automatic recloser along with the three DG units in the specified location, the system's SAIDI would decrease from 927.25 minutes (obtained in the IEEE original feeder) to 425.52 minutes, which correspond to over 54% in SAIDI improvements. For SAIFI index, the modified system has reduced the frequency of interruptions from 5.352 to 3.532 per customer, marking a 34% reduction in SAIFI. However, we noticed again that there is no significant change applied to CAIDI even in the case of the DG, which shows that we cannot consider CAIDI a real measuring for system reliability improvements. Figure 7 shows the savings obtained by installing the three DG units on the distribution feeder. In the case when adding the automatic recloser between 834-842 or 842-844, it is projected that the utility will experience the greatest savings among other options, which can be proven by the fact that the system will be able to recon figure in order to maintain service to a substantial number of customers during different outage scenarios. Thus, it is recommended, in case of this feeder only, that one automatic recloser would be enough to achieve the targeted reliability goal, and to isolate a proper portion of the network as a small islanded microgrid during outages, to maintain service to a substantial number of customers.

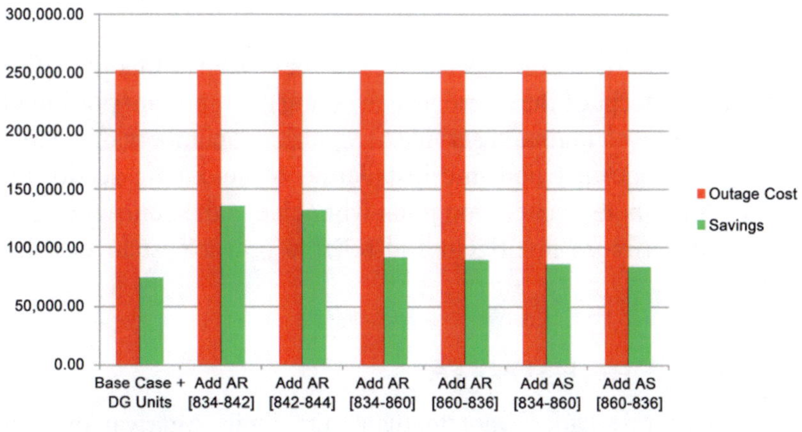

Figure 7: The savings vs outage costs for case 3 considering 3 DG units.

The reduction in the SAIDI and SAIFI indices when adding multiple reclosers could be attributable to the fact that the DG units have another virtue in improving reliability by taking the form of peak shaving, where the DG units can generate more on-site power than the demand on the feeder, allowing more power to support the grid during normal operation.

CONCLUSION

This paper analyzes and examines the effects of smart grid applications on the reliability of power distribution systems. Unlike the generation and transmission sectors, the distribution of power systems did not receive much attention until recent years, where many have emphasized the great potential that can be achieved in this field of the electrical systems. The concept of the smart grid is very broad and difficult to summarize. However, the effects of two main applications of the smart grid have been examined in this work, which are the optimal use of the automatic sectionalizing devices, such as the automatic recloser and the automatic switches, and the accommodation of distributed generation. To reach the goal of this work, several case studies were applied. The results show that the optimal installation of the automatic reclosers will enable the concept of a self-healing power distribution grid that can recover quickly and automatically from major disturbance events, and restore power to as many customers as possible, resulting in significant improvement in the system indices, for instance in SAIDI and SAIFI. This has resulted also in great savings to the power providers, since the utility will be able, using the automatic reclosers and/or switches, to reduce the interruption duration which reduces the expected unserved kWh to the consumers. In addition, we simulate the effect of the installation of the distributed generators on the system. Results have shown that the DG units can apply the concept of the microgrid, isolating an important portion of the distribution feeder to maintain service to significant numbers of customers. In all, we conclude that both the installation of automatic sectionalizing devices and distributed generation units will achieve the concept of a smart grid, providing a more intelligent and reliable power system distribution network.

ACKNOWLEDGEMENTS

The authors of this paper want to thank Dr. Sudir Agarwal of General Reliability for all valuable comments and help he provided in this work. We recognize the great virtues of DISREL, the intelligent-based software that was used while conducting this work. For more details on the work presented in this paper, you can review the work presented by Aljohani (2014) in reference [14] .

REFERENCES

1. Brown, R.E. (2009) Electric Power Distribution Reliability. 2nd Edition, Marcel Dekker, New York.
2. United States Department of Energy (2014) Office of Electricity Delivery and Energy Reliability. http://energy.gov/oe/technology-development/smart-grid
3. Billinton, R. and Allan, R.N. (1996) Reliability Evaluation of Power Systems. 2nd Edition, Plenum Press, New York. https://doi.org/10.1007/978-1-4899-1860-4
4. Short, T.A. (2004) Electric Power Distribution Handbook. CRC Press, Boca Raton.
5. Munasinghe, M., Scott, W.G. and Gellerson, M. (1979) The Economics of Power System Reliability and Planning: Theory and Case Study. John Hopkins University Press, Baltimore.
6. Li, W. (2011) Probabilistic Transmission System Planning. John Wily & Sons, Inc., Hoboken. https://doi.org/10.1002/9780470932117
7. Arritt, R.F. and Dungan, R.C. (2011) Distribution System Analysis and Future Smart Grid. IEEE Transactions on Industry Applications, 47, 2343-2350. https://doi.org/10.1109/repcon.2011.5756725
8. Werner, V., Hall, D., Robinson, R. and Warren, C. (2006) Collecting and Categorizing Information Related to Electric Power Distribution Interruption Events: Data Consistency and Categorization for Benchmarking Surveys. IEEE Transactions on Power Delivery, 21, 480-483. https://doi.org/10.1109/TPWRD.2005.852303
9. Lasseter, R.H. (2011) Smart Distribution: Coupled Microgrids. Proceedings of the IEEE, 99, 1074-1082. https://doi.org/10.1109/JPROC.2011.2114630
10. Piagi, P. and Lasseter, R.H. (2006). Autonomous Control of Microgrids. Proceedings of the IEEE Power Engineering Society Meeting, Montreal, Quebec, 8 June 2006, 8. https://doi.org/10.1109/pes.2006.1708993
11. Zheng, H., Cheng, Y., Gou, B., Frank, D., Bern, A. and Muston, W.E. (2012) Impact of Automatic Switches on Power Distribution Reliability. Electric Power Systems Research, 83, 51-57. https://doi.org/10.1016/j.epsr.2011.08.018
12. Billinton, R. and Jonnavithula, S. (1996) Optimal Switching Device Placement in Radial Distribution Systems. IEEE Transactions on Power Delivery, 11, 1646-1651. https://doi.org/10.1109/61.517529

13. Lim, I., Sidhu, T.S., Choi, M.S., Lee, S.J. and Ha, B.N. (2013) An Optimal Composition and Placement of Automatic Switches in DAS. IEEE Transactions on Power Delivery, 28, 1474-1482. https://doi.org/10.1109/tpwrd.2013.2239314
14. Aljohani, T.M. (2014) Distribution System Reliability Analysis for Smart Grid Applications. ProQuest Dissertations Publishing, Southern California.
15. Williams, J (2013) Zero Carbon Homes—A Road Map. Earthscan and Routledge Press, London, 42.
16. Lawrence Berkeley National Laboratory. Environmental Energy Technologies Division (2003) A Framework and Review of Customer Outage Costs: Integration and Analysis of Electric Utility Outage Cost Surveys. Survey, Lawrence Berkeley National Laboratory, Berkeley.
17. Chowdhury, A. and Koval, D. (2009) Power Distribution System Reliability: Practical Methods and Applications. John Wiley and Sons, Hoboken. https://doi.org/10.1002/9780470459355
18. Florida Electric Cooperatives Association (2012) Improving Network Reliability with Reclosers. http://www.feca.com/RecloserTechnology.pdf
19. Chowdhury, A., Agarwal, S.K. and Koval, D. (2003) Reliability Modeling of Distributed Generators in Conventional Distribution Systems Planning and Analysis. IEEE Transactions on Industry Applications, 39, 1493-1498.
20. Al-muhaini, M. and Heydt, G. (2013) Evaluating Future Power Distribution System Reliability Modeling Including Distributed Generation. IEEE Transactions on Power Delivery, 28, 2264-2272.
21. Mozina, C.J. (2013) Impact of Smart Grids and Green Power Generation on Distribution Systems. IEEE Transactions on Industry Applications, 49, 1079-1090.

Chapter 14

An Approach to Assess the Resiliency of Electric Power Grids

Navin Shenoy, R. Ramakumar

School of Electrical and Computer Engineering, Oklahoma State University, Stillwater, USA

ABSTRACT

Modern electric power grids face a variety of new challenges and there is an urgent need to improve grid resilience more than ever before. The best approach would be to focus primarily on the grid intelligence rather than implementing redundant preventive measures. This paper presents the foundation for an intelligent operational strategy so as to enable the grid to assess its current dynamic state instantaneously. Traditional forms of real-time power system security assessment consist mainly of methods based on power flow analyses and hence, are static in nature. For dynamic security assessment, it is necessary to carry out time-domain simulations (TDS) that

Citation: Shenoy, N. and Ramakumar, R. (2015), "An Approach to Assess the Resiliency of Electric Power Grids". *Journal of Power and Energy Engineering*, **3**, 1-13. doi: 10.4236/jpee.2015.311001.

are computationally too involved to be performed in real-time. The paper employs machine learning (ML) techniques for real-time assessment of grid resiliency. ML techniques have the capability to organize large amounts of data gathered from such time-domain simulations and thereby extract useful information in order to better assess the system security instantaneously. Further, this paper develops an approach to show that a few operating points of the system called as landmark points contain enough information to capture the nonlinear dynamics present in the system. The proposed approach shows improvement in comparison to the case without landmark points.

Keywords: Grid Resilience, Machine Learning, Smart Grids, Time-domain Analysis, Dynamic Security Assessment

INTRODUCTION

In the wake of new vulnerabilities such as those arising from severe weather events and cyber-attacks, current electric grids can no longer be allowed to operate as they did in the past. It is becoming increasingly difficult to analyze different combinations of contingencies under changing scenarios. Grid resilience and improved situational awareness will form the basis of future electric grids in order to tackle these new challenges. The most cost effective way to meet such stringent requirements is through intelligent operation of the grid by employing data driven models that are both informational and analytical in nature. The key attribute involved here is the ability to assess the current state of the power system in real-time in terms of its security. Power system security is defined as its ability to survive imminent disturbances (contingencies) without interruption of customer service. Historically, it has been recognized that for a power system to be secure, it must be stable against all types of disturbances [1] [2] . Hence, stability analysis is an important component that can facilitate the assessment of power system security and thus, its resiliency.

Security in terms of operational requirements implies that following a sudden disturbance, power system would be secure if and only if: 1) it could survive the transient swings and reach an acceptable steady state condition, and 2) there are no limit violations in the new steady state condition. The first requirement can be met by carrying out time-domain simulations in order to investigate the instability phenomena such as loss of synchronism or voltage collapse in the post-contingency transient phase. The second requirement is met by using power- flow based methods in order to assess the new steady state condition for voltage and current limit violations.

Time-domain simulations (TDS) are computationally involved and too complex to be performed in real-time. Therefore, for many years in the past, the electric utility industry's framework for real-time security assessment mainly consisted of solution methods that would meet only the second requirement stated earlier. Such a type of real-time security analysis is prevalent even today and is commonly referred to as "Static Security Assessment (SSA)". On the other hand, a "Dynamic Security Assessment (DSA)" procedure would strive to meet both the requirements (as stated earlier) in real-time in order to assess power system security.

Different forms of DSA practices have existed in North America since the late 1980s [3] . Modern DSA implementations are able to complete a computation cycle within 5 - 20 minutes after a real-time snapshot (base case) of the system is available [4] . Real-time snapshots are provided by existing SCADA-based state estimators every few seconds or minutes depending on the size of the system [5] -[9] . Thus, these modern DSA implementations can be termed as "near real-time" and not "real-time". However, the latest PMU-based data collection technology can provide much better snapshots wherein the measurements are transmitted to the main control center at rates as fast as 60 samples/second [10] . Thus, DSA implementations of the future will be required to handle large amounts of data and complete the computation cycles much faster in order to assess the system security in true "real-time". Mathematically, such an instantaneous assessment would be possible only if grid resilience against any contingency can be expressed as a function of the state estimator output. In other words, input to the data-driven models must consist of only steady-state (static) quantities namely bus voltages and bus angles derived from power-flow based methods.

Machine learning (ML) techniques have the ability to assimilate and reason with knowledge the way human brain does. Such techniques are primarily driven by data that could be in the form of various power system parameters such as [11] -[13] : voltage, current, power, frequency, power angles etc. ML techniques can capture the nonlinear dynamics of power systems by extracting useful information from such large amounts of data. DSA tools employing such ML techniques will have the ability to determine stability limits in real-time. Such sophisticated tools will be able to analyze the current and future dynamics of power systems without carrying out extensive time-domain simulations. Additionally, these tools would also benefit the system operators by providing them with real-time information on trends in system security, thereby facilitating faster decision-making during crucial times. Also, as the entry of renewable energy systems further increases grid complexity, it is possible to extend the proposed work in order to accommodate online training, thereby resulting in a smart tool that can very effectively assess the system security in real-time.

This paper presents a framework that would enable implementation of such powerful machine learning techniques for real-time assessment of grid resilience. A standard IEEE 14-bus system is used in this paper for simulation purposes [14]. Firstly, a set of multiple steady-state operating points is generated by performing a SSA on the base case. Secondly, a TDS is performed on each operating point to assess the grid resilience against a specific contingency, thus generating a dataset for this work. The paper highlights the importance of selecting a few cases as landmark points in the operational space under consideration. Further, it presents a procedure to select the best landmark points in order to improve the prediction accuracy on the original dataset, thereby enhancing the ability to assess grid resilience instantaneously.

STATIC SECURITY ASSESSMENT (SSA)

Static security assessment (SSA) provides a mathematical framework to compute stability limits for individual buses and lines based on power flow based methods. This involves checking for steady state voltage violations at every bus in the system. Power-Voltage (PV) curves are plotted for each bus by systematically loading the base case of the power system under consideration. This is achieved by means of an algorithm called as "Continuation Power Flow (CPF)" [15].

CPF is a "case worsening" procedure where the power system is loaded in steps as follows:

$$P_G = \lambda P_{G0}$$
$$P_L = \lambda P_{L0}$$
$$Q_L = \lambda Q_{L0} \qquad (1)$$

where P_{G0}, P_{L0}, Q_{L0} are the base case generator and load powers (in per-unit) and λ is the loading parameter (in per-unit). CPF facilitates plotting of voltage curves as a function of loading parameter λ, for each bus.

As stated earlier, such a framework can be used to generate a dataset consisting of multiple steady-state operating points. For an n-bus system, every such operating point can be represented by a feature vector x of dimension 2n consisting of n bus voltages and n bus angles as features. A set S containing such objects is given by,

$$\left\{x \in \Re^{2n} : x = \left[V_1, \cdots, V_n, \delta_1, \cdots, \delta_n\right]^{\mathrm{T}}\right\} \tag{2}$$

where V_i's are bus voltages (in per unit) and δ_i's are bus angles (in degrees).

SSA is performed on the standard IEEE 14-bus system for the following voltage stability criteria at each bus: $V_{max} = 1.2$ pu and $V_{min} = 0.8$ pu. Generators are represented by machine models along with automatic voltage regulators and turbine governors. A CPF routine is performed for each line outage of this power system. Thus, a maximum loading parameter λ_{maxi} is calculated for each line outage i, taking voltage stability criteria into account. The set represented by Equation (2) is generated only for values of λ given by,

$$1 \leq \lambda \leq \lambda_{\max i}, \forall i \tag{3}$$

It has to be noted that these λ_{maxi} values account for only steady-state voltage violations and hence, do not provide any information about dynamic system security. In order to account for dynamic stability, time-domain simulations are performed for each operating point, as described in the next section. All routines are carried out using the PSAT toolbox for Matlab [16] . Figure 1 shows V-λ curves for a particular line outage.

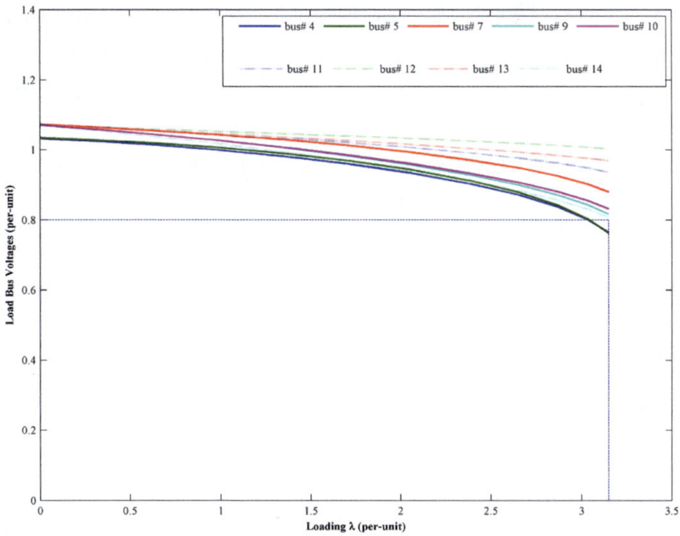

Figure 1: V-λ curves of PQ buses for outage of line#16 (bus 2 to bus 4), bus 5 voltage reaches V_{min} at λ = 3.1487.

DYNAMIC SECURITY ASSESSMENT (DSA)

The goal of a DSA is to classify different cases based on their dynamic security severity. Dynamic security depends on the time responses of various system variables for the contingency under consideration. As mentioned earlier, it is not possible to perform computationally intensive time-domain simulations in real-time. Nonetheless, machine learning techniques have the ability to extract information from offline time-domain simulations. Subsequently, such useful information can be used to predict dynamic system security for new configurations in order to avoid lengthy time-domain simulations. To implement such an application, detailed time-domain simulations are required to be conducted for different operating points. Thus, a database, on which ML techniques can operate, needs to be generated in offline mode.

The database is generated in the form of a feature matrix X and an output vector y. Each row of the feature matrix X represents a steady state operating point in the form of object $x \in \mathfrak{R}^{2n}$ from set S as defined in equation (2). Matrix X contains total number of 'm' such objects and hence, its size is (m × 2n). A time-domain simulation for a specific contingency is performed on each of these m objects. These simulations are tagged as "stable" or "unstable" depending on the time responses of system variables. Output vector y is a binary column vector with m rows wherein each row represents whether the corresponding TDS is stable(1) or unstable(0). For the IEEE 14-bus test system considered in this paper, a load disturbance of 0.2 per-unit (increase) is applied to every steady state operating point generated in the previous section. Stability is decided based on the average values of voltage violations over the entire simulation period (V_{max} = 1.2 pu and V_{min} = 0.8 pu). Figure 2(a) and Figure 2(b) show voltage dynamics at all buses for a stable and unstable case respectively.

Essentially, DSA is a mapping between each object x and its resiliency against the contingency under consideration, expressed by function f such that,

$$f(x) = \begin{cases} 1, & \text{if TDS is stable} \\ 0, & \text{if TDS is unstable} \end{cases} \qquad (4)$$

The next section describes the application of machine learning techniques in order to arrive at this unknown function f.

APPLICATION OF MACHINE LEARNING TECHNIQUES

Machine learning techniques can be applied to the database as generated in the previous section in the form of feature matrix X (size m × 2n) and output vector y (size m × 1). Each row i of matrix X is in the form of object $x \in \Re^{2n}$ from set S as defined in Equation (2) and is referred to as the ith training example: $x^{(i)}$. Similarly, the ith

row from vector y represents the output of the ith training example and is represented by a bit $y^{(i)}$ (either 0 or 1). Therefore, we have,

$x^{(i)}$ = i^{th} training example

$y^{(i)}$ = output (stability) of the ith training example

For "2n" features and "m" training examples, matrix X and vector y are given as follows,

$$X = \begin{bmatrix} \cdots & x^{(1)^T} & \cdots \\ \cdots & x^{(2)^T} & \cdots \\ \cdots & \vdots & \cdots \\ \cdots & x^{(m)^T} & \cdots \end{bmatrix}_{m\times 2n} \qquad y = \begin{bmatrix} y^{(1)} \\ y^{(2)} \\ \vdots \\ y^{(m)} \end{bmatrix}_{m\times 1} \tag{5}$$

Next, a prediction/hypothesis function h in terms of parameter vector q (column vector) of size 2n is proposed as follows,

$$h_\theta(x) = g\left(\theta^T x\right) \tag{6}$$

where x is any training example vector and g depends on the machine learning algorithm being employed.

The cost function J for machine learning algorithms is generally of the form [17] ,

$$J(\theta) = \frac{1}{2m}\sum_{i=1}^{m}\left[h_\theta\left(x^{(i)}\right) - y^{(i)}\right]^2 \tag{7}$$

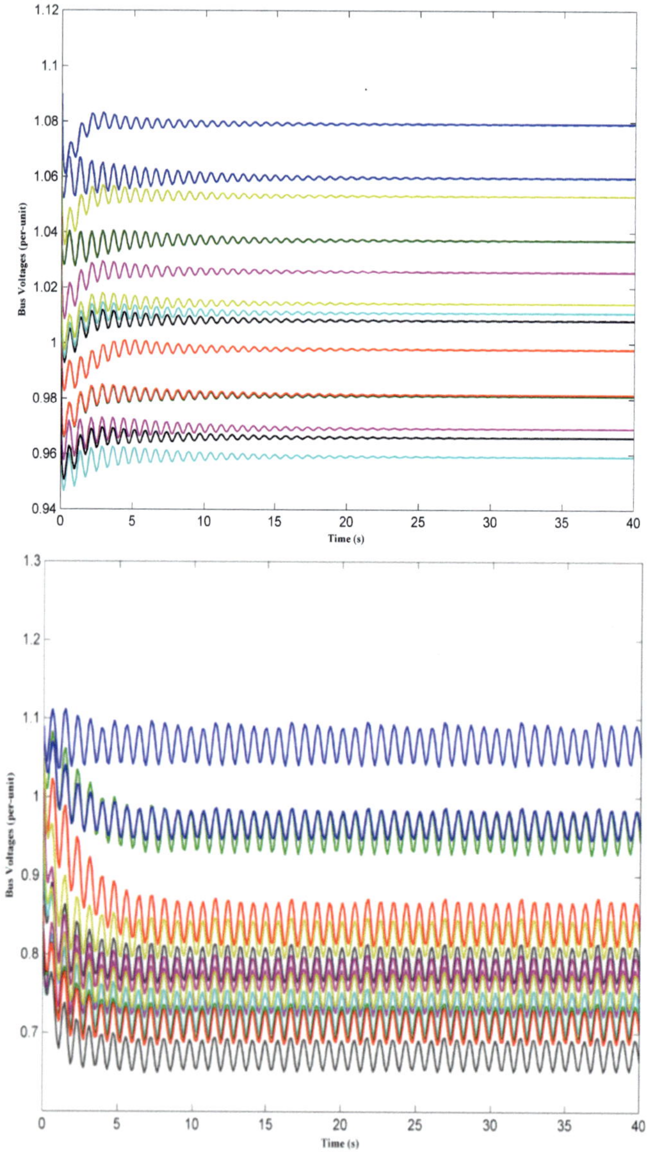

(b)

Figure 2: (a) Bus voltages (stable case); (b) Bus voltages (unstable case).

The above cost function is the mean of the sum of squared errors in predicting the outputs of m training examples. Such a cost function can be minimized by using analytical method or batch gradient descent method. The optimal parameter vector q thus derived can be used for predicting the stability of future cases in real-time.

The problem presented in this paper is to classify a TDS as stable (1) or unstable (0). For such classification problems, logistic regression can be used, in which case functions g and J are given as follows [18] ,

$$g(z) = \frac{1}{1+e^{-z}} \tag{8}$$

and

$$J(\theta) = -\frac{1}{m}\sum_{i=1}^{m}\left[y^{(i)}\log\left(h_\theta\left(x^{(i)}\right)\right) + \left(1-y^{(i)}\right)\log\left(1-h_\theta\left(x^{(i)}\right)\right)\right] \tag{9}$$

The function g(z) given in equation (8) is a sigmoid function and its value lies between 0 and 1. For classification purposes, TDS cases for which g(z) is greater than 0.5 can be considered as stable and the rest as unstable. At this point, it should be noted that function h given in equation (6) approximates the unknown function f of the previous section, when the parameter q is optimal. The approximated function f_{apprx} can be given by,

$$f_{apprx}(x) = \begin{cases} 1, & \text{if } h_\theta(x) \geq 0.5 \\ 0, & \text{if } h_\theta(x) < 0.5 \end{cases} \tag{10}$$

In order to test the algorithm, the 14-bus dataset represented by matrix X and vector y (as generated in the previous section) can be divided into a training set (75%) and a test set (25%), which is a normal practice in ML domain. We may also delete the constant feature columns from X such as those containing PV bus voltages and reference angles, since such constant feature values do not add any valuable information. Therefore, an original matrix X with 22 columns (features) is used in this paper. Figure 3(a) and Figure 3(b) show the learning curves for the training and test sets respectively. Learning curves are plotted by varying the number of examples m in the training set. As highlighted in these figures, the average prediction errors on the training and test sets are calculated as 1.245% and 2.599% respectively. For m objects, prediction error is the percentage of examples that are classified incorrectly by the function f_{apprx} given in Equation (10) and it is calculated as follows,

$$\% \text{ Error} = \frac{100}{m}\sum_{i=1}^{m} err\left(f_{apprx}\left(x^{(i)}\right), y^{(i)}\right)$$

where

$$err\left(f_{apprx}\left(x^{(i)}\right), y^{(i)}\right) = \begin{cases} 1, & \text{if } f_{apprx}\left(x^{(i)}\right) = 1 \,\&\, y^{(i)} = 0 \text{ or} \\ & \text{if } f_{apprx}\left(x^{(i)}\right) = 0 \,\&\, y^{(i)} = 1 \\ 0, & \text{otherwise} \end{cases} \quad (11)$$

The next section of this paper introduces the concept of "landmark points" and "linear kernel". Further, this paper presents a strategy to select best landmark points in order to improve the prediction accuracy.

LANDMARK POINTS AND LINEAR KERNEL

The concept of selecting landmark points gains importance from the fact that a few training examples may contain the most relevant information about the inherent dynamics present in the dataset [19] . This section investigates the possibility of selecting such landmark points within the operational space under consideration in order to improve prediction accuracy without compromising computational efficiency. Essentially, these landmark points are 2n-dimensional objects belonging to the same set S given by Equation (2). In order to demonstrate the effectiveness of this concept, L number of landmark points are drawn at random from the rows of matrix X and then, every (training example, landmark) pair is compared using a linear kernel [20] . A linear kernel measures the similarity between training example $x^{(i)}$ and landmark $l^{(j)}$using the dot product and is given by,

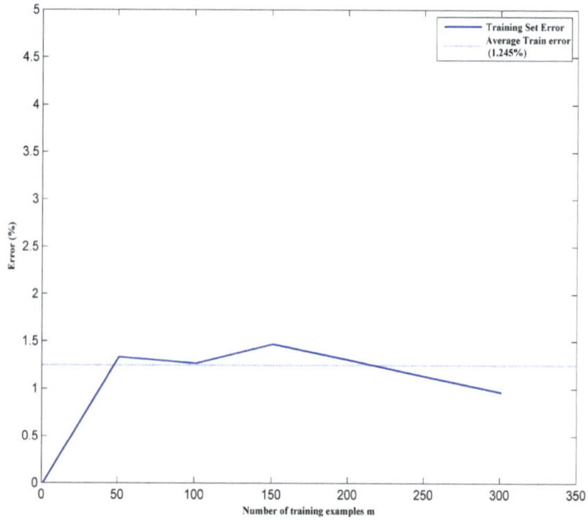

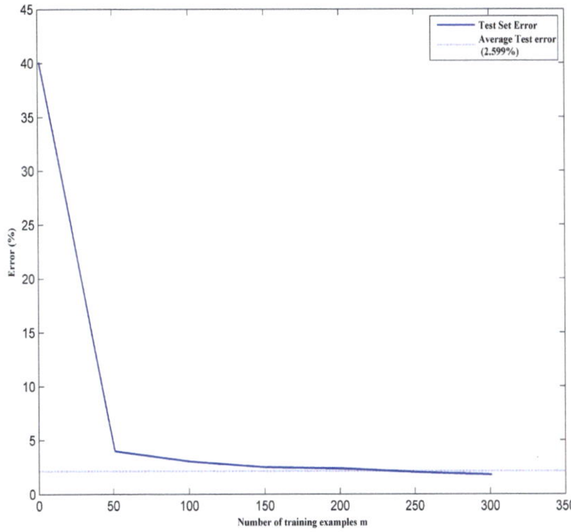

Figure 3: (a) Learning curve (training set); (b) Learning curve (test set).

$$sim\left(x^{(i)}, l^{(j)}\right) = x^{(i)\mathrm{T}} l^{(j)} \tag{12}$$

Similarity is calculated between all training examples i: 1 < i < m and landmark points j: 1 < j < L. The original feature matrix X (size m × 2n) gets transformed into a new matrix X'_{random} (size m × L) which can be now used for training and testing purposes. Computational efficiency is maintained by enforcing the following constraint,

$$L \leq 2n \tag{13}$$

As shown in Figure 4(a) and Figure 4(b), prediction errors on the training and test sets decrease as the number of landmark points increase. However, it should be noted that such a random selection of landmark points does not guarantee better performance when compared with the average training and test set errors calculated in the previous section.

STRATEGY TO SELECT BEST LANDMARK POINTS

Choosing the most appropriate set of landmark points for a given dataset is not an easy task. In this section, the k-means algorithm is used to derive better landmark points as compared to the random ones selected in the previous section [21] . Using k-means algorithm, centroids can be calculated for any feature matrix X. A total number of L such centroids are generated from X for use as landmark points and then, using linear kernel a new matrix $X'_{centroids}$ is formed like in the previous section.

In an attempt to find the best landmark points, the original matrix X is divided into 2 matrices X_{stable} and $X_{unstable}$ consisting of only stable and unstable cases respectively. Using k-means, a total number of L centroids are generated for each of these matrices separately and again using linear kernel, two new matrices X'_{stable} and $X'_{unstable}$ are formed.

The strategy for selecting best landmark points can be stated as follows,

- Select L random examples from original matrix X as landmarks and generate X'_{random}
- Select L centroids from original matrix X as landmarks and generate $X'_{centroids}$
- Select L centroids from X_{stable} as landmarks and generate X'_{stable}
- Select L centroids from $X_{unstable}$ as landmarks and generate $X'_{unstable}$
- Plot learning curves using X'_{random}, $X'_{centroids}$, X'_{stable}, $X'_{unstable}$
- Compare the training and test set errors and select the best L landmarks

Figure 5(a) and Figure 5(b) show the learning curves for each of the above matrices with L = 22 landmarks for the IEEE 14 bus dataset generated earlier. From these figures we can conclude that centroids selected from the unstable cases are the best landmark points for this dataset. Moreover, it has to be noted that computational efficiency is not compromised since the total number of landmarks used here (L = 22) is not greater than the total number of columns in the original matrix X (=22). Figure 6(a) and Figure 6(b) again compare the learning with increasing number of land marks for the case of random landmarks against landmarks selected as centroids from only unstable cases. Figure 7(a) and Figure 7(b) plot the learning curves for the original matrix X (without any landmarks) and $X'_{unstable}$ (best landmarks). These plots confirm that when best landmarks are employed, prediction accuracy improves on both, the training set and the test set.

CONCLUDING REMARKS

The ability to assess the current state of the power system instantaneously is the key attribute needed for enhanced grid resilience. Electric power entities carry out large number of offline studies on power system models of different sizes, thus generating tons of data. Machine learning techniques can be employed to use such huge databases in order to learn the inherent non-linear relationships that exist among different power system parameters. Such useful information can be later used online for real-time security analysis.

This paper presents a framework to apply machine learning techniques for real-time assessment of the grid resilience against any contingency with respect to its static and dynamic stability using offline databases. Further, this paper demonstrates a strategy to select best landmark points in order to improve prediction accuracy without compromising computational efficiency. Moreover, ML algorithms are easily scalable and hence, the proposed approach can be extended for analyzing grid resilience against multiple contingencies. Metrics for grid resilience can be developed based on such multi-contingency analyses. With large-scale penetration of renewable energy in to the current grid and emergence of microgrids, future grid applications would require real-time training in order to extract useful information on a continuous basis.

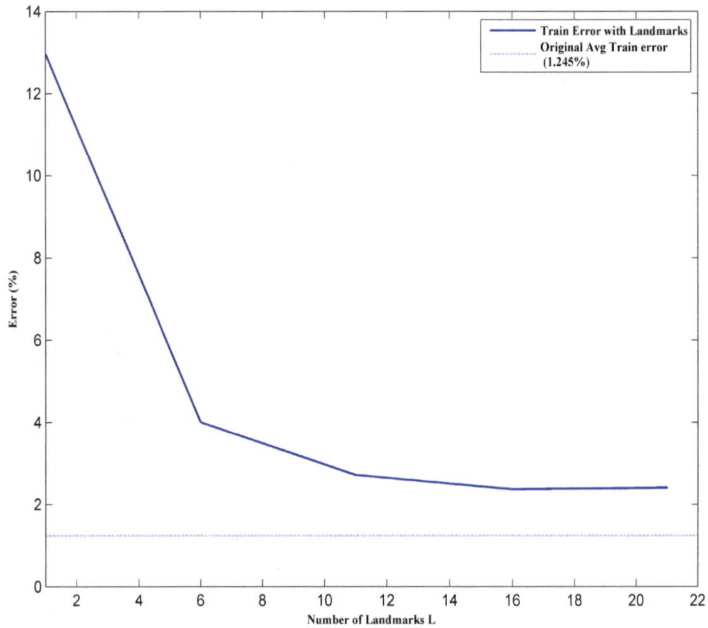

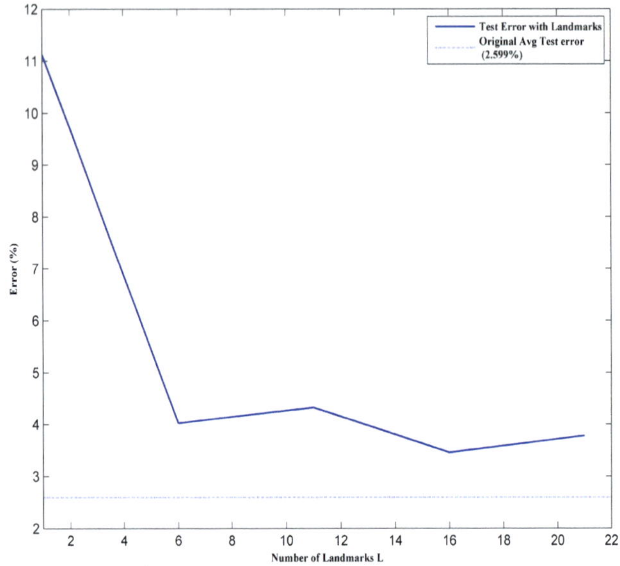

Figure 4: (a) % Error vs num. of landmarks (training set); (b) % Error vs num. of landmarks (test set).

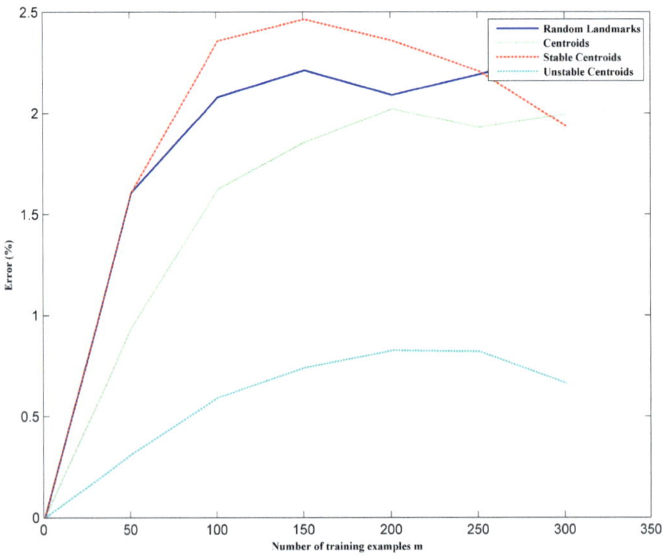

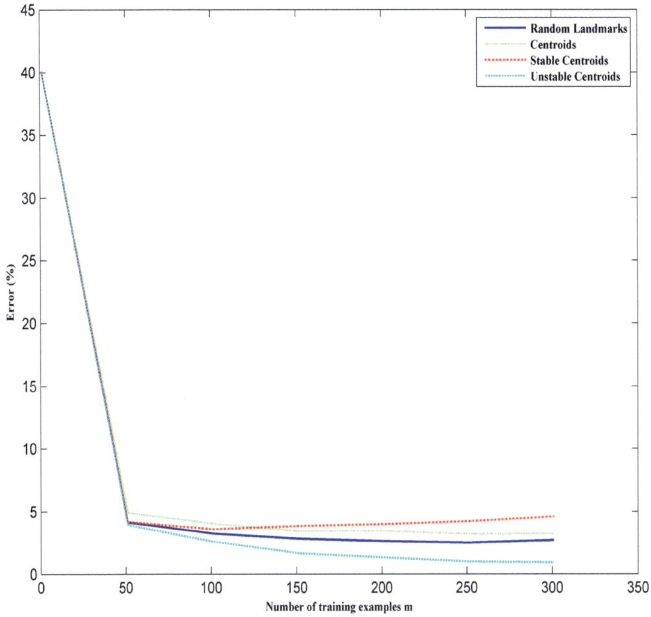

Figure 5: (a) Learning curves (training set); (b) Learning curves (test set).

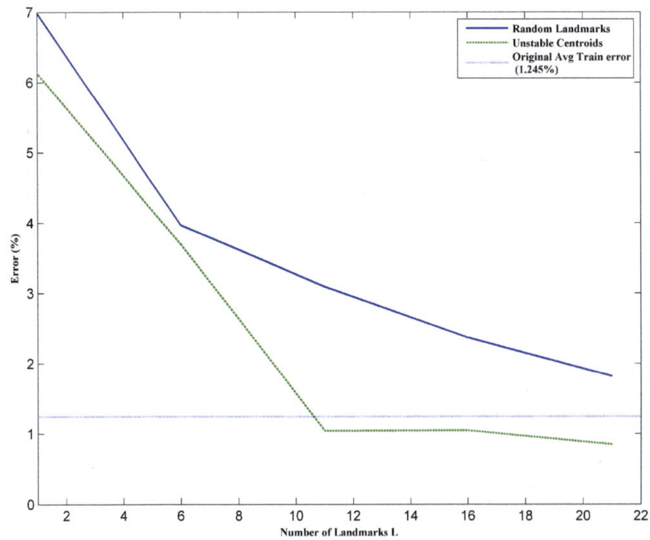

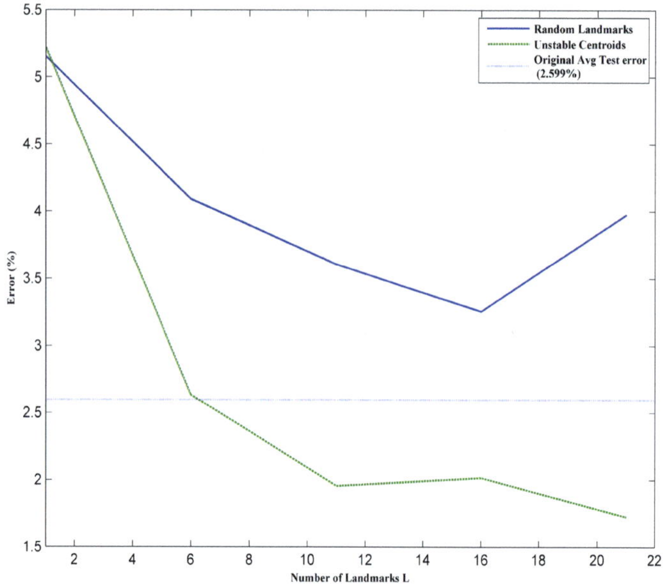

Figure 6: (a) % Error vs num. of landmarks (training set); (b) % Error vs num. of landmarks (test set).

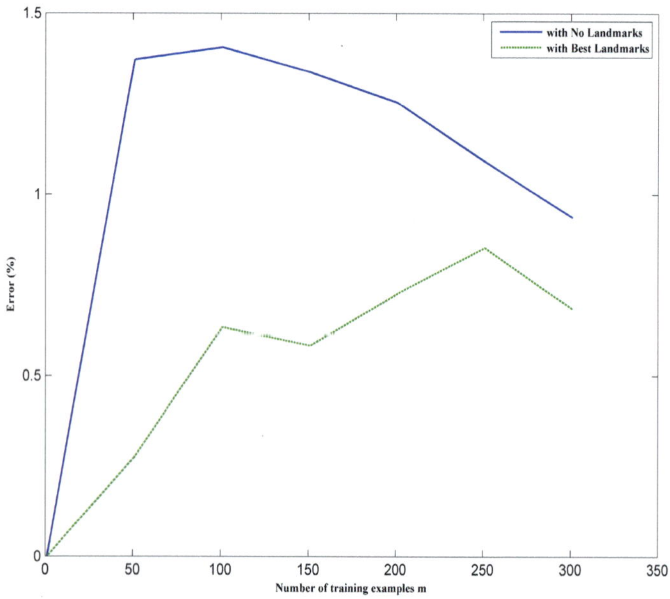

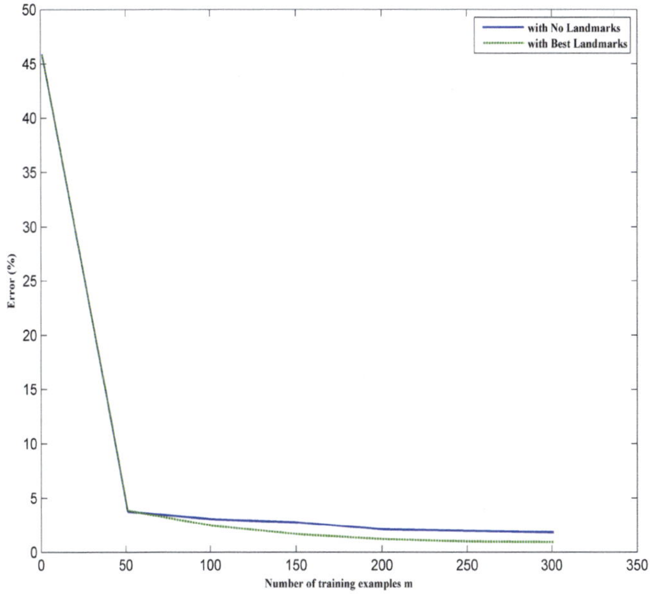

Figure 7: (a) Learning curves (training set); (b) Learning curves (test set).

Machine learning techniques can accommodate such complex requirements posed by the continually changing electric grid and hence, would definitely play an important role in realizing next-gen real-time applications.

ACKNOWLEDGEMENTS

This work was supported by the OSU Engineering Energy Laboratory and the PSO/Albrecht Naeter Professorship in the School of Electrical and Computer Engineering.

REFERENCES

1. Kundur, P., et al. (2004) Definition and Classification of Power System Stability IEEE/CIGRE Joint Task Force on Stability Terms and Definitions. IEEE Transactions on Power Systems, 19, 1387-1401. http://dx.doi.org/10.1109/TPWRS.2004.825981
2. Wang, L. and Morison, K. (2006) Implementation of Online Security Assessment. IEEE Power and Energy Magazine, 4, 46-59. http://dx.doi.org/10.1109/MPAE.2006.1687817
3. Fouad, A., Aboytes, F. and Carvalho, V.F. (1988) Dynamic Security Assessment Practices in North America. IEEE Transactions on Power Systems, 3, 1310-1321. http://dx.doi.org/10.1109/59.14597
4. Grigsby, L.L. (2012) Power System Stability and Control. 3rd Edition, CRC Press, Boca Raton. http://dx.doi.org/10.1201/b12113
5. Jardim, J., Neto, C. and dos Santos, M.G. (2006) Brazilian System Operator Online Security Assessment System. IEEE Power Systems Conference and Exposition, Minneapolis, 25-29 July 2010, 7-12.
6. Tong, J. and Wang, L. (2006) Design of a DSA Tool for Real Time System Operations. International Conference on Power System Technology, Chongqing, 22-26 October 2006, 1-5. http://dx.doi.org/10.1109/icpst.2006.321419
7. Savulescu, S.C. (2009) Real-Time Stability Assessment in Modern Power System Control Centers. John Wiley & Sons, Hoboken. http://dx.doi.org/10.1002/9780470423912
8. Chiang, H.-D., Tong, J. and Tada, Y. (2010) On-Line Transient Stability Screening of 14,000-Bus Models Using TEPCO-BCU: Evaluations and Methods. IEEE Power and Energy Society General Meeting, Minneapolis, 25-29 July 2010, 1-8.
9. Yao, Z. and Atanackovic, D. (2010) Issues on Security Region Search by Online DSA. IEEE Power and Energy Society General Meeting, Minneapolis, 25-29 July 2010, 1-4.
10. Ekanayake, J., Jenkins, N., Liyanage, K., Wu, J. and Yokoyama, A. (2012) Smart grid: Technology and Applications. John Wiley & Sons, Hoboken. http://dx.doi.org/10.1002/9781119968696
11. Ongsakul, W. and Dieu, V.N. (2013) Artificial Intelligence in Power System Optimization. CRC Press, Hoboken.
12. Warwick, K., Ekwue, A. and Aggarwal, R. (1997) Artificial Intelligence Techniques in Power Systems. IEE Press, London.

13. Song, Y.-H., Johns, A. and Aggarwal, R. (1996) Computational Intelligence Applications to Power Systems, Vol. 15. Springer Science & Business Media, Berlin, Heidelberg.
14. Power Systems Test Case Archive. http://www.ee.washington.edu/research/pstca/
15. Ajjarapu, V. and Christy, C. (1992) The Continuation Power Flow: A Tool for Steady State Voltage Stability Analysis. IEEE Transactions on Power Systems, 7, 416-423. http://dx.doi.org/10.1109/59.141737
16. Milano, F. (n.d.) PSAT, Matlab-Based Power System Analysis Toolbox. http://faraday1.ucd.ie/psat.html
17. Ng, A. (2015) Machine Learning. http://cs229.stanford.edu/
18. Barber, D. (2012) Bayesian Reasoning and Machine Learning. Cambridge University Press, Cambridge.
19. Tipping, M.E. (2001) Sparse Bayesian Learning and the Relevance Vector Machine. The Journal of Machine Learning Research, 1, 211-244.
20. Murphy, K.P. (2012) Machine Learning: A Probabilistic Perspective. MIT Press, Cambridge, MA.
21. Smola, A. and Vishwanathan, S. (2008) Introduction to Machine Learning. Cambridge University Press, Cambridge, UK.

SECTION 4:
INTELLIGENT IMPLEMENTATION OF SMART GRIDS

Chapter 15

Intelligent Load Shedding Using TCP/IP for Smart Grids*

Muhammad Qamar Raza, Muhammad Ali, Nauman Tareen, Waheed ur Rehman, Asadullah Khan, Azam Ul Asar

Department of Electrical Engineering COMSATS, Institute of Information Technology, Abbottabad, Pakistan

ABSTRACT

Computerized power management system with fast and optimal communication network overcomes all major discrepancies of undue or inadequate load relief that were present in old conventional systems. This paper presents the basic perception and methodology of modern and true intelligent load shedding scheme in micro grids topology by employing TCP/IP protocol for fast and intelligent switching. The network understudy performs load management and power distribution intelligently in a unified network. Generated power is efficiently distributed among local loads

Citation: M. Raza, M. Ali, N. Tareen, W. Rehman, A. Khan and A. Asar, "Intelligent Load Shedding Using TCP/IP for Smart Grids," *Energy and Power Engineering*, Vol. 4 No. 6, 2012, pp. 398-403. doi: 10.4236/epe.2012.46053.

through fast communication system of server in the form of source and clients in the form of loads through TCP/IP. The efficient use of information between server and clients enables to astutely control the load shedding in a power system of micro grids system. The processing time of above stated system comes out to be 10 ms faster than others which ensure very less delay as compared to conventional methods. The Micro Grids system operating through TCP/IP control has been implemented in MATLAB/SIMULINK and results have been verified.

Keywords: Intelligent Load Shedding (ILS); TCP/IP; Micro Grids (MGs); Server and Clients

INTRODUCTION AND BACKGROUND

In power system terminology, when some interdisturbances are leading the overall system towards instability, the final unwanted solution is load shedding. During load shedding some loads from interconnected network are disconnected for fast and quick recovery of an overall system to its initial settling state. Every grid station has its separate control that controls only the loads in terms of source management. For good management the power generation must be greater than the power consumption then system remains stable. Problems will occur if generation is unable to meet the demands of loads. Then small grids stations are not able to provide energy to their consumers and the only solution is load shedding, to remove power supply from the loads for the system stability.

Disturbances of any nature like hurricanes, faults, transients etc leads a system towards a fatal scenario and these unwanted actions may take place either on generation side, load side or transmission side depending upon the case. Damir proposed artificial neural network approach for intelligent detection [1,2] in power systems for stable operation on transmission lines and some issues were also discussed in a report [3] of NERC. Some methods were also suggested for transmission lines control [4] operation. Schemes of load shedding that has been adopted in past have been discussed in power system literature [5-8]. In breaker interlock scheme, when mains breaker was sending an emergency signal to connected loads breakers all loads were shed without any sequence or priority. However, high priority loads are left untouched during any disturbance and low priority loads are shed from supply. In second method called as under frequency relay scheme that was blind for detecting disturbances but reacted to changes. Slow response time of frequency relays and false dropping out of loads without check were its major discrepancies. Later on, PLCs System was introduced in an interconnected network but was lacking intelligence due to its blindness to transient shifts and slow response time for breakers to trip in emergency condition. Also, more loads were shed than required. Methods of checking voltage instability and actual loss of transmission capacity [9,10] have not been found.

The need for local load shedding [11] was proposed by Adibi for an over loaded equipment under certain time constraints. Conventional schemes were replaced by "Intelligent Load Shedding" [1] scheme which not only has fast response time by introducing TCP/IP for data transfer but manages dropping of specific loads based upon priority and accurately predicts shifts in available generation.

The paper under study has been divided into four main sections illustrating out the work in detail. The first section discusses ILS design using Transmission Control Protocol (TCP/IP), a link that provides intelligent network relation to fast data transfer while the second section relates a LV system (MG) with intelligent control and third section explains the implemented network in MATLAB/Simulink. Finally results are discussed and conclusions with future work have been presented.

ILS DESIGN USING TCP/IP

In power system each connected network plays an important role either on supply side or generation side or load side .If one connected device becomes problematic than overall system goes towards instability. This is the very important and key feature in power system operation and control because the control and stability are related to each other uncontrolled system is basically unstable. And if that uncontrollable scenario in a power system is not tackled by quick shutdown than heavy blackouts and big disaster may occur in a system. This scene should be avoided because it really affects economic conditions of the organizations that are responsible for its functioning. Here reliability plays a very notable role in a sense that reliable systems are effective and unreliable systems are not stable for very long time. For good and efficient system stability should persists for an unlimited period of time until and unless operating conditions varies drastically. Quick and controlled actions are needed in any system to uproot the changes in controlling parameters which otherwise destroys the systems performance. Otherwise faults occur in a huge amount and connected equipment becomes faulty. It is the scenario where intelligent networks plays important role as traffic control and data transfer is very vital since jamming of information signals will prove fatal.

In Figure 1, block diagram shows the step by step events that will take place in a system quickly. Each block represents an event that has been taking place in a system. First of all power will be produced that will be constant and supplied to the central controller which will make a quick action and send information in form of bits to the local controllers that connected to the respective clients using computational engine. After receiving the useful information by local controllers it has been sent to clients for switching load breakers for an intelligent load shedding. Here all the information has been exchanged through TCP/IP platform because TCP/IP controls transmission, demand, distribution, load management and power in Power systems in sensible way and has much more recommendations proposed by Holland [12] for shifting electrical networks from SCADA to it. As numbers of system may be interconnected and they might sent data and information to each other so the system should be strong to hold such a heavy traffic and should be manage from not to get mal functional. As controller handles large amount of data and update it rapidly after mille second so it not easy for it to make decisions of each control and emergency signal at that specific time intervals. And delay can make Power system gets unstable. Therefore for the data transferring reliable medium should be used, having large band width and speed. Channel response should be fast too. For this to be accomplished transmission control protocol is prior for interconnected power system which not only provides high band width but also much quick in response. Secondly, it is connection oriented protocol and data traffic remains secure and safe.

Figure 2 shows that, server client—client model and information that has been transmitted on TCP platform. Server is the central controller having a data base and clients are basically loads here. Local Ethernet networking has been extended there. Communication between both parties is really fast and controlled intelligently by server.

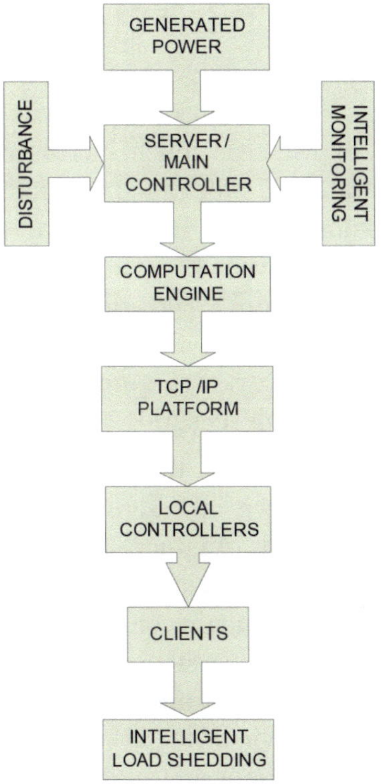

Figure 1: Block diagram of ILS design on TCP/IP platform.

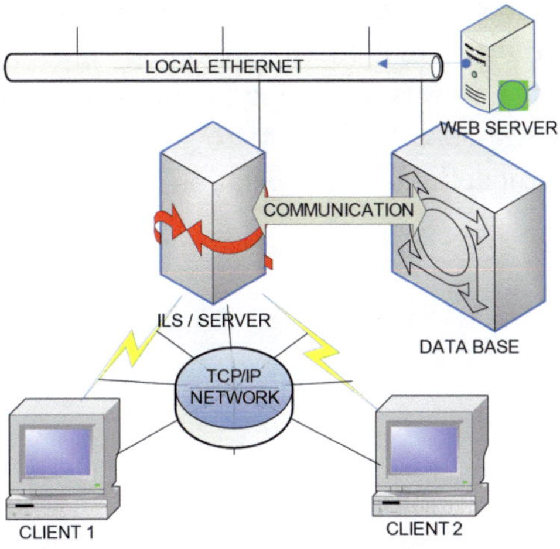

Figure 2: Server—Client intelligent network model design.

MG'S NETWORK WITH INTELLIGENT CONTROL

Micro Grids are the small networks [13] with either renewable energy resources or generating sources connected to it in a controlled fashion. The non-autonomous form of MG has much increased reliability than stand alone MG. It has a central controller called as micro grid central controller which has a strong communication network with its local controllers as clients. Useful information between both the parties is carried through TCP platform. MG connected with utility performs stable operations and quality of power is delivered to consumers. Several issues regarding micro grids [13,14] have been discussed in detail. Figure 3 shows an intelligent control network of MG's in server-client scenario.

Simple model of a MG network that has taken as base for intelligent control implementation is shown in Figure 4 with one generation source and two clients as constant loads. As micro grids contains many generation sources and different kinds of loads connected to it so each load has its own controller which sends signals to main controller for making of decisions in quick succession. In such type of systems, a LAN is mainly obtained and if many micro grids are connected to each other than wide area network is obtained to control all the connected system working well. Nowadays, different power system applications are following this protocol as a leading way of communication between local controllers and main controller, e.g., wind farms, local factories, industries, hydal generation systems etc all are trying to shift previous systems to TCP/IP networking embedded inside each operating instrument. In modern terminology if path communication is between controllers and every operating load or instrument than it is called intelligent control response [15] in micro grids because stable operation requires quick control actions in its operation. A simplified model is shown in Figure 5.

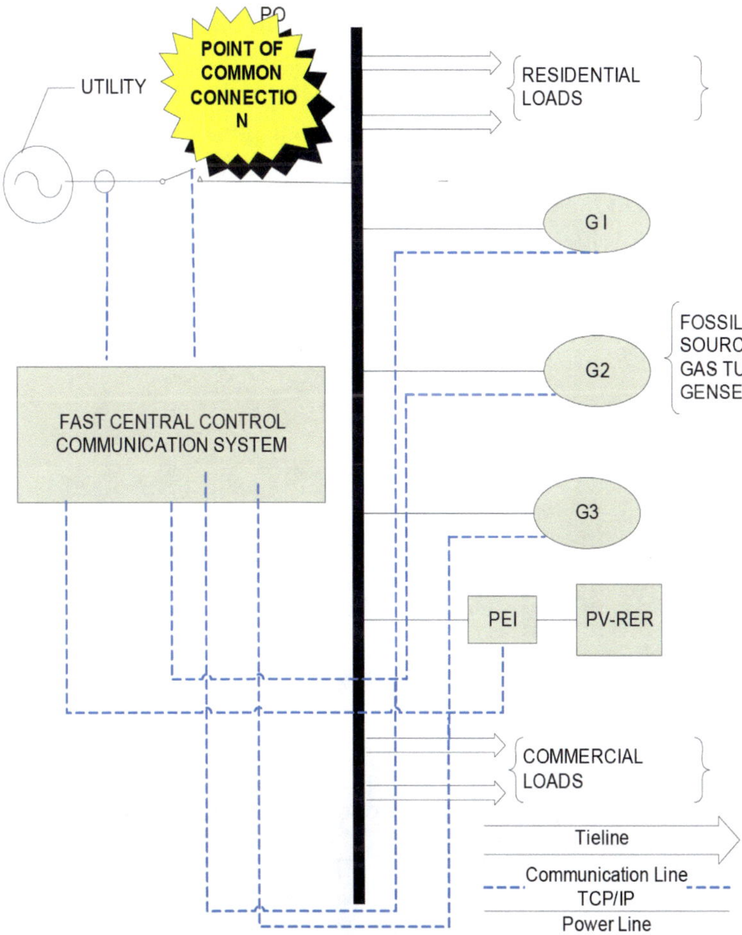

Figure 3: MG in intelligent control scenario.

IMPLEMENTED NETWORK IN MATLAB

In this article, two controllers have been defined, one is main controller acting as a server which is primary one and the other one is secondary which acts as client. TCP/IP networking channel is wide spread on the whole network. Loads to be shed are decided by server and this useful information is carried to clients which has its own local controllers through TCP as shown in Figure 6. Priority of loads has been defined in the server and quick decisions of ILS are carried out with the help of TCP/IP [12].

Final system that has been implemented in MATLAB/Simulink is shown in Figure 7. To precede that very basic idea of ILS, the work basically comprises of four major systems. And these networks were made in MATLAB/SIMULINK. First of all the very basic system was a manual control network that has been developed in SIMULINK and two controllers with defined programming distributing generated power actually between two loads was discussed. Generated power was manually controlled and loads were connected and disconnected according to defined priority. Switching of loads with power available and defined priority is shown in Table 1.

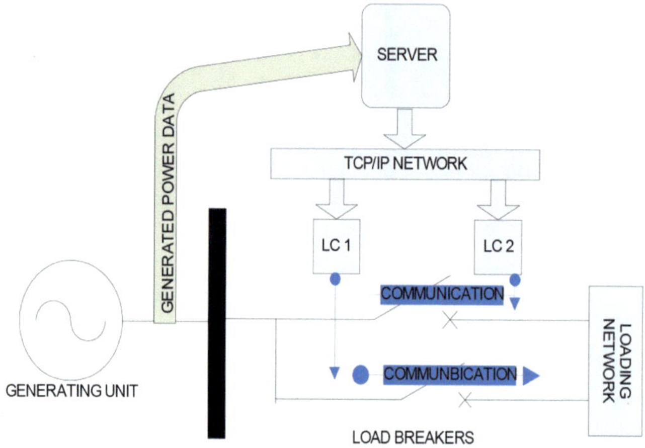

Figure 4: Power system implemented network architecture.

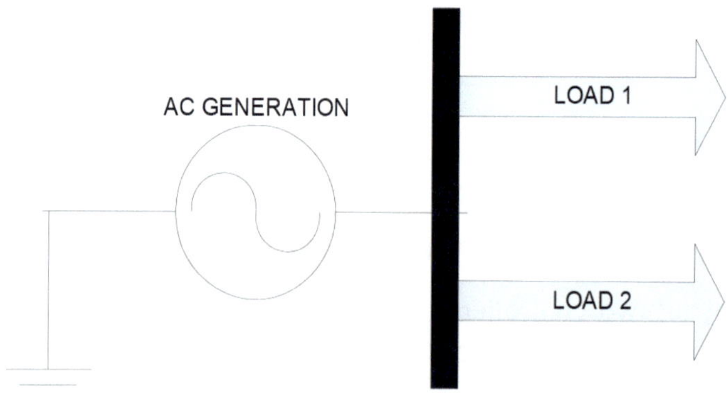

Figure 5: MG simplified model.

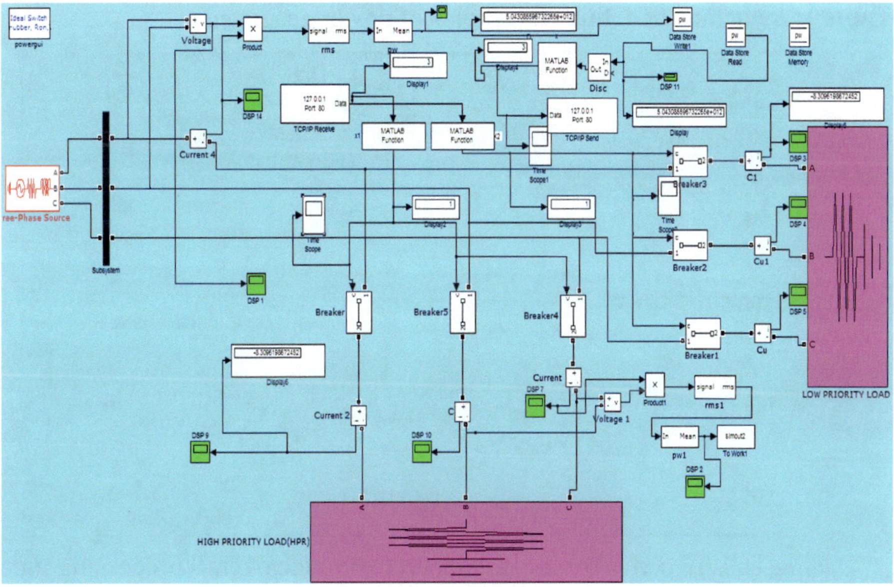

Figure 6: Implemented network in MATLAB.

First system was replaced by second one in the automated control topology. In this case the produce power was automated made constant in a steady state conditions and then managed according to supply. In that particular scheme, constant power was handled through automated system. Using binary number system, loads were operated. Third system was built on the bases of automation with improved concept of data read and data write with same controller options using temporary memory storage option. Then finally the fourth system was very improved form of all of them. There were three controllers that were built. One was acting as server whiles other two as local load controller acting as clients.

The useful and important information was shared dynamically using transmission control protocol. Using this platform, for fast traffic delivery, control decisions were taken. As the load management was done automated through TCP/IP plat form through intelligent controllers, so scheme of ILS was developed in MATLAB. Previous system that power system engineers developed like breaker interlock technique, under frequency decay technique and PLC's operating systems have major disadvantage of slow response time for the breakers to operate.

Table 1: Switching at defined thresholds of power generated

Supply "Volts"	Generated Power "Watts"	Decision Criteria of Server	Data Sent from Sever	Data Received by Clients (HP) (LP)	Load Breakers Position (HP) (LP)
25 KV	5E12 W	Pw ≥ 5E12	3 (11)	1 1	On On Both Closed
2500 V	2.5E4 W	Pw < 5E12 & pw ≥ 2.5E4	2 (10)	1 0	On Off Closed/Open
250 V	2.5E2 W	Pw < 2.5E4 & pw ≥ 2.5E2	1 (01)	0 1	Off On Open/Closed
250 V			0 (00)	0 0	Off Off Both Open

Using TCP/IP plat form for fast data transfer, the overall processing time that came out was 10 mille second.

RESULTS AND CONCLUSIONS

Figure 7 explains the nature of power produced from the generating source which is constant since the system can only accept the constant output because it will not make accurate decisions on the basis of varying power. It also shows that it takes a very small time to for constant output being produced. If the source voltage magnitudes are to be varied than output will also vary in the same manner. Figure 8 shows that relationship data send by sever and time. -Figure 9 shows the information generated by server to clients.

Figure 9 shows that 10 ms is the total processing time taken for the data to reach from the server to clients and finally to load breakers through TCP/IP platform. One "1" signal is received at 10 ms second so breaker at that position will remain closed and load connected to it will function normally.

Figure 10 shows a current signal flowing through the load breaker after it receives a control signal at 10 ms and after that connected transmission line will be live for that connected breaker. So the systems become intelligent when the responses are quick, fast but accurate according to the desired condition. Intelligence lies in the fact that system responds only to true directed states. Intelligent systems are needed every where either in power systems, communication systems etc. because without them sometimes there are always a chance that information is miss interpreted and unwanted events occur.

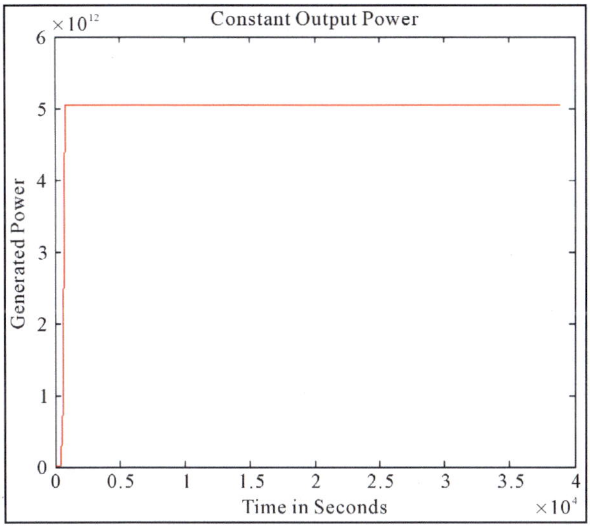

Figure 7: MG power delivered to server.

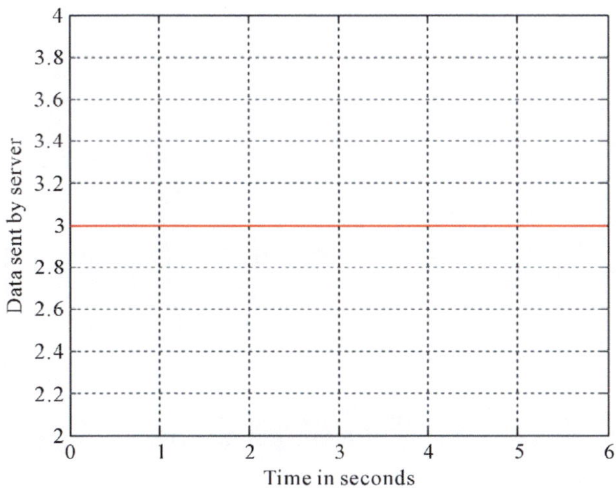

Figure 8: Information delivered by Server.

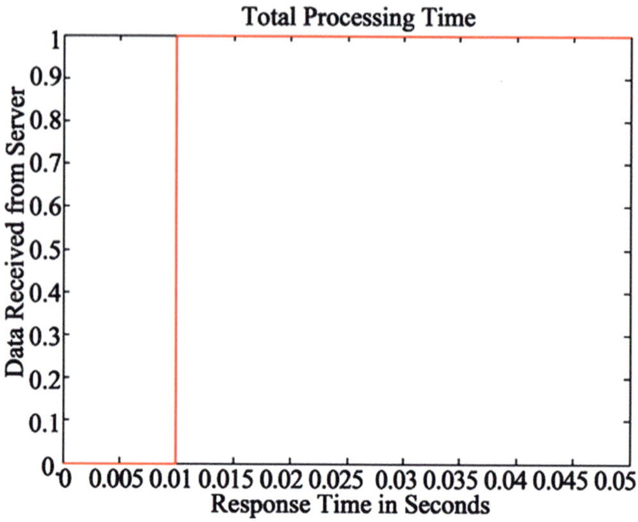

Figure 9: Total processing time.

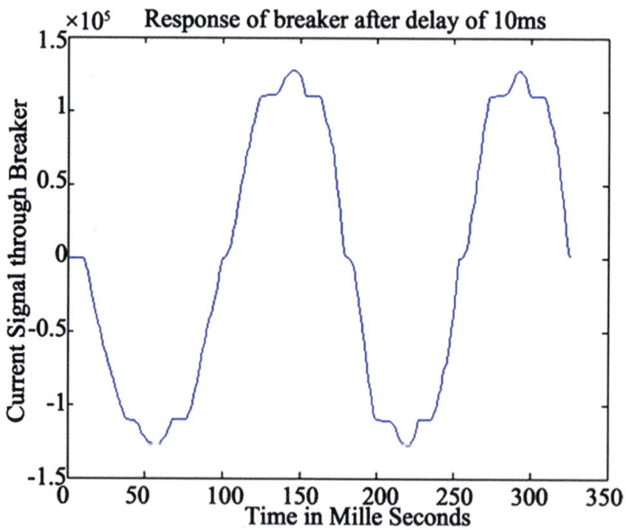

Figure 10: Load breakers response after receiving control signal.

Intelligence of a system basically lies in a controller which is the brain of a system which monitors and controls the whole network. In power systems especially when supply is insufficient to meet the demand of connected loads the intelligence lies in a fact that which loads are shifted either in on or off position.

FUTURE WORK

Muhammad Qamar Raza, Muhammad Ali, Nauman Tareen, Waheed ur Rehman, Asadullah Khan, Azam Ul Asar Since the latest trends are now from micro grids to smart grids which is a big change and development in power industry. As the smart grid concept includes control up to the instrument level, so if the instrument corresponds and communicates intelligently and accurately so all previous problems will be smoothened out in near future. The concept of TCP/IP brings evolution in a system whenever there is a huge interconnected traffic involved. Also Smart grids technology will be flourished and polished by it. A lot of work is still going on in this new area and will continue in new future.

REFERENCES

1. F. Shookh and J. J. Dai, "An ILS System Application in Large Industrial Facility," Industry Applications Conference, 2-6 October 2005, pp. 417-425.
2. L. H. Fink, D. E. Badley, J. E. Koehler, D. A. McInnis and J. J. Redmond, "Emergency Control Practices," Transactions on PAS, Vol. 104, No. 9, 1985, pp. 2336-2441.
3. North American Electric Reliability Council, "1988: System Disturbance," NERC Report, North American Electric Reliability Council, July 1989.
4. L. O. Barthold, "Technical Limits to Transmission System Operation," EPRI EL-5859, Final Report, Electric Power Research Institute, June 1988.
5. A. Maiorano, R. Sbrizzai, F. Torelli and M. Trovato, "Intelligent Load Shedding Schemes for Industrial Customers with Cogeneration Facilities," IEEE Transactions on Power Engineering Society, Vol. 2, No. 31, 1999, pp. 925-930.
6. D. Novosel and R. L. King, "Using Artificial Neural Networks for Load Shedding to Alleviate Overloaded Lines," IEEE Transactions on Power Delivery, Vol. 9, No. 1, 1994, pp. 425-433.
7. L. J. Shih, W. J. Lee, J. C. Gu and Y. H. Moon, "Application of df/dt in Power System Protection and Its Implementation in Micro Controller Based Intelligent Load Shedding Relay," Industrial and Commercial Power Systems Technical Conference, Memphis, 6-9 May 1991, pp. 11-17.
8. W. J. Lee and J. C. Gu, "A Micro Computer-Based Intelligent Load Shedding Relay," IEEE Transactions on Power Delivery, Vol. 4, No. 4, 1989, pp. 2018-2024.
9. P. Crossley, F. Ilar and D. Karlsson, "System Protection Schemes in Power Networks," CIGRE Technical Report No. 187, 2001.
10. C. W. Taylor, "Power System Voltage Stability," McGrawHill, New York, 1994.
11. M. M. Adibi and D. K. Thome, "Local Load Shedding," IEEE Transaction on PWRS, Vol. 3, No. 3, 1988, pp. 1220-1229. doi:10.1109/59.14585
12. Kwok-Hong, B. Holland, "Migrating Electric Power Networks from SCADA to TCP/IP," Power Engineering Journal, Vol. 16, No. 6, 2002, pp. 305-311.

13. Micro Grids Workgroup, "Large Scale Integration of Micro Generation to Low Voltage Grids Target Action 1: Annex 1, Description of Work," Athens, May 2002.
14. D. Pudjianto and G. Strbac, "Investigation of Regulatory, Commercial, Economic and Environmental Issues in Micro Grids," 2005 International Conference on Future Power Systems, Amsterdam, 18 November 2005, 6 p.
15. R. H. Lasseter, "Control and Design of Micro Grid Components," PSERC Final Report, 2007.

Chapter 16

Intelligent Load Management Scheme for a Residential Community in Smart Grids Network Using Fair Emergency Demand Response Programs

Muhammad Ali[1], Zulfiqar Ali Zaidi[2], Qamar Zia[3], Kamal Haider[4], Amjad Ullah[3], Muhammad Asif[5]

[1]Electrical Engineering Department, COMSATS Institute of IT, Abbottabad, Pakistan

[2]Department of Mathematics, COMSATS Institute of IT, Abbottabad, Pakistan

[3]Electrical Engineering Department, NWFP University of Engineering & Technology, Peshawar, Pakistan

[4]Electrical Engineering Department, Gandhara University, Peshawar, Pakistan

[5]Electrical Engineering Department, CECOS University of IT, Peshawar, Pakistan

ABSTRACT

In the framework of liberalized deregulated electricity market, dynamic competitive environment exists between wholesale and retail dealers for energy supplying and management. Smart Grids topology in form of energy management has forced power supplying agencies to become globally

Citation: M. Ali, Z. Zaidi, Q. Zia, K. Haider, A. Ullah and M. Asif, "Intelligent Load Management Scheme for a Residential Community in Smart Grids Network Using Fair Emergency Demand Response Programs," *Energy and Power Engineering*, Vol. 4 No. 5, 2012, pp. 339-348. doi: 10.4236/epe.2012.45044.

competitive. Demand Response (DR) Programs in context with smart energy network have influenced prosumers and consumers towards it. In this paper Fair Emergency Demand Response Program (FEDRP) is integrated for managing the loads intelligently by using the platform of Smart Grids for Residential Setup. The paper also provides detailed modelling and analysis of respective demands of residential consumers in relation with economic load model for FEDRP. Due to increased customer's partaking in this program the load on the utility is reduced and managed intelligently during emergency hours by providing fair and attractive incentives to residential clients, thus shifting peak load to off peak hours. The numerical and graphical results are matched for intelligent load management scenario.

Keywords: Demand Response (DR); Fair Emergency Demand Response Program (FEDRP); Intelligent Load Management (ILM); Residential Area Networks (RAN); Smart Grids

INTRODUCTION

The electric industry is poised to make the renovation from a centralized, producer controlled net-work to one that is less centralized and more consumers interactive. The advancement to smarter grid promises to change the industry's intact business model and it will be beneficent to all i.e. utilities, energy service providers, technology automation vendors and all consumers of electric power.

SG brings improvement in the existing electric grid by incorporating intelligence to each single grid component and the grid architecture. In Residential Area Network (RAN), there is energy manager called REM communicates with Home Energy Manager (HEM) through wireless technology IEEE 802.16. The REM updates the customers about demand response programs, the peak hours, off peak hours etc. through Smart Meters (SM). In [1], author mentioned that in Home Area Network (HAN), home appliance including electric vehicle chargers, security products, refrigerators, microwave, and air conditioners etc. communicates with each other and HEM using Zigbee technology. In [2], authors suggested Zigbee for home automation due to its low power consumption, low cost, a lot of network nodes and reliability.

In [3] author describes that in smart grid topology end user are facilitated by offering different demand response (DR) programs either incentive based or price based. In [4] Demand Response is defined as Changes in electric usage by end-use customers from their normal consumption patterns in response to changes in the price of electricity over time, or to incentive payments designed to induce lower electricity use at times of high wholesale market prices or when system reliability is jeopardized. DR programs are classified into two main categories i.e. Incentive Based and Price Based Programs (PBP). Incentive Based Programs (IBP) is further divided into Classical Programs and Market Based Programs. Classical IBP further sub categories into Direct load control programs and interruptible programs. Market based IBP includes EDRP, Demand Bidding, Capacity Market, Ancillary Services Market. PBP contains Time of Use, Critical Peak Pricing, Extreme Day CPP, Extreme Day Pricing, Real Time Pricing. In market based programs, participation in the programs are given cash for the load reduction during critical hours.

The paper is divided into six broad sections in which second section highlights related work and third section focuses on problem formulation. Fourth section shows detailed model and analysis of FEDRP under the concept of residential area networks and is subdivided into five sub sections. Fifth section shows all numerical and graphical analysis in detail and last section shows conclusion of the work.

RELATED WORK

Recently energy management is an active topic due to continuous rise in global energy consumption continuously [5]. As a result, the existing electricity grid is expected to experience difficulties in generating the necessary power for large amounts of increasing load, distributing the required power and keeping the generated power and the load balanced. As in [6], the participation in DR programs is helpful in customer bill reduction as they reduce load during peak hours as their normal consumption is less than their class average.

During the peak hours, the load on the grid increases than the base load. As mentioned in [7], it is not possible for a power plant to generate adequate power at peak load level and store it when the load is lower, backup plants are used to accommodate the peak loads. Thus these plants incur extra cost for the utility due to extra generation to convene load demands. To compensate the cost, the prosumer has to increase the cost of unit which reduces customer participation during peak hours. The load management techniques are used to reduced peak demands in order to reduce the burden from the grid [8].

For REM different appliances scheduling schemes have also been proposed to reduce the load in SG. In [9], the authors use the particle optimization technique to schedule demands in an automated way. In [10], authors reduce the peak to average electricity usage ratio by Optimal Consumption Schedule (OCS) for the customers in a neighbourhood. The authors in [11], an optimized REM algorithm is proposed that is helpful in reducing the peak load in which appliance start period is scheduled. In [12], an automatic controller design is suggested that schedule appliances to provide an optimal cost. A neural network base prediction approach has been proposed in [13], to optimize the schedule of micro CHP devices. In [14], an energy management protocol is proposed in which consumer sets maximum consumption value and the residential gateway can turn off the device in standby mode.

In [15], Emergency Demand Response Program, is used as a method for Available Transfer Capability enhancement, and this implementation is evaluated from both economical and reliability view points. For this aim, the Emergency Demand Response Program is implemented for specific loads which are chosen according to a sensitivity analysis.

PROBLEM FORMULATION

Smart grid advent brings challenges with opportunities for the end-users and utility. EDRP is one of incentive based program which is offered to consumer to reduce their loads during peak hours [16-18] by giving them incentive payments. In residential area networks (RAN) customer response is totally dependent upon program associated cost; if price is high it must effect customer participation in certain program. In EDRP end user are charged at high prices during peak hours than off-peak hours for any load either "must run" load or "optional" loads, it results in less consumer participation which ultimately cause less revenue generation for utility. Although incentives attract users to cut down their demand during peak hours, but in any case user has to pay high price for must run and variable loads during peak hours.

The scope of this paper is to incorporate fairness in existing EDRP to sustain stability between utility and enduser. Fairness means to make DR programs more reliable and viable, the author in [19] gives idea about clustering based on different categories and shows how customer's participation can be enhance in DR programs, also different schemes are described in [20] for load management This article mainly focuses on how fairness can be amalgamate in RAN for this article suggests the concept of

Fair emergency demand response program (FEDRP). In residential setup load can be categorized by author in as fixed or "must run" load and variable or "optional" load. In existing EDRP fixed and variable loads are charged at same price during peak and off peak hours which cause less consumers participation and satisfaction. In proposed FEDRP customer must be provided same prices in peak and off peak hours for fixed or "must run" loads and only variable or optional loads prices are time variant from peak to off-peak; and also incentive will be offered end users for cutting down their loads during peak hours. This article provides best possible solution for achieving maximum end user participation and to reduce the loads during peak hours.

MODELING AND ANALYSIS OF FEDRP

Fixed and Variable Load Economic Model for RAN

Demand management is most important technique to maximize benefit of both client and utility. To augment the utility revenue, maximum customer involvement is crucial. This simple and widely used model is based on an assumption in which demand will change linearly in respect to the elasticity. To formulate maximum customer involvement, demand of customer is to be analyze against change in prices for must run and optional loads. The price elasticity of demand is defined as the proportion of change in demand to the change in price.

$$e = \partial d / \partial p = p_a / d_a \cdot \frac{\Delta d}{\Delta p} \qquad (1)$$

$$e_{tt} = \frac{\partial d_y(i) / d_x(i)}{\partial \mathrm{p}_t / \mathrm{p}_t} \qquad (2)$$

$$e_{tt'} = \frac{\partial d_y(i) / d_x(i)}{\partial \mathrm{p}_{t'} / \mathrm{p}_{t'}} \qquad (3)$$

Logarithmic modeling of elastic load:

If customer demand changes based on incentive offered by utility so;

$$\Delta d(i) = d_y(i) - d_x(i) \qquad (4)$$

The prize incentive attracts the consumer so total incentive INC (Δd_t) function is given as;

$$\mathrm{INC}(\Delta d(i)) = Q(i)(d_x(i) - d_y(i)) \tag{5}$$

Customer benefit for participating in DR program will be;

$$\mathrm{B}[d_y(i)] - d_y(i) \cdot p_t + \mathrm{INC}(\Delta d(i)) \tag{6}$$

$$\frac{\partial \mathrm{B}[d_y(i)]}{\partial d_y(i)} - p_t + \frac{\partial \mathrm{INC}(\Delta d(i))}{\partial d_y(i)} = 0$$

$$\frac{\partial \mathrm{B}[d_y(i)]}{\partial d_y(i)} = p_t + Q(i)$$

$$\partial d_y(i)/d_x(i) = e_{tt'} \frac{\partial(p_{t'} + Q(i)')}{(p_{t'} + Q(i)')}$$

By assuming constant elasticity for NT hours period, $e_{tt'}$ is equal to constant for t' and $t' \in \mathrm{NT}$ integration of each term we obtain following relation.

$$\int_{\mathrm{d}_x(i)}^{\mathrm{d}_y(i)} \partial \mathrm{d}_y(i)/\mathrm{d}_x(i)$$

$$= \sum_{t'=1}^{\mathrm{NT}} e_{tt'} \cdot \left[\int_{p_{t'}^0}^{p_{t'}} \frac{\partial(p_{t'})}{(p_{t'} + Q(i)')} + \int_0^{Q(i)'} \frac{\partial(Q(i)')}{(p_{t'} + Q(i)')} \right] \tag{7}$$

Combining the optimum customer behavior that leads to

$$d_y(i)$$

$$= d_x(i) + d_x(i) \prod_{t'=1}^{\mathrm{NT}} \mathrm{Ln}\left[\left(p_{t'} + Q(i)'\right)^2 / p_{t'}\left(p_{t'}^0 + Q(i)'\right) \right]^{e_{tt'}} \tag{8}$$

Parameter η is DR potential which can be entered to model as follows:

$$d_y(i)$$

$$= d_x(i) + \eta d_x(i) \prod_{t'=1}^{\mathrm{NT}} \mathrm{Ln}\left[\left(p_{t'} + Q(i)'\right)^2 / p_{t'}\left(p_{t'}^0 + Q(i)'\right) \right]^{e_{tt'}} \tag{9}$$

Larger the value of η means the more customer tendency to reduce or shift consumption from one hour to the other.

For Fixed Loads

For non shift able or must run loads we have elasticity known as "self elasticity" as [10] describes it.

$$e_s = \frac{\Delta d_s}{\Delta p_s} \leq 0 \tag{10}$$

As in our FEDRP must run loads price remain fixed and invariant of peak and off-peak hours so demand at any time for must run (base) loads will be given as

$$d_y(i) = d_x(i) \tag{11}$$

So, demand for base loads will remain same through peak and off-peak hours and eventually customer's participation increases in FEDRP.

Cost of Customer Participation

$$C_{FEDRP} = \left(d_x(i) - d_y(i)\right) \cdot Q(i) \tag{12}$$

$$C_{FEDRP} = \frac{\left(Q(i)\right)^2 \cdot \sum_{t'=1}^{T} e(t,t')}{p(t)}$$

Demand Modeling of RAN

In a residential setup demand of electricity vary with consumer level, for a same price at some definite interval demand of a home may be different from other home. In our proposed FEDRP demand (D_h) of consumer is alienated into two portions, must run loads demand (d_f) and optional load demand (d_v);

$$D_h = d_f + d_v \tag{13}$$

In Figure 1(a) price of electricity is changing with the demand for variable loads with change in price in such a way that $p_1 > p_2 > p_3 > p_4$. d_1 is demand of customer during peak hours where price increases which ultimately result in less customer's demand and for d_2, d_3 and d_4 end user's demand is increased due to decreased prices, and for fixed loads as described in Figure 1(b) price remain same during peak and off peak hours.

In Figure 2 residential area setup is modeled such that demand of consumers is varying such that:

$$D_{hd} > D_{hc} > D_{hb} > D_{ha} \quad (14)$$

$$D_{ha} = d_1^a + d_2^a + d_3^a + d_4^a \quad (15)$$

D_{ha} at different intervals will be:

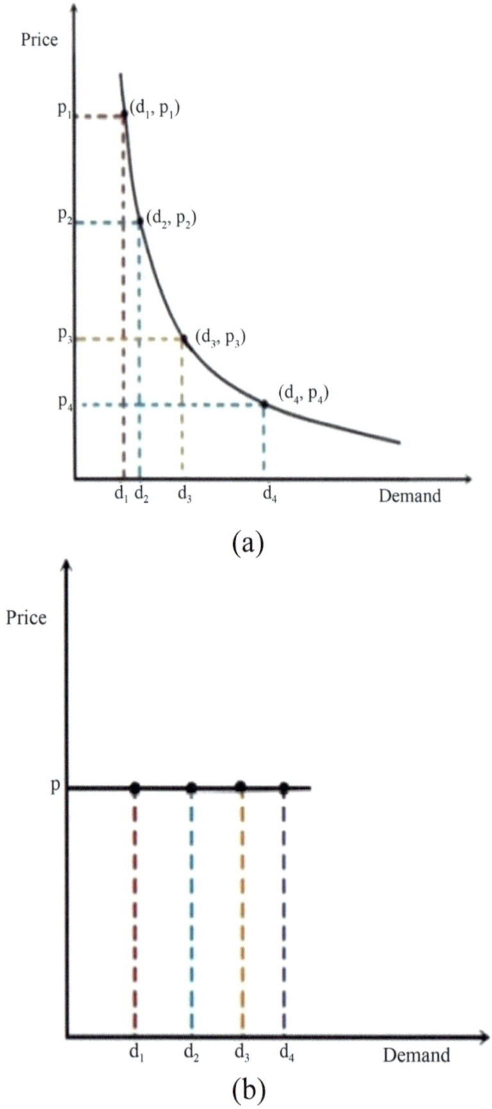

Figure 1: (a) Demand and price curve for optional (variable) loads; (b) Demand and price curve for fixed (must run) loads.

$$D_{hb} = \sum_{i=1}^{6} d_1^a + \sum_{i=7}^{14} d_2^a + \sum_{i=15}^{20} d_3^a + \sum_{i=21}^{24} d_4^a \tag{16}$$

where "i" determine the period of time that customer demand during interval d_1, d_2, d_3, and d_4.

Now for second home demand considering both optional and must run load will be:

$$D_{hb} = \sum_{i=1}^{6} d_1^b + \sum_{i=7}^{14} d_2^b + \sum_{i=15}^{20} d_3^b + \sum_{i=21}^{24} d_4^b \tag{17}$$

Similarly demands of home three and four are given as:

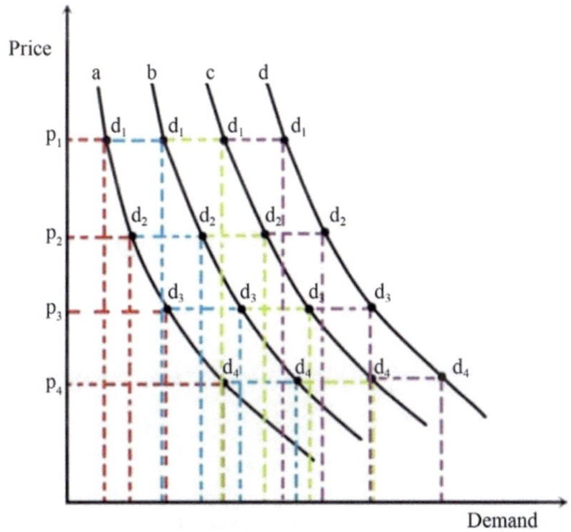

Figure 2: Demand and price curve for variable loads four homes in RAN.

$$D_{hb} = \sum_{i=1}^{6} d_1^c + \sum_{i=7}^{14} d_2^c + \sum_{i=15}^{20} d_3^c + \sum_{i=21}^{24} d_4^c \tag{18}$$

$$D_{hb} = \sum_{i=1}^{6} d_1^d + \sum_{i=7}^{14} d_2^d + \sum_{i=15}^{20} d_3^d + \sum_{i=21}^{24} d_4^d \tag{19}$$

so the total fixed demand and variable demand for RAN during 24 hours is given as;

$$d_f = \begin{bmatrix} \sum_{i=1}^{6}\left(d_{1f}^{a} + d_{1f}^{b} + d_{1f}^{c} + d_{1f}^{d}\right) \\ +\sum_{i=7}^{14}\left(d_{2f}^{a} + d_{2f}^{b} + d_{2f}^{c} + d_{2f}^{d}\right) \\ +\sum_{15}^{20}\left(d_{3f}^{a} + d_{3f}^{b} + d_{3f}^{c} + d_{3f}^{d}\right) \\ +\sum_{21}^{24}\left(d_{4f}^{a} + d_{4f}^{b} + d_{4f}^{c} + d_{4f}^{d}\right) \end{bmatrix} \quad (20)$$

Above equation describes the fixed demand in RAN during 24 hours similarly variable demand during a day is expressed as;

$$d_v = \begin{bmatrix} \sum_{i=1}^{6}\left(d_{1v}^{a} + d_{1v}^{b} + d_{1v}^{c} + d_{1v}^{d}\right) \\ +\sum_{i=7}^{14}\left(d_{2v}^{a} + d_{2v}^{b} + d_{2v}^{c} + d_{2v}^{d}\right) \\ +\sum_{15}^{20}\left(d_{3v}^{a} + d_{3v}^{b} + d_{3v}^{c} + d_{3v}^{d}\right) \\ +\sum_{21}^{24}\left(d_{4v}^{a} + d_{4v}^{b} + d_{4v}^{c} + d_{4v}^{d}\right) \end{bmatrix} \quad (21)$$

So total demand of these homes in residential setup from (16), (17), (18), and (19) will be:

$$D_t = D_{ha} + D_{hb} + D_{hc} + D_{hd} \quad (22)$$

As total demand is equal to total fixed and total variable demand

$$D_h = d_f + d_v$$

$$D_t = \begin{bmatrix} \sum_{i=1}^{6}\left(\left(d_{1f}^{a} + d_{1v}^{a}\right) + \left(d_{1f}^{b} + d_{1v}^{b}\right) + \left(d_{1f}^{c} + d_{1v}^{c}\right)\right. \\ +\left(d_{1f}^{d} + d_{1v}^{d}\right) + \sum_{i=7}^{14}\left(d_{2f}^{a} + d_{2v}^{a}\right) + \left(d_{2f}^{b} + d_{2v}^{b}\right) \\ +\left(d_{2f}^{c} + d_{2v}^{c}\right) + \left(d_{2f}^{d} + d_{2v}^{d}\right) + \sum_{15}^{20}\left(d_{3f}^{a} + d_{3v}^{a}\right) \\ +\left(d_{3f}^{b} + d_{3v}^{b}\right) + \left(d_{3f}^{c} + d_{3v}^{c}\right) + \left(d_{3f}^{d} + d_{3v}^{d}\right) \\ +\sum_{21}^{24}\left(d_{4f}^{a} + d_{4v}^{a}\right) + \left(d_{4f}^{b} + d_{4v}^{b}\right) + \left(d_{4f}^{c} + d_{4v}^{c}\right) \\ \left.+\left(d_{4f}^{d} + d_{4v}^{d}\right)\right) \end{bmatrix} \quad (23)$$

By simplifying above equation, so total demand of these homes in residential setup from (16), (17), (18), and (19) will be:

$$D_t = D_{ha} + D_{hb} + D_{hc} + D_{hd}$$

$$D_t = \begin{bmatrix} \sum_{i=1}^{6}\left(d_1^a + d_1^b + d_1^c + d_1^d\right) + \sum_{i=7}^{14}\left(d_2^a + d_2^b + d_2^c + d_2^d\right) \\ + \sum_{15}^{20}\left(d_3^a + d_3^b + d_3^c + d_3^d\right) + \sum_{21}^{24}\left(d_4^a + d_4^b + d_4^c + d_4^d\right) \end{bmatrix} \quad (24)$$

Considering total demand for residential setup of "m" homes with different demands during "4" intervals in 24 hours is described as

$$\begin{aligned} D_t = & \sum_{i=1}^{6}\left(d_1^a + d_1^b + d_1^c + d_1^d + \cdots + d_1^m\right) \\ & + \sum_{i=7}^{14}\left(d_2^a + d_2^b + d_2^c + d_2^d + \cdots + d_2^m\right) \\ & + \sum_{15}^{20}\left(d_3^a + d_3^b + d_3^c + d_3^d + \cdots d_3^m\right) \\ & + \sum_{21}^{24}\left(d_4^a + d_4^b + d_4^c + \cdots + d_4^m\right) \end{aligned} \quad (25)$$

Equation (22) shows the total demand of "m" homes during 24 hours at four different intervals.

Utility Revenue

Scope of this article is to increase utility revenue along with the maximum customer satisfaction. In FEDRP utility revenue is of two types, first that utility obtain from fixed or "must run" loads and other revenue from variable or "optional" loads. This means that revenue and benefit of utility will be given as:

$$R = f(p, d)$$

In proposed model company revenue is of two types, first that utility obtain from fixed or "must run" loads and other revenue from variable or "optional" loads. This means that revenue will be given as:

$$R = R_f + R_v \tag{26}$$

Taking into consideration the demands of four homes and their respective revenues at four intervals is calculated as:

Revenue from fixed (must run) loads will be:

$$R_f = \sum_{i=0}^{3} p \left[\int_0^{d_{i+1}^a} \mathrm{d}D + \int_0^{d_{i+1}^b} \mathrm{d}D + \int_0^{d_{i+1}^c} \mathrm{d}D + \int_0^{d_{i+1}^d} \mathrm{d}D \right] \tag{27}$$

Revenue from variable (optional) loads will be:

$$R_v = \sum_{i=0}^{3} p_{i+1} \left[\int_0^{d_{i+1}^a} \mathrm{d}D + \int_0^{d_{i+1}^b} \mathrm{d}D + \int_0^{d_{i+1}^c} \mathrm{d}D + \int_0^{d_{i+1}^d} \mathrm{d}D \right] \tag{28}$$

Using R_f, R_v in (23):

$$R = \sum_{i=0}^{3} \left[(p + p_{i+1}) \cdot \left\{ \int_0^{d_{i+1}^a} \mathrm{d}D + \int_0^{d_{i+1}^b} \mathrm{d}D + \int_0^{d_{i+1}^c} \mathrm{d}D + \int_0^{d_{i+1}^d} \mathrm{d}D \right\} \right] \tag{29}$$

For "m" homes revenue to utility at four intervals is described as (see Equation (30), below)

Benefit function for utility is given as:

Profit = Revenue – Total operating cost

$$y = R - \sum_{k=1}^{u} G_k \cdot C \tag{31}$$

$$y = \sum_{i=0}^{3} \left[(p + p_{i+1}) \cdot \left\{ \int_0^{d_{i+1}^a} \mathrm{d}D + \int_0^{d_{i+1}^b} \mathrm{d}D + \int_0^{d_{i+1}^c} \mathrm{d}D + \int_0^{d_{i+1}^d} \mathrm{d}D \right\} \right] - \sum_{k=1}^{u} G_k \cdot C \tag{32}$$

NUMERICAL RESULTS AND SIMULATION

In this section numerical and graphically study has been evaluated considering FEDRP. For this purpose daily load curve of Pakistani local grid has been taken for this simulation studies. The curve is divided into three sections as low load, off load and peak load periods as shown in Table 1 and energy prices are taken in rupees in different periods.

$$R=\sum_{i=0}^{3}\left[\left(p+p_{i+1}\right)\cdot\left\{\int_{0}^{d_{i+1}^{a}} dD+\int_{0}^{d_{i+1}^{b}} dD+\int_{0}^{d_{i+1}^{c}} dD+\int_{0}^{d_{i+1}^{d}} dD+\cdots+\int_{0}^{d_{i+1}^{m}} dD\right\}\right] \quad (30)$$

The selected values of self and cross elasticity's have been shown in **Table 2**.

- Fairness index has been calculated for fair emergency demand response program as follows;
- Fairness index (FI) in [4] is described as ratio of customers whose demand is satisfied to total number of customers. Utility should effectively manage the customers demand weather for must run or optional loads. Fairness index given as:

$$FI=\frac{\left(\sum_{i=1}^{a} d_f\right)\cdot\beta+\left(\sum_{i=1}^{a} d_v\right)\cdot\gamma}{\alpha\cdot\left(\gamma+\beta\right)} \quad (33)$$

where β, γ are the priority of different loads and α is number of total customers. Considering the case for example in peak hours fixed load demand of 40 customers are satisfied and optional needs for 30 customers satisfied, and also priority of fixed load is twice of optional load [6], then FI = 0.91. The client contentment for no mandatory load depends on price variation for this load during peak and off-peak hours. Higher the prices less will be the demand of customer, for satisfying its optional load.

Total numerical analysis of daily load curve has been evaluated in Table 3 which shows utility fixed and variable revenues (R_f and R_v) under different cases of incentives given to clients for program participation. Initial fixed and variable demands (D_{fi} and D_{vi}) for the utility is shown with zero dollar incentives but the important thing is that as giving 25 dollar incentive for peak reduction to clients, D_v shifts tremendously as compared to 10 and 20 dollar incentives.

Total energy and peak reductions have been calculated in Table 4.

Figures 3-5 show the total variable demand analysis when different cases of consumer participation have been taken under different scenarios of incentives given by the utility company to those customers who sign up the contract for FEDRP. Incentives in the form of 10, 20 and 25 dollars are given to the participants to reduce their variable load during emergency or peak hour. It has been analyzed in these figures that energy of daily load during peak hours has been sufficiently reduced and is shifted to off peak periods.

Table 1: Energy prices of daily load curve

	Low-Load	Off-Period	Peak
Period (HRS)	00:00 am to 9:00 am	09:00 am to 5:00 pm	5:00 pm to 00:00 am
Energy Prices (per KWH)	6 Rps	8 Rps	10 Rps

Table 2: Self and cross elasticity's

	Low Load	Off-Peak	Peak
Low Load	–0.1	0.01	0.012
Off-Peak	0.01	–0.1	0.016
Peak	0.012	0.016	–0.1

Giving more incentives has permitted consumers to shift their load to off peak period. Similarly, Figures 6-8 show fixed revenues generated for utility company offering the program under different cases of incentives. The cases of variable revenues generated to utility from customer's participation using their variable load has been calculated numerically and analyzed graphically as demonstrated in Figures 9-11. In all the analysis cases of incentives are compared clearly with non incentive cases which prove the whole scenario.

Table 3: Total numerical analysis of load curve

D_i (MW)	D_{fi} (MW)	D_{vi} (MW)	D_v (10$) Inc	D_v (20$) Inc	D_v (25$) Inc	R_f (10$) Inc	R_f (20$) Inc	R_f (25$) Inc
22.08	4.338	17.57	17.7	17.9	18	130.559	131.23	133.23
22.08	4.44	17.764	17.8	18	18.16	136.047	139.86	141
22.5	4.49	17.89	18	18.2	18.31	148.25	150.804	155.804
22.75	4.54	18.228	18.2	18.3	18.32	157.65	159.67	161.67
23	4.62	18.515	18.4	18.4	18.38	163.791	168	169.8
23.5	4.75	18.8	18.5	18.42	18.4	166.368	169.04	171.04
23.75	4.82	18.9	18.625	18.45	18.4	167.348	168.83	169.83
24	4.825	18.68	18.6	18.43	18.41	168.034	168	168.89
24.14	4.75	18.468	18.458	18.4	18.4	168.712	167.978	167.778
23.92	4.71	18.18	18.28	18.3	18.35	169.89	168	166.68
23.72	4.73	18	18.1	18.2	18.25	170.671	171	170.5
23.5	4.74	17.82	17.9	18.1	18.17	171.228	173	174.4
23.65	4.78	17.73	17.83	18.07	18.16	172.45	173.68	174.5
23.8	4.8	17.86	17.92	18.1	18.18	173.02	174	174.3
24	4.84	18.2	18.28	18.41	18.6	173.765	173.8	173.8
24.2	4.91	18.59	18.6	18.8	18.8	174.87	172.78	172
24.7	4.92	19.279	19.13	18.95	18.9	175.176	172.359	171.89
24.5	4.94	19.65	19.25	19	18.93	175.368	172.616	171.616
25	4.92	19.785	19.3	19.02	18.94	175.341	172.621	171.621
24.7	4.92	19.5	19.28	19.018	18.95	175.29	173.6491	172.1
24.5	4.87	19.305	19.24	19	18.93	174.813	176	176
24.3	4.85	18.815	18.915	18.99	18.92	174.54	175.92	176.92
24.15	4.78	18.64	18.84	18.876	18.88	174.22	175.86	177.4
23.7	4.75	18.275	18.375	18.5	18.6	174.162	175.78	177.6

R_v (10\$) Inc	R_v (20\$) Inc	R_v (25\$) Inc
143	147	148
144.612	148.612	150.612
147.48	153	154
154.2	156.2	157.2
156.4	158.3	158.3
159.8	160	160.5
162.2	161	161.7
162.12	162.52	161.92
161.212	161.8212	161.8212
160	161	161
161	168	172
180	183	183
184.76	188.76	190
189.32	193	193
208.4	200.4	199.4
214.08	206.08	201.08
216.9	207	203
217.8	207.6	202.6
215.065	205.065	203.065
210.5	204.5	201.5
180.745	200.0012	200.0012
174	180	185
170	175	178
168.575	170.575	172.575

Table 4: Energy and peak reductions in % with different scenarios

Different Inc Cases	Total Energy MWhr	Energy Reduction (%)	Peak (MW)	Peak Reduction (%)
(0$) Inc	580	0	24.7	0
(10$) Inc	575	0.862069	24.5	0.809717
(20$)Inc	568	2.068966	24.09	2.469636
(25$) Inc	566.89	2.260345	23.9	3.238866

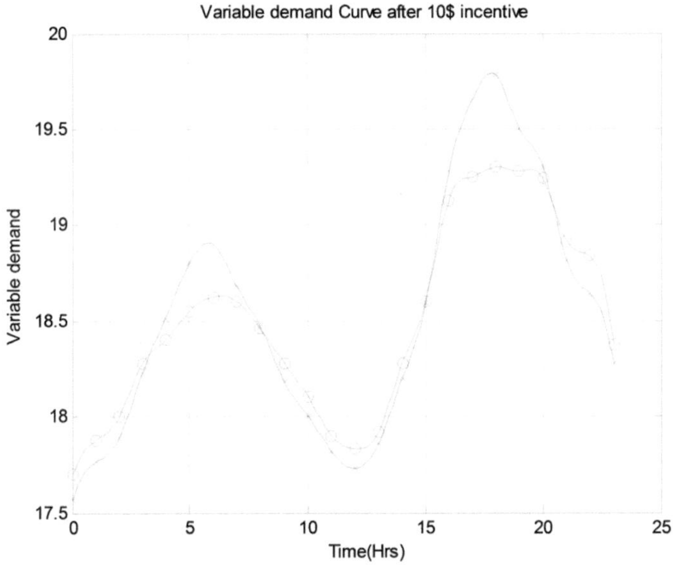

Figure 3: Variable demand for 10$ incentive.

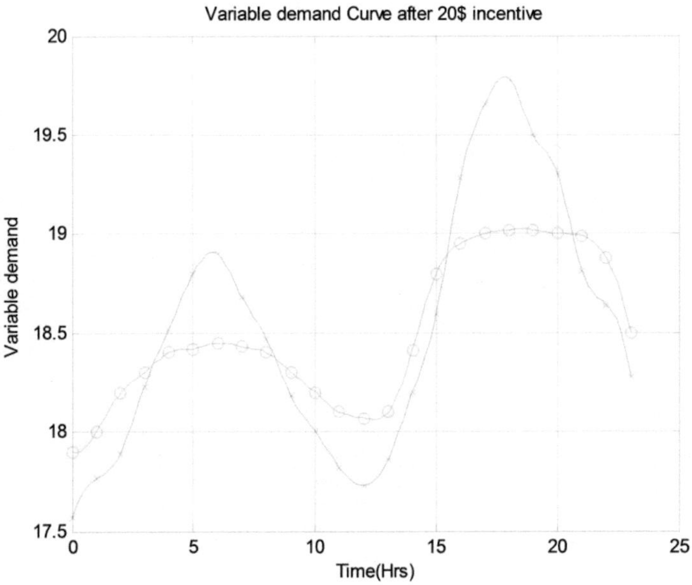

Figure 4: Variable demand for 20$ incentive.

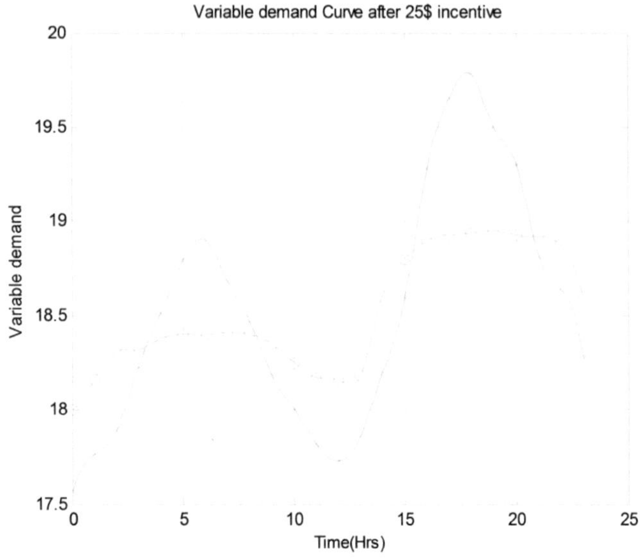

Figure 5: Variable demand for 25$ incentive.

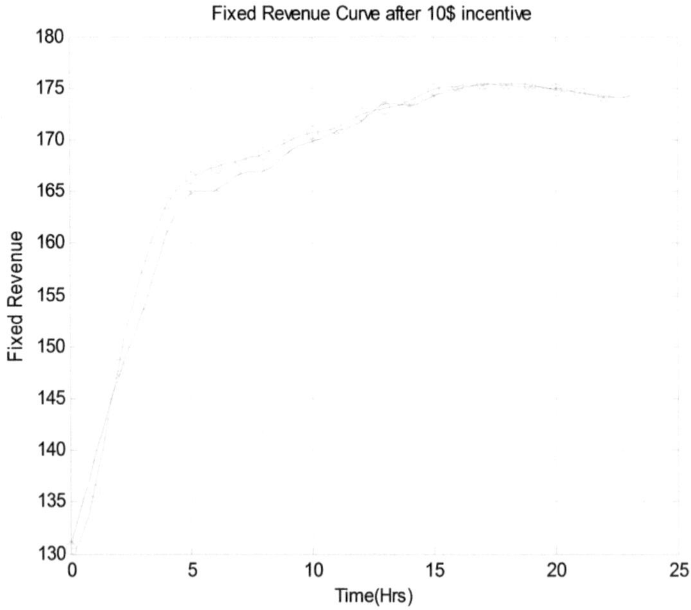

Figure 6: Fixed revenue for 10$ incentive.

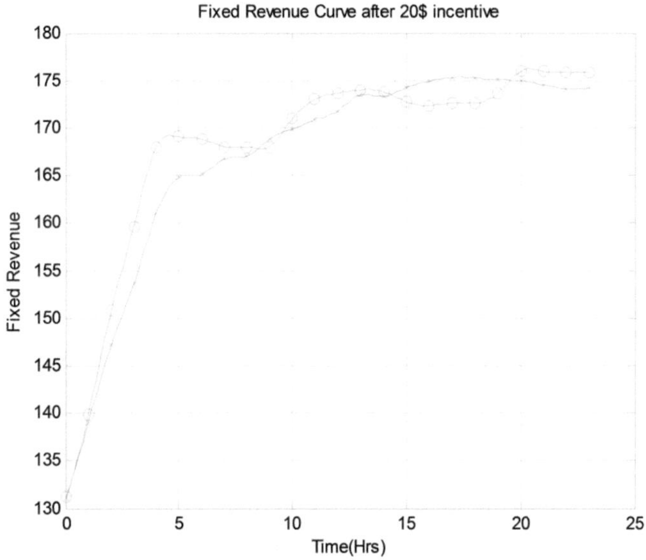

Figure 7: Fixed revenue for 20$ incentive.

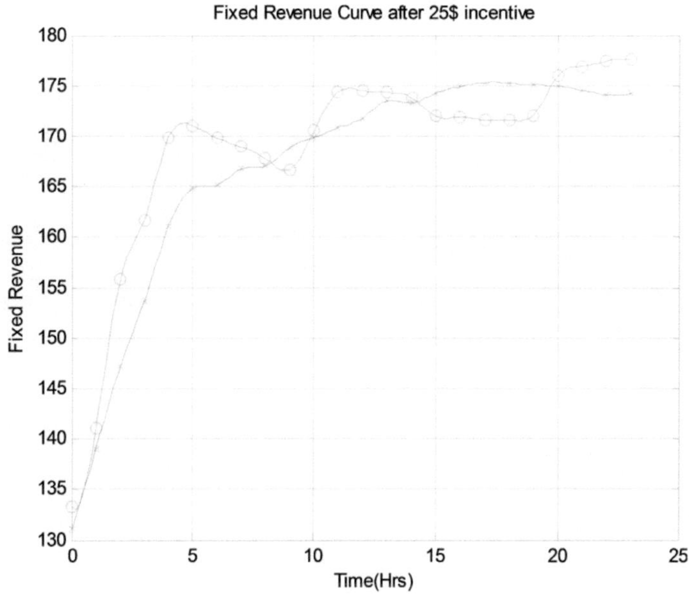

Figure 8: Fixed revenue for 25$ incentive.

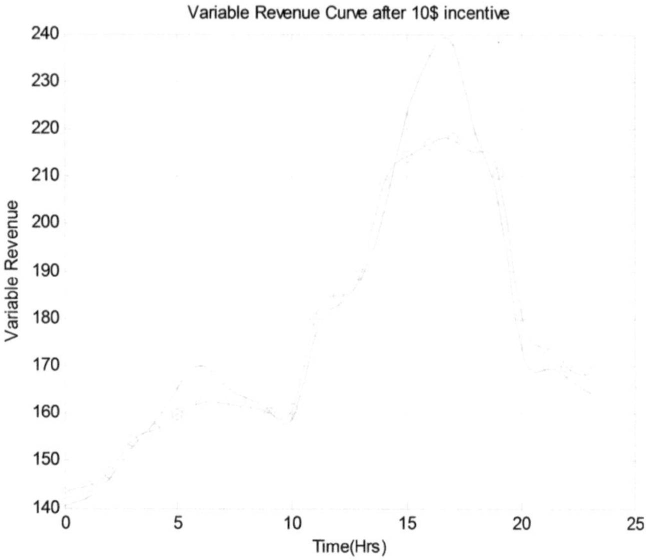

Figure 9: Variable revenue for 10$ incentive.

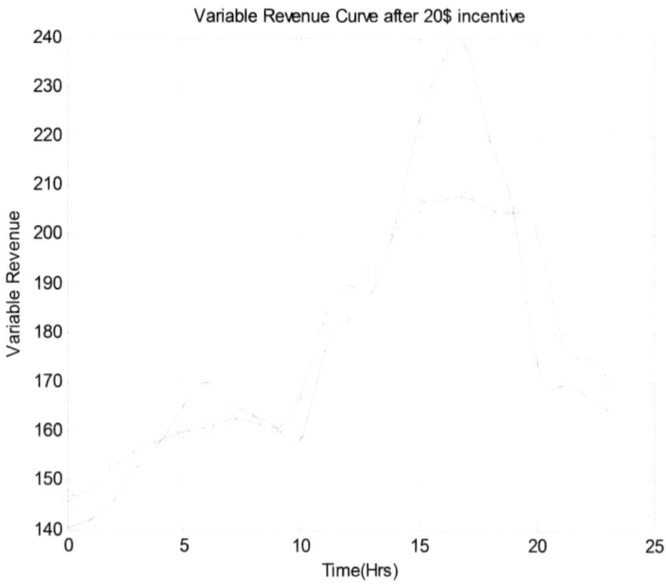

Figure 10: Variable revenue for 20$ incentive.

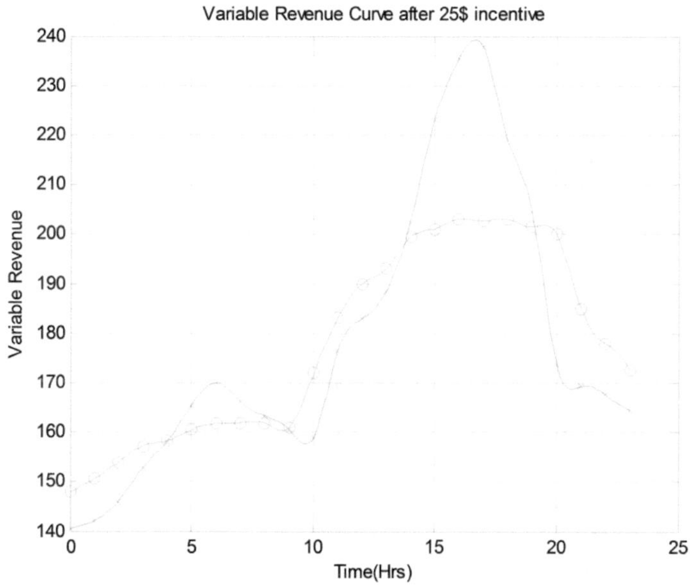

Figure 11: Variable revenue for 25$ incentive.

CONCLUSION

In this paper utility revenue and profit is modeled considering RAN for different user levels consumption, also demand of consumer is modeled mathematically and graphically. As consumer participation and satisfaction in DR program is basic tool to measure competitiveness for any DR program in market, so in this article end user participation is represented graphically with the comparison of initial demand before FEDRP. For customer contentment fairness index of the FEDRP is also calculated. Demand curve of FEDRP is plotted and also modeled numerically and compared with the existing EDRP.

ACKNOWLEDGEMENTS

Muhammad Ali, Zulfiqar Ali Zaidi, Qamar Zia, Kamal Haider, Amjad Ullah, Muhammad Asif The special thank goes to Dr. Imdad and Mr. Waheed-urRehaman who supported us in providing help in the preparation regarding to this article. We are also thankful to Mr. Istiaq Khan and Mr. Abdul Rehman as they have made a great contribution throughout our work especially in editing the text and paper formatting.

REFERENCES

1. J. Pelletier, "ZigBee Powered Smart Grids Coming to a Home near You?" 2009.
2. Z. Md. Fadlullah, M. M. Fouda, N. Kato, A. Takeuchi, N. Iwasaki and Y. Nozaki, "Toward Intelligent Machine- to-Machine Communications in Smart Grid," IEEE Communications Magazine, vol. 49, no. 4, 2011, pp. 60-65.
3. M. H. Albadi and E. F. El-Saadany, "A Summary of Demand Response in Electricity Markets," Electric Power Systems Research, vol. 78, 2008, pp. 1989-1996. Hdoi:10.1016/j.epsr.2008.04.002
4. US Department of Energy, "Benefits of Demand Response in Electricity Markets and Recommendations for Achieving Them: A Report to the United States Congress Pursuant to Section 1252 of the Energy Policy Act of 2005," 2006.
5. J. J. Conti, P. D. Holtberg, J. A. Beamon, A. M. Schaal, G. E. Sweetnam and A. S. Kydes, "Annual Energy Outlook with Projections to 2035, Report of US Energy Information Administration (EIA)," 2010. http://www.eia.doe.gov
6. Charles River Associate, "Primes on Demand Side Management with an emphasis on Price Responsive Programs," Report Prepared for the World Bank, Washington DC. http://www.worldbank.org
7. G. T. Bellarmine, "Load Management Techniques," Proceedings of the IEEE, Nashville, 7-9 April 2000, p. 139.
8. G. T. Bellarmine and N. S. S. Arokiaswamy, "Load Management," Wiley Encyclopedia of Electrical and Electronics Engineering, Vol. 11, 1999, 482-494.
9. M. A. A. Pedrasa, T. D. Spooner and I. F. MacGill, "Coordinated Scheduling of Residential Distributed Energy Resources to Optimize Smart Home Energy Services," IEEE Transactions on Smart Grid, vol. 1, no. 2, 2010, pp. 134-143. Hdoi:10.1109/TSG.2010.2053053
10. A.-H. Mohsenian-Rad, V. W. S. Wong, J. Jatskevich and R. Schober, "Optimal and Autonomous Incentive-Based Energy Consumption Scheduling Algorithm for Smart Grid," IEEE Transactions on Smart Grid, Gaithersburg, 19-21 January 2010, pp. 1-6.
11. M. Erol-Kantarci, "Wireless Sensor Networks for Cost- Efficient Residential Energy Management in the Smart Grid," IEEE Transactions on Smart Grid, vol. 2, no. 2, 2011, pp. 314-325.

12. A.-H. Mohsenian-Rad and A. Leon-Garcia, "Optimal Residential Load Control with Price Prediction in RealTime Electricity Pricing Environments," IEEE Transactions on Smart Grid, vol. 1, no. 2, 2010, pp. 120-133. Hdoi:10.1109/TSG.2010.2055903
13. A. Molderink, V. Bakker, M. Bosman Johann, L. Hurink and G. J. M. Smit, "Management and Control of Domestic Smart Grid Technology," IEEE Transactions on Smart Grid, vol. 1, no. 2, 2010, pp. 109-119. Hdoi:10.1109/TSG.2010.2055904
14. S. Tompros, N. Mouratidis, M. Draaijer, A. Foglar and H. Hrasnica, "Enabling Applicability of Energy Saving Applications on the Appliances of the Home Environment," IEEE Network, vol. 23, no. 6, 2009, pp. 8-16. Hdoi:10.1109/MNET.2009.5350347
15. E. Shayesteh, A. Yousefi, M. Parsa Moghaddam and M. K. Sheikh-El-Eslami, "ATC Enhancement Using Emergency Demand Response Program," IEEE Power Systems Conference and Exposition, Seattle, 15-18 March 2009, pp. 1-7.
16. M. H. Albadi and E. F. El-Saadany, "Demand Response in Electricity Markets: An Overview," IEEE PES General Meeting, Tampa, 24-28 June 2007, pp. 1-5.
17. US Department of Energy, "Benefits of Demand Response in Electricity Markets and Recommendation for Achieving Them," A Report to the US Congress, 2006.
18. R. Tyagi and J. W. Black, "Emergency Demand Response for Distribution System Contingencies," IEEE Transmission and Distribution Conference and Exposition, New Orleans, 19-22 April 2010, pp. 1-4.
19. S. Valero and M. Ortiz, "Methods for Customer and Demand Response Policies Selection in New Electricity Markets, Generation, Transmission & Distribution," IET Generation, Transmission & Distribution, Vol. 1, No. 1, 2007, pp. 104-110.
20. P. Moses and M. S. Moasum, "Load Management in Smart Grids Considering Harmonic Distortion and Transformer Detering," IEEE Innovative Smart Grid Technologies, Gaithersburg, 19-21 January 2010, pp. 1-7.

Chapter 17

Towards Implementation of Smart Grid: An Updated Review on Electrical Energy Storage Systems

Md Multan Biswas[1], Md Shafiul Azim[2], Tonmoy Kumar Saha[2], Umama Zobayer[3], Monalisa Chowdhury Urmi[3]

[1]Department of Electrical and Electronic Engineering, Bangladesh University of Engineering and Technology, Dhaka, Bangladesh

[2]Department of Electrical and Electronic Engineering, Khulna University of Engineering and Technology, Khulna, Bangladesh

[3]Department of Electrical and Electronic Engineering, Stamford University Bangladesh, Dhaka, Bangladesh

ABSTRACT

A smart grid will require, to greater or lesser degrees, advanced tools for planning and operation, broadly accepted communications platforms, smart sensors and controls, and real-time pricing. The smart grid has been described as something of an ecosystem with constantly communication, proactive,

Citation: M. Biswas, M. Azim, T. Saha, U. Zobayer and M. Urmi, "Towards Implementation of Smart Grid: An Updated Review on Electrical Energy Storage Systems," *Smart Grid and Renewable Energy*, Vol. 4 No. 1, 2013, pp. 122-132. doi: 10.4236/sgre.2013.41015.

and virtually self-aware. The use of smart grid has a lot of economical and environmental advantages; however it has a downside of instability and unpredictability introduced by distributed generation (DG) from renewable energy into the public electric systems. Variable energies such as solar and wind power have a lack of stability and to avoid short-term fluctuations in power supplied to the grid, a local storage subsystem could be used to provide higher quality and stability in the fed energy. Energy storage systems (ESSs) would be a facilitator of smart grid deployment and a "small amount" of storage would have a "great impact" on the future power grid. The smart grid, with its various superior communications and control features, would make it possible to integrate the potential application of widely dispersed battery storage systems as well other ESSs. This work deals with a detailed updated review on available ESSs applications in future smart power grids. It also highlights latest projects carried out on different ESSs throughout all around the world.

Keywords: Battery; Distributed Generation; Hybrid Energy Storage Systems; Power Quality; Smart Grid

INTRODUCTION

The world's electricity systems face a number of challenges, including aging infrastructure, continuous increase in demand, the integration of growing numbers of variable renewable energy DGs and electric and hybrid electric vehicles, the need to improve the security of power supply, and the need to lower carbon dioxide (CO_2) emissions [1-3]. These challenges must be addressed also with regard to each region's unique technical, economic, and commercial regulatory environment [4]. Smart grid technologies offer ways not just to meet these challenges but also to develop a sustainable energy supply that is more energy efficient and more affordable.

Compared to other industries, our electrical grid has been largely bypassed by technological innovation until relatively recently, owing to the fact that historically it has been heavily regulated and modeled to keep the lights on and costs low. Partly for this reason, its modernization by means of information-technology tools and techniques has been somewhat of a back-burner priority. Like the telecom and internet revolutions that preceded it, technology holds the key to the smart grid and its realization. The smart grid and the technologies embodied within it are an essential set of investments that will help bring our electric grid into the 21st century using megabytes of data to move megawatts of electricity more efficiently, reliably, and affordably [5-9].

The smart grid will:

- Provide power quality for the digital economy;
- Accommodate all generation and storage options;
- Enable new products, services, and markets;
- Optimize asset utilization and operate efficiently;
- Enable active participation by consumers;
- Anticipate and respond to system disturbances;
- Operate resiliently against attack and natural disaster.

Storage is perhaps the most important smart grid advanced component because of its key role in complementing renewable generation. With the proper amount and type of storage broadly deployed and optimally controlled, renewable generation can be transformed from an energy source into a dispatchable generation source [10-12]. And with the addition of energy storage, more wind and solar generation can be added to a typical power system that employs a large percentage of slow-response fossil and nuclear generation. It is feasible that, the penetration of renewables can be significantly above 20 percent with the addition of sufficient energy storage technologies [13].

This paper is organized as follows. Section 2 presents a brief description of smart grid technology with its essential features. The necessity and prospect of energy storage systems in future smart power grid are broadly discussed in Section 3. Finally, Section 4 elaborately reviews the available energy storage technologies that are considered for use in smart grid applications with updated technologies, which is followed by some concluding remarks in Section 5.

SMART GRID: TECHNOLOGY DESCRIPTION

Though there has been much debate over the exact definition of smart grid, it actually comprises a broad range of technology solutions that optimize the energy value chain. Depending on where and how a specific utility operates across that chain, it can benefit from deploying certain parts of a smart grid solution set. Smart grid is a large electricity network that uses digital and other advanced technologies to improve efficiency, reliability, and security of the electric system: from large generation, through the delivery systems to electricity consumers and a growing number of distributed-generation and energy storage resources [14-16]. Smart grids co-ordinate the needs and capabilities of all power generators, grid operators, end-users, and electricity market stakeholders to operate all parts of the system as efficiently as possible, minimising costs and environmental impacts while maximising system stability, reliability, and resilience. Smart grids are an evolving set of technologies that will be deployed at different rates in a variety of settings around the world, depending on local commercial attractiveness, compatibility with existing technologies, regulatory developments, and investment frameworks. Figure 1 demonstrates the evolutionary character of smart power network [4].

Advanced metering infrastructure (AMI) is an approach to integrating consumers based upon the development of open standards. It provides consumers with the ability to use electricity more efficiently and provides utilities with the ability to detect problems on their systems and operate them more effectively [17,18]. AMI enables consumer-friendly efficiency concepts like "Prices to Devices" to work like this: Assuming that energy is priced on what it costs in near real-time—price signals are relayed to "smart" home controllers or end-consumer devices like thermostats, washer/dryers, and refrigerators—the home's major energy-consumers [19-21]. The devices, in turn, process the information based on consumers' learned wishes and power accordingly. The house or office responds to the occupants, rather than vice-versa. Because this interaction occurs largely in the background with minimal human intervention, there's a dramatic savings on energy that would otherwise be consumed [14]. Far more than "smart meters," a fully-functioning smart grid will feature sensors throughout the transmission and distribution grid to collect data, realtime two-way communications to move that data between utilities and consumers, and the computing power necessary to make that intelligence actionable and transactive as shown in Figure 2. Indeed, only by bringing the tools, techniques, and technologies that enabled the internet to the utility and the electric grid is such a transformation possible [8].

Smart Grid's Principal Characteristics

A smart power grid brings the power of networked, interactive technologies into an electricity system, giving utilities and consumer's unprecedented control over energy use, improving power grid operations, and ultimately reducing costs to consumers. In brief, the main features of smart grids are [9,22-26]:

1) A smart grid accommodates not only large, centralised power plants, but also the growing array of customer-sited distributed energy resources.

2) Not all commercial enterprises, and certainly not all residential customers, need the same quality of power. A smart grid supplies varying grades (and prices) of power.

3) Correctly designed and operated markets efficiently create an opportunity for consumers to choose among competing services.

4) A smart grid applies the latest technologies to optimize the use of its assets. For example, optimized capacity can be attainable with dynamic ratings, which allow assets to be used at greater loads by continuously sensing and rating their capacities.

5) Provides resiliency to disturbances, attacks, and natural disasters. Self-healing actions result in reduced interruption of service to consumers and help service providers better manage the delivery infrastructure.

6) Consumers help balance supply and demand, and ensure reliability by modifying the way they use and purchase electricity.

ENERGY STORAGE—A KEY ENABLER OF SMART GRID

Liberalization of the power market and widespread use of distributed energy resources (DERs), in particular DG and ESSs, could enable smart grids to have a significant influence on electricity market prices and ancillary services [27]. As focused in [27], the charge/discharge of various ESSs can be controlled in order to guarantying proper applications.

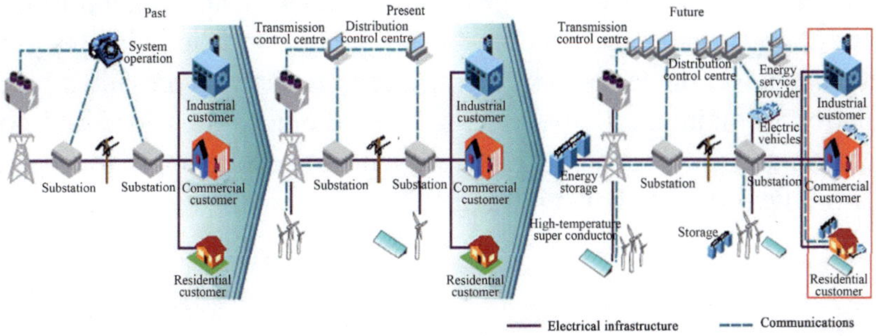

Figure 1: Smarter electricity systems.

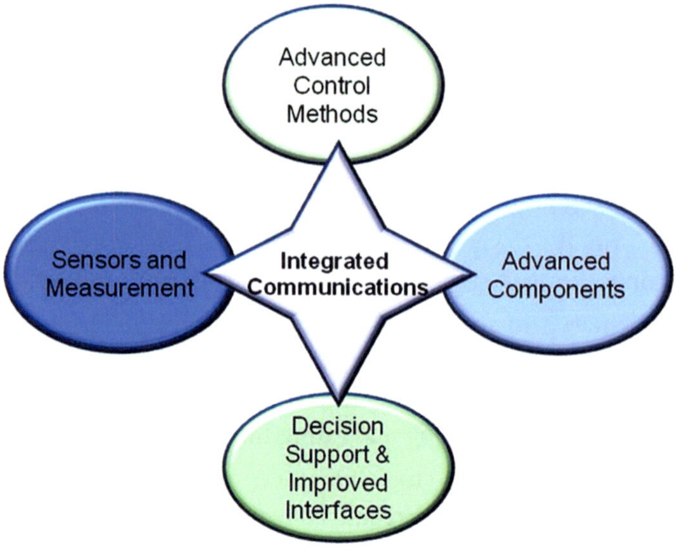

Figure 2: Advanced communication and control in smart grids.

Further, in modern power distribution systems, where a significant amount of the total electricity demand is met by renewable generation, ESSs can mitigate the uncertainties of energy sources (such as solar and wind) and can store the energy during high renewable production and/or low price periods, and deliver when either necessary or convenient [28-31]. Based on the ESSs technologies, in [27] the applications of ESSs are classified in instantaneous, short-, midand long-term. Instantaneous and short-term applications are involved in real time regulations, for example aiming at ancillary services provision or integration of electric drive vehicles batteries in the networks [32,33].

Energy storage systems improve the efficiency and reliability of the electric supply system by reducing the requirements for spinning reserves to meet peak power demands [34]. Storage devices can also provide frequency regulation to maintain the balance between the network's load and power generated, and they can achieve a more reliable power supply for high tech industrial facilities as shown in Figure 3.

Figure 3: Distributed energy storage in a Windfarm.

High voltage power electronics, such as switches, rectifiers, inverters, and controllers, allow electric power to be precisely and rapidly controlled to support long distance transmission [35,36]. Thus, energy storage and power electronics hold substantial promise for transforming the electric power industry [28]. This capability will allow the system to respond effectively to disturbances and operate more efficiently, thereby reducing the need for additional infrastructure.

AVAILABLE ENERGY STORAGE SYSTEMS

The adoption of smart power grid devices throughout utility networks will effect tremendous change in grid operations and usage of electricity over the next two decades. Increased deployment of energy storage devices in the distribution grid will help make this process happen more effectively and improve system performance. The energy storage systems (ESSs) applicable in power networks can be divided into two major categories [29] as shown in **Figure 4**. The first category being largescale storage systems that can be used in utility transmission applications. The second group includes small-scale storage systems sited at the consumer's premises. Some of the available energy storage systems are potentially discussed in the following sub-sections.

Battery

Battery is one or more electrochemical cells that convert chemical energy into electrical energy. Batteries have been using for energy storage purpose for over one-hundred years and possess some very important, unique, and desirable features. Battery energy storage systems (BESSs) are modular, quiet, and non-polluting [37]. Batteries are manufactured in a wide variety of capacities ranging from less than 100 watts to modular configurations of several megawatts. As a result, batteries can be used for various utility applications in the areas of generation, transmission and distribution, and customer service and they can be installed relatively quickly. Battery has convenient size and voltage characteristics. The operating principle of a typical battery is shown in Figure 5.

A grid-scale BESS consists of a battery bank, control system, power electronics interfacing circuit for ac-dc power conversion, protective circuitry, and a transformer to convert the BESS output to the desired transmission or distribution system voltage level [38]. The battery bank consists of numerous batteries connected in a combination series-parallel configuration to provide the desired energy and power capabilities for the application. The use of Lead Acid batteries for energy storage dates back to mid-1800s. There have been developed several new battery technologies to store more energy, last longer and less cost than the Lead Acid battery.

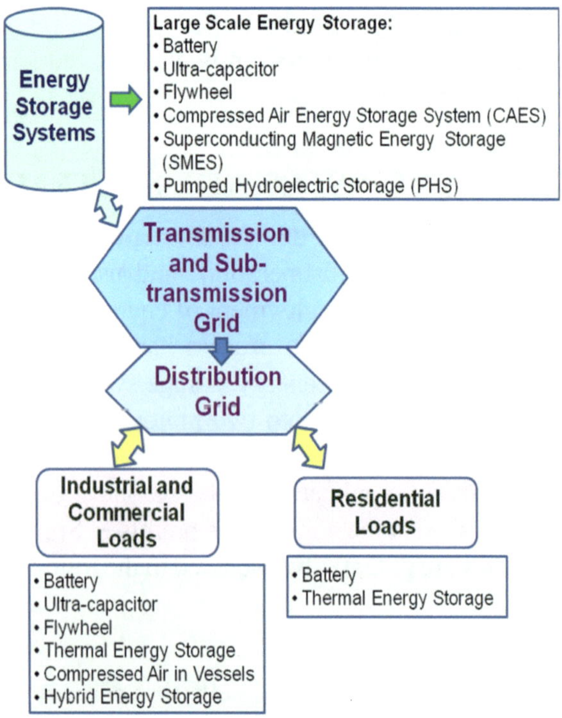

Figure 4: Energy storage systems in smart power networks.

Battery technologies differ widely in terms of their energy and power densities, energy efficiencies, cycle-life, availability, and operating conditions. Some of these new battery technologies are Lithium Ion, Lithium Polymer, Nickel Metal Hydride (Ni-MH), Vanadium Redox (VRB), Nickel Cadmium (Ni-Cd), Sodium Sulfur (NaS), and Zinc Bromide [39-41]. Table **1** summarizes the characteristic parameters of different batteries [27,28, 42-44].

Ultra-Capacitors

Ultra-capacitors (UCs), also known as super-capacitors, or electric double-layer capacitors (EDLCs), store energy in the electrical double layer at an electrode/electrolyte interface. There are no chemical reactions involved in the UC's energy storage mechanism. The main parts of an electrochemical capacitor are electrodes and electrolyte. The electrical energy E accumulated in ultracapacitors is related to the capacitance C or the stored charges Q and voltage V by following formula:

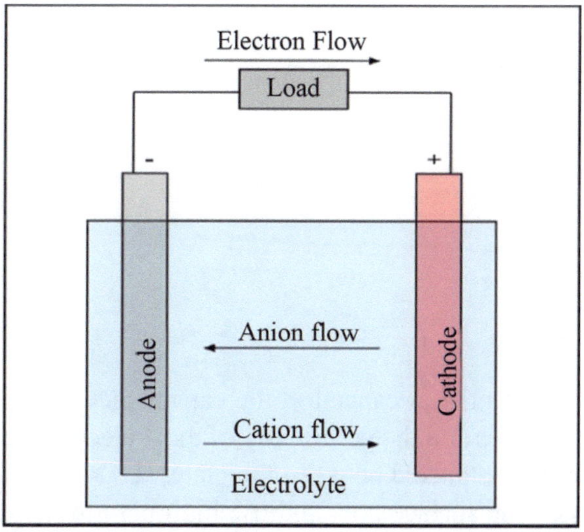

Figure 5: Working principle of a battery.

Table 1: Comparison of different types of batteries

Type	Energy Density (Wh/kg)	Energy Efficiency (%)	Power Density (W/Kg)	Cycle Life (Cycles)	Self Discharge (%/Month)
Lead-Acid	30 - 40	70 - 90	180	200 - 2000	3 - 4
Li-Ion	100 - 250	75 - 90	1800	500 - 2000	5 - 10
Li Polymer	130 - 200	70	3000	>1200	4 - 8
Ni-MH	30 - 80	70	250 - 1000	500 - 100	30
Ni-Cd	40 - 60	60 - 90	140 - 180	500 - 2000	10 - 15
NaS	150	80 - 90	120 - 150	2500	-
VRB	25 - 40	80	100 - 150	>16,000	<1
Zinc Bromide	70	70	-	1000	-

$$E = \frac{1}{2}CV^2 = \frac{1}{2}QV \quad (1)$$

And,

$$Q = CV = \frac{A\varepsilon}{d}V \quad (2)$$

As for the conventional capacitor, the capacitance C is proportional to the area A of the plates and the permitivity of the dielectric ε and is inversely proportional to the distance d between the plates. UCs are designed to have a very high electrode surface area and use high permitivity dielectric. Due to the high permeability and close proximity of the electrodes, UCs have a low voltage withstand capability (usually 2 - 3 V) [30]. The electrode surface area is maximized by using porous carbon as the current collector, allowing a relatively large amount of energy to be stored at the collector surface. Therefore, UCs attain very high capacitance ratings. Larger UCs have capacities up to 5000 farads [45].

Ultracapacitors store energy by physically separating unlike charges. UCs have a long cycle life due to the fact that there are no chemical changes on the electrodes ideally in normal operation and UCs have superior efficiency. UCs also provide exceptional power density, since the charges are physically stored on the electrodes. Conversely, energy density is low since the electrons are not bound by chemical reactions [30]. The UC is temperature resistant with an operating range between −40°C to +65°C and is also shock and vibration resistant [46, 47].

In order to be a viable alternative in a large scale energy storage system they will need to be able to handle multiple kV. The ability of modular, non-polluting, quiet, quick charge and discharge capability, long life (10 to 12 years), and very high cycle life makes the UC a very desirable energy storage device. They can be use for short term ride through capabilities as well as voltage regulation, frequency control and other power quality issues [47,48]. UCs are currently available in many sizes. Two 3000 F capacitors are shown in Figure 6. There is even a startup company that claims to be able to create ultra capacitors with higher energy densities than lead-acid, Nickel metal hydride and even lithium ion batteries [49]. There have also been advances in the design of the ultracapacitors using nano-tube technology to improve the surface area of the capacitor. This "nano-tube ultracapacitor" would improve the ultracapacitor's energy density to be compatible with again that of a chemical battery [46,50,51].

Flywheel

During the past decade, flywheel energy storage systems (FESSs) have been rediscovered by the power industry due to their advantages in comparison with other energy storage systems. FESSs have found an important technical role on the application of enhancing the electric power quality, grid voltage and frequency support, and unbalanced load compensation. By virtue of their high dynamics, long lifetime, and good efficiency, FESSs are well suited for short-term storage systems [52-54].

FESS consists of a flywheel coupled to permanent magnets synchronous machine (PMSM) as shown in Figure 7. Flywheels store energy in the form of momentum in a rotating wheel or cylinder. A FESS stores energy through accelerating a rotor up to a high rate of speed and maintaining the energy in the system as inertial energy. Advanced composite materials are sometimes used for the rotor to lower its weight while allowing for the extremely high speeds. The flywheel releases the energy by reversing the process and using the motor as a generator. As the flywheel releases its stored energy, the flywheel's rotor slows until it is fully discharged.

Figure 6: 3000 farad ultracapacitors.

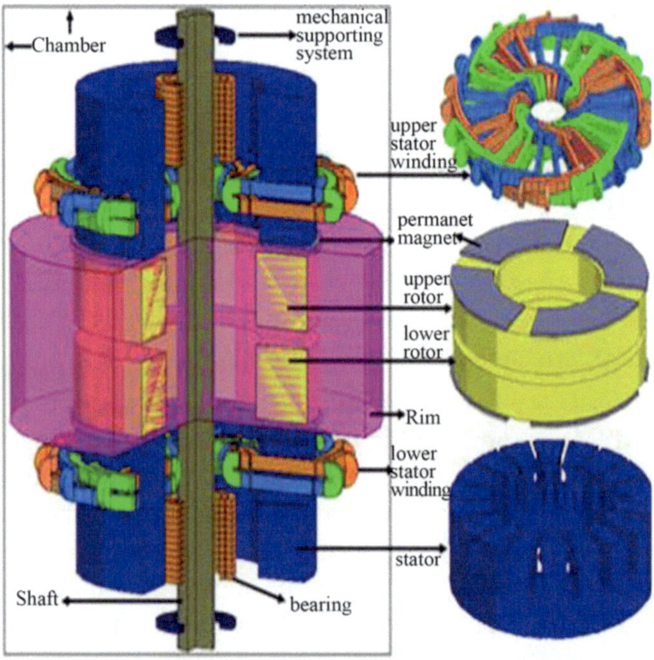

Figure 7: Flywheel energy storage system.

The energy that can be stored depends on its rotational velocity ω and moment of inertia I, as in (3). They have high power density and more energy can be stored if the flywheel rotates at higher rotational speed [55]. Power electronics are used to ensure that output voltage has appropriate amplitude and frequency characteristics.

$$E = \frac{1}{2} I \omega^2 \qquad (3)$$

Flywheel energy storage systems using superconducting magnetic bearing (SMB) together with a permanent magnet bearing (PMB) is one of the most promising electro-mechanical energy storage systems consisting long life, high energy density, high efficiency, with no pollution or toxic material disposal problems, and low rotational loss by non-contact superconductor bearing [55- 57].

Flywheels have seen most commercial success targeted for power delivery capabilities typically in the 150 kW-1MW range [50,58]. Recently, a 20 MW FESS plant has been successfully established in Stephentown, New York by Beacon Power Corp. under a pilot project of the Department of Energy (DOE), New York State Energy Research and Development Authority (NYSERDA) and currently 40 MW FESS plant project is under development [59].

Superconducting Magnetic Energy Storage (SMES)

Superconducting magnetic energy storage (SMES) systems store energy in the magnetic field produced by current flowing through a superconducting coil. The SMES principle is based on inductive energy storage in the magnetic field produced by current flowing through a superconducting coil.

A SMES system consists of four major subsystems [60], plus miscellaneous equipment for system control, data collection, and so on. The major subsystems are: 1) superconducting coil with magnet (SCM); 2) power conditioning system (PCS) that controls the flow of current into and out of the coil to charge and discharge the SMES; 3) cryogenic system (CS) that maintains the coil at a low enough temperature to maintain superconductivity; and 4) the control unit (CU), as shown in Figure 8.

For a SMES system, the inductively stored energy (E) and the rated power (P) are commonly the given specifications for SMES devices, and can be expressed as follows:

$$E = \frac{1}{2} L I^2 \; ; P = \frac{dE}{dt} = L I \frac{dI}{dt} = V I \tag{4}$$

where L is the inductance of the coil, I is the dc current flowing through the coil, and V is the voltage across the coil. During SMES operation, the magnet coils have to remain in the superconducting status. A refrigerator in the cryogenic system maintains the required temperature for proper superconducting operation [61].

Among the different variants of flexible alternating current transmission system (FACTS) devices and ESSs currently available, static synchronous compensators (STATCOM) integrated with SMES has been proposed as the most adequate for participating of the primary frequency control because of SMES has high efficiency (95% to 98% [61]) and rapid response to power demand [62-64]. Depending on the control loop of its power conversion unit and switching characteristics, the SMES system can respond very rapidly (MWs/milliseconds). The ability of injecting/absorbing real or reactive power can increase the effectiveness of the control, and enhance system reliability and availability. Comparing with other storage technologies, the SMES technology has a unique advantage in two types of applications: Power system transmission control and stabilization, and power quality improvement [65].

Several SMESs in the range of kWh to MWh scale have been already implemented for compensation of load/ generation fluctuation as well as energy storage [66]. Recently, 10 MVA/20 MJ SMES prototype has been tested at an actual power system including hydro power generators in order to compensate the fluctuating power load for a metal rolling factory [67].

Compressed Air Energy Storage System (CAES)

Compressed air energy storage system (CAES) is a hybrid technology of long term power storage and generation. A CAES plant mainly consists of 1) compressor train; 2) motor-generator unit; 3) gas turbine; and 4) underground compressed air storage tank (see Figure 9). During low-cost off-peak load periods, air is compressed and stored in large underground salt caverns [68]. Upon demand, the process is reversed; the compressed air is returned to the surface; this air is used to burn natural gas in the combustion chambers. The resulting combustion gas is then expanded in the two-stage gas turbine to spin the generator and produce electricity.

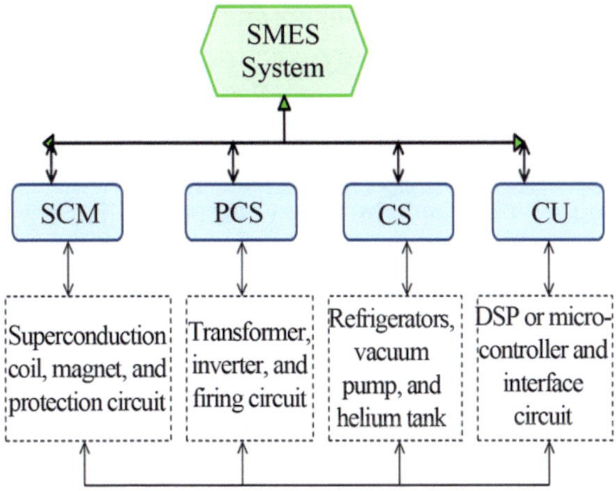

Figure 8: Block diagram of an SMES system.

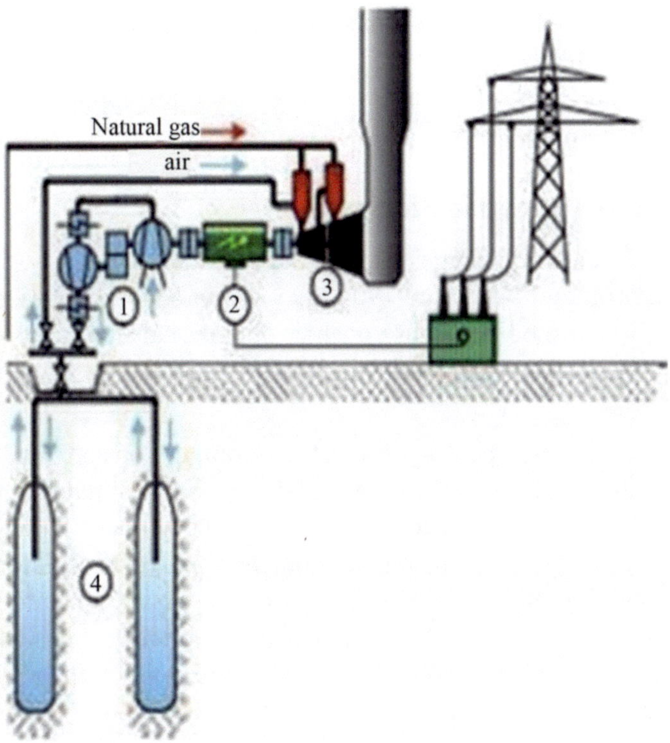

Figure 9: Major components of a CAES plant.

CAES can be placed near consumers and in large, small, and micro scale smart power networks. As in many countries, the potentials of installing new hydrostorage plants with pumping facilities are limited and many other energy storage devices are far from being economic, CAES may be an attractive investment opportunity for such purposes. The first such plant was built in Huntorf, Germany, in 1978 with a capacity of 290 MW [68]. In 1991, the first U.S. CAES facility was built in McIntosh, Alabama, by the Alabama Electric Cooperative and EPRI, and has a capacity rating of 110 MW [69]. Liu et al. [70] has proposed a novel hybridfuel CAES system for China. The design is based on using standard, industry proven equipment components to deliver a reliable and economic compressed air energy cycle. The resulting CAES plant consists of 410 MW generation, 205 MW of compression, and 2050 MW of storage.

In the past few years, research has been conducted to improve the efficiency of the turbines and heat transfer mechanisms used to pump and retrieve compressed air [71-73]. In an adiabatic CAES, the air's heat energy is stored separately and recovered before the compressed air is expanded in an air turbine. Such plants are currently under development and promise higher efficiencies and zero direct CO_2-emissions [71].

Pumped Hydroelectric Storage

Pumped hydroelectric storage (PHS) is the oldest kind of large-scale energy storage technology [74]. Since 1904, they are in active operation and new ones are still being built because of their operational flexibility and ability to provide rapid response to changes in system loading or spot price of electricity.

Conventional pumped hydroelectric storage consists of two large reservoirs, one is located at base level and the other is situated at a different elevation. Water is pumped to the upper reservoir where it can be stored as potential energy. Upon demand, water is released back into the lower reservoir, passing through hydraulic turbines which generate electrical power as high as 1000 MW. Pumped hydroelectric storage has huge energy and power capacity. Recently [75-77] examined the impact of pumped storage units together with large renewable penetration. They focused on reducing system operating costs and maximizing usage of renewable energy based on unit commitments and dispatches. Currently, efforts aimed at increasing the use of pumped hydro storage are focused on the development of underground facilities. Modern pumped storage plants are often designed to have fast start, load ramping and unloading capabilities. They can respond to load changes within seconds.

Hybrid Energy Storage System

The selection of the energy storage system for a particular application in power grid sometimes depends on a suitable combination of the power and energy ratings, energy, power density, cost, weight, volume, and operating temperature etc of ESSs. These requirements may not be achievable from a single energy storage source. To implement such optimal applications, hybrid energy storage devices (HESDs) have been proposed. In a HESD, two or more different energy storage systems with complementary characteristics can be combined together electronically. For future grid application proposed HESDs are listed next, with the energy-supplying device listed first, followed by the power-supplying device:

- battery and ultracapacitor [78-81];
- battery and flywheels [82];
- CAES and battery or ultracapacitor [83];
- battery and SMES [84].

Figure 10 demonstrates the relationship between power operational range and discharge time at rated power for various energy storage systems such as battery, ultracapacitors, flywheel, SMES, CAES, and pumped hydro. Table 2 summarizes the characteristic parameters of different energy storage technologies [30,42-44,59,83].

CONCLUDING REMARKS

In this paper the potential application of different ESSs and their prospect have been discussed and analyzed in detail for future smart power grids. Energy can be stored both for short and long durations. Various storage systems are available for these purposes. The importance of storage systems in electricity grids is finally receiving the attention of system planners as more storage options participation of storage is increasing. The design of smart Md Multan Biswas, Md Shafiul Azim, Tonmoy Kumar Saha, Umama Zobayer, Monalisa Chowdhury Urmi grids in the future will take advantage of storage to deal with more dynamic loads and sources.

Table 2: Comparison of various energy storage systems

Type	Energy Density (Wh/kg)	Energy Efficiency (%)	Power Density (W/Kg)	Life Time (Cycles)	Self Discharge (%/Month)	Environmental Effect
Battery	25 - 250	60 - 90	100 - 3000	150 - 3000	3 - 30	Toxic
Ultra Capacitor	<50	95	4000	10 k - 100 k	High	Benign
Flywheel	100 - 130	95	1000	>125,000	High	Benign
SMES	30 - 80	95	Very High	-	Negligible	Benign
CAES	10 - 30	50	Fair	40 Years	Negligible	Benign
PHS	0.3	65 - 80	-	75 Years	Negligible	Benign

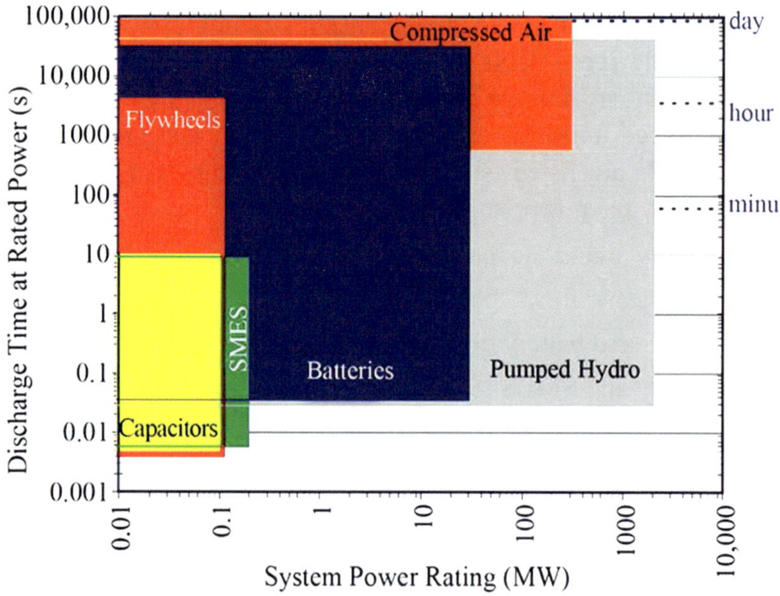

Figure 10: The relationship between power rating and discharge time for different storage technologies.

As market rules are adjusted to take advantage of the benefits of bulk and distributed storage devices, the overall capabilities and reliability of more complex electricity networks should continue to improve as fully integrated smart grids.

REFERENCES

1. K. Moslehi and R. Kumar, "A Reliability Perspective of the Smart Grid," IEEE Transactions on Smart Grid, Vol. 1, No. 1, 2010, pp. 57-64. doi:10.1109/TSG.2010.2046346
2. P. Zhang, F. Li and N. Bhatt, "Next-Generation Monitoring, Analysis, and Control for the Future Smart Control Center," IEEE Transactions on Smart Grid, Vol. 1, No. 2, 2010, pp. 186-192. doi:10.1109/TSG.2010.2053855
3. S. M. Amin and B. F. Wollenberg, "Toward a Smart Grid: Power Delivery for the 21st Century," IEEE Power Energy Magazine, Vol. 3, No. 5, 2005, pp. 34-41. doi:10.1109/MPAE.2005.1507024
4. International Energy Agency, "Technology Roadmap: Smart Grids," 2011. http://www.iea.org/publications/freepublications/publication/name,3972,en.html
5. US Department of Energy, "How the Smart Grid Promotes a Greener Future," 2010. http://energy.gov/oe/downloads/how-smart-grid-promotes-greener-future
6. H. Gharavi and R. Ghafurian, "Smart Grid: The Electric Energy System of the Future," IEEE Proceedings, Vol. 99, No. 6, 2011, pp. 917-921.
7. J. Xia and Y. Wang, "Secure Key Distribution for the Smart Grid," IEEE Transactions on Smart Grid, Vol. 3, No. 3, 2012, pp. 1437-1443.
8. A. Molderink, V. Bakker, M. G. C. Bosman, J. L. Hurink and G. J. M. Smit, "Management and Control of Domestic Smart Grid Technology," IEEE Transactions on Smart Grid, Vol. 1, No. 2, 2010, pp. 109-119. doi :10.1109/TSG.2010.2055904
9. A. Bose, "Smart Transmission Grid Applications and Their Supporting Infrastructure," IEEE Transactions on Smart Grid, Vol. 1, No. 1, 2010, pp. 11-19. doi:10.1109/TSG.2010.2044899
10. B. P. Roberts and C. Sandberg, "The Role of Energy Storage in Development of Smart Grids," IEEE Proceedings, Vol. 99, No. 6, 2011, pp. 1139-1144. doi:10.1109/JPROC.2011.2116752
11. M. S. Whittingham, "History, Evolution, and Future Status of Energy Storage," IEEE Proceedings, Vol. 100, No. , 2012, pp. 1518-1534.
12. G. D. Rodriguez, "A Utility Perspective of the Role of Energy Storage in the Smart Grid," 2010 IEEE Power and Energy Society General Meeting, Minneapolis, 25-29 July 2010, pp. 1-2. doi:10.1109/PES.2010.5589870

13. US Department of Energy, "Energy Storage—A Key Enabler of the Smart Grid," 2009. http://www.netl.doe.gov/smartgrid/refshelf.html#White%20Papers
14. US Department of Energy, "The Smart Grid: An Introduction, 2008. http://energy.gov/oe/downloads/smart-grid-introduction-0
15. F. Li, W. Qiao, H. Sun, H. Wan, J. Wang, et al., "Smart Transmission Grid: Vision and Framework," IEEE Transactions on Smart Grid, Vol. 1, No. 2, 2010, pp. 168-177. doi:10.1109/TSG.2010.2053726
16. M. D. Hopkins, A. Pahwa and T. Easton, "Intelligent Dispatch for Distributed Renewable Resources," IEEE Transactions on Smart Grid, Vol. 3, No. 2, 2012, pp. 1047- 1054. doi:10.1109/TSG.2012.2190946
17. S. S. S. R. Depuru, L. Wang, V. Devabhaktuni and N. Gudi, "Smart Meters for Power Grid—Challenges, Issues, Advantages and Status," 2011 IEEE/PES Power Systems Conference and Exposition (PSCE), Phoenix, 20-23 March 2011, pp. 1-7. doi:10.1109/PSCE.2011.5772451
18. D. Rua, D. Issicaba, F. J. Soares, P. M. R. Almeida, R. J. Rei and J. A. P. Lopes, "Advanced Metering Infrastructure Functionalities for Electric Mobility," 2010 IEEE PES Innovative Smart Grid Technologies Conference Europe (ISGT Europe), Porto, 11-13 October 2010, pp. 1-7.
19. H. Suleiman, K. A. Ahmed, N. Zafar, et al., "Inter-Domain Analysis of Smart Grid Domain Dependencies Using Domain-Link Matrices," IEEE Transactions on Smart Grid, Vol. 3, No. 2, 2012, pp. 692-709. doi:10.1109/TSG.2011.2176151
20. Microsoft Power and Utilities, "Smart Energy Reference Architecture (SERA)," 2009.
21. A. P. S. Meliopoulos, G. Cokkinides, R. Huang, E. Farantatos, S. Choi, et al., "Smart Grid Technologies for Autonomous Operation and Control," IEEE Transactions on Smart Grid, Vol. 2, No. 1, 2011, pp. 1-10. doi:10.1109/TSG.2010.2091656
22. A. Ipakchi and F. Albuyeh, "Grid of the Future," IEEE Power Energy Magazine, Vol. 7, No. 2, 2009, pp. 52-62. doi:10.1109/MPE.2008.931384
23. R. A. F. Currie, G. W. Ault, C. E. T. Foote, N. M. McNeill, et al., "Smarter Ways to Provide Grid Connections for Renewable Generators," IEEE PES General Meeting, Minneapolis, 25-29 July 2010, pp. 1-6.

24. Q. Pang, H. Gao and M. Xiang, "Design of Intelligent Terminal Unit for Smart Distribution Grid," Proceedings of the IEEE China International Conference on Electricity Distribution (CICED), Nanjing, 13-16 September 2010, pp. 1-6.
25. F. Rahimi and A. Ipakchi, "Demand Response as a Market Resource under the Smart Grid Paradigm," IEEE Transactions on Smart Grid, Vol. 1, No. 1, 2010, pp. 82-88. doi:10.1109/TSG.2010.2045906
26. E. Santacana, G. Rackliffe, L. Tang and X. Feng, "Getting Smart," IEEE Power Energy Magazine, Vol. 8, No. 2, 2010, pp. 41-48. doi:10.1109/MPE.2009.935557
27. K. C. Divya and J. Østergaard, "Battery Energy Storage Technology for Power Systems: An Overview," Electric Power System Research, Vol. 79, No. 4, 2009, pp. 511- 520. doi:10.1016/j.epsr.2008.09.017
28. M. M. Chowdhury, M. E. Haque, M. Aktarujjaman, M. Negnevitsky and A. Gargoom, "Grid Integration Impacts and Energy Storage Systems for Wind Energy Applications—A Review," Proceedings of the IEEE PES General Meeting, San Diego, 24-29 July 2011, pp. 1-8.
29. S. C. Smith, P. K. Sen and B. Kroposki, "Advancement of Energy Storage Devices and Applications in Electrical Power System," IEEE PES General Meeting, Pittsburgh, 20-24 July 2008, pp. 1-8.
30. S. Vazquez, S. M. Lukic, E. Galvan, L. G. Franquelo, et al., "Energy Storage Systems for Transport and Grid Applications," IEEE Transactions on Industrial Electronics, Vol. 57, No. 12, 2010, pp. 3881-3895. doi:10.1109/TIE.2010.2076414
31. A. A. Thatte and L. Xie, "Towards a Unified Operational Value Index of Energy Storage in Smart Grid Environment," IEEE Transactions on Smart Grid, Vol. 3, No. 3, 2012, pp. 1418-1426.
32. W. Kempton and J. Tomic, "Vehicle-to-Grid Power Fundamentals: Calculating Capacity and Net Revenue," Journal of Power Sources, Vol. 144, No. 1, 2005, pp. 268-279. doi:10.1016/j.jpowsour.2004.12.025
33. A. Bracale, P. Caramia and D. Proto, "Optimal Operation of Smart Grids Including Distributed Generation Units and Plug in Vehicles," International Conference on Renewable Energies Power Quality (ICREPQ'11), Las Palmas de Gran Canaria, 13-15 April 2011, pp. 553-558.
34. M. M. Biswas, K. K. Das, I. A. Baqee, M. A. H. Sadi and H. M. S. Forhad, "Prospects of Renewable Energy and Energy Storage Systems

in Bangladesh and Developing Economics," Global Journal of Researches in Engineering (GJRE), Vol. 11, No. 5, 2011, pp. 23-31.

35. Y. M. Cheng, Y. C. Liu, S. C. Hung and C. S. Cheng, "Multi-Input Inverter for Grid-Connected Hybrid PV/ Wind Power System," IEEE Transactions on Power Electronics, Vol. 22, No. 3, 2007, pp. 1070-1076. doi:10.1109/TPEL.2007.897117
36. J. M. Carrasco, L. G. Franquelo, J. T. Bialasiewicz, E. Galvan, R. C. P. Guisado, et al., "Power Electronic Systems for the Grid Integration of Renewable Energy Sources: A Survey," IEEE Transactions on Power Electronics, Vol. 53, No 4, 2006, pp. 1002-1016. doi:10.1109/TIE.2006.878356
37. R. B. Schainker, "Executive Overview: Energy Storage Options for a Sustainable Energy Future," IEEE PES General Meeting, Palo Alto, June 2004, Vol. 2, pp. 2309- 2314.
38. C. A. Hill, M. C. Such, D. Chen, J. Gonzalez, et al., "Battery Energy Storage for Enabling Integration of Distributed Solar Power Generation," IEEE Transactions on Smart Grid, Vol. 3, No. 2, 2012, pp. 850-857. doi:10.1109/TSG.2012.2190113
39. J. McDowall, "Nickel-Cadmium Batteries for Energy Storage Applications," The 14th Annual Battery Conference on Applications and Advances, Long Beach, 12-15 January 1999, pp. 303-308.
40. J. P. Barton and D. G. Infield, "Energy Storage and Its Use with Intermittent Renewable Energy," IEEE Transactions on Energy Conversion, Vol. 19, No. 2, 2004, pp. 441-448. doi:10.1109/TEC.2003.822305
41. P. C. Symons, "Opportunities for Energy Storage in Stressed Electrical Supply Systems," IEEE PES Summer Meeting, Vancouver, July 2001, Vol. 1, pp. 448-449.
42. D. Y. Goswami and F. Kreith, "Energy Conversion," CRC Press, Boca Raton, 2007. doi:10.1201/9781420044324
43. A. A. H. Hussein, S. Harb, N. Kutkut, J. Shen and I. Batarseh, "Design Considerations for Distributed MicroSto- rage Systems in Residential Applications," 32nd International Telecommunications Energy Conference (INTELEC), Orlando, 6-10 June 2010, pp. 1-6.
44. B. R. Williams and T. Hennessy, "Energy Oasis [Vanadium Redox Battery System in Power Distribution Application]," IEEE Power Engineer, Vol. 19, No. 1, 2005, pp. 28-31. doi:10.1049/pe:20050105

45. Nesscap Ultracapacitors Products, "(EDLC) Electric Double-Layer Capacitor," 2012. http://www.nesscap.com/product/edlc_large1.jsp
46. D. Halber, "Researchers Fired Up over New Battery," MIT Tech Talk, Vol. 50, No. 16, 2006, pp. 1-5. http://web.mit.edu/newsoffice/2006/batteries-0208.html
47. Maxwell Technologies, "An Overview of Ultracapacitor Technology," 2012. http://www.maxwell.com/ultracapacitors/
48. R. D. Weir and C. W. Nelson, "Utilization of Poly (Ethylene Terephthalate) Plastic and Composition-Modified Barium Titanate Powders in a Matrix that Allows Polarization and the Use of Integrated-Circuit Technologies for the Production of Lightweight Ultrahigh Electrical Energy Storage Units (EESU)," US Patent No. 7466536, 2008.
49. H. Pan, J. Li and Y. P. Feng, "Carbon Nanotubes for Supercapacitor," Nanoscale Research Letters, Vol. 5, No. 3, 2010, pp. 654-668. doi:10.1007/s11671-009-9508-2
50. H. Zhang, G. P. Cao, Y. S. Yang and Z. N. Gu, "Comparison between Electrochemical Properties of Aligned Carbon Nanotube Array and Entangled Carbon Nanotube Electrodes," Journal of Electrochemical Society, Vol. 155, No. 2, 2008, pp. 19-22. doi:10.1149/1.2811864
51. R. Hebner, J. Beno and A. Walls, "Flywheel Batteries Come Around Again," IEEE Spectrum, Vol. 39, No. 4, 2002, pp. 46-51. doi:10.1109/6.993788
52. G. Cimuca, C. Saudemont, B. Robyns and M. M. Radulescu, "Control and Performance Evaluation of a Flywheel Energy Storage System Associated to a VariableSpeed Wind Generator," IEEE Transactions on Industrial Electronics, Vol. 53, No. 4, 2006, pp. 1074-1085. doi:10.1109/TIE.2006.878326
53. G. Cimuca, S. Breban, M. Radulescu, C. Saudemont, et al., "Design and Control Strategies of an Induction-Machine-Based Flywheel Energy Storage System Associated to a Variable-Speed Wind Generator," IEEE Transactions on Energy Conversion, Vol. 25, No. 2, 2010, pp. 526-534. doi:10.1109/TEC.2010.2045925
54. M. Subkhan and M. Komori, "New Concept for Flywheel Energy Storage System Using SMB and PMB," IEEE Transactions on Applied Superconductivity, Vol. 21, No. 3, 2011, pp. 1485-1488. doi:10.1109/TASC.2010.2098470

55. J. Lee, S. Jeong, Y. H. Han and B. J. Park, "Concept of Cold Energy Storage for Superconducting Flywheel Energy Storage System," IEEE Transactions on Applied Superconductivity, Vol. 21, No. 3, 2011, pp. 2221-2224. doi:10.1109/TASC.2010.2094177
56. T. Ichihara, K. Matsunaga, M. Kita, I. Hirabayashi, et al., "Application of Superconducting Magnetic Bearings to a 10 kWh-Class Flywheel Energy Storage System," IEEE Transactions on Applied Superconductivity, Vol. 15, No. 2, 2005, pp. 2245-2248. doi:10.1109/TASC.2005.849622
57. N. Hamsic, A. Schmelter, A. Mohd, et al., "Stabilising the Grid Voltage and Frequency in Isolated Power Systems Using a Flywheel Energy Storage System," The Great Wall World Renewable Energy Forum, Beijing, 23-27 October 2006, pp. 1-6.
58. E. Ortjohann, N. Hamsic, A. Schmelter, et al., "Increasing Renewable Energy Penetration in Isolated Grids Using a Flywheel Energy Storage System," International Conference on Power Engineering, Energy and Electrical Drives, Setubal, 12-14 April 2007, pp. 195-200.
59. M. Lazarewicz and J. Judson, "Performance of First 20 MW Commercial Flywheel Frequency Regulation Plant," ESA 2011 Annual Meeting, Beacon Power Corporation, San Jose, June 2011.
60. D. Sutanto and K. W. E. Cheng, "Superconducting Magnetic Energy Storage Systems for Power System Applications," International Conference on Applied Superconductivity and Electromagnetic Devices, Chengdu, 25-27 September 2009, pp. 377-380.
61. IEEE Task Force and T&D Committee, "Detailed Modeling of Superconducting Magnetic Energy Storage (SMES) System," IEEE Transactions on Power Delivery, Vol. 21, No. 2, 2006, pp. 699-710. doi:10.1109/TPWRD.2005.864075
62. V. Karasik, K. Dixon, C. Weber, B. Batchelder, G. Campbell and P. Ribeiro, "SMES for Power Utility Applications: A Review of Technical and Cost Considerations," IEEE Transactions on Applied Superconductivity, Vol. 9, No. 2, 1999, pp. 541-546. doi:10.1109/77.783354
63. M. G. Molina and P. E. Mercado, "Comparative Evaluation of Performance of a STATCOM and SSSC both Integrated with SMES for Controlling the Power System Frequency," 2004 IEEE/PES Transmission and Distribution Conference and Exposition: Latin America, São Paulo, 8-11 November 2004, pp. 535-541.

64. M. G. Molina and P. E. Mercado, "New Energy Storage Devices for Applications on Frequency Control of the Power System Using FACTS Controllers," Proceedings of X ERLAC, Iguazú, 18-22 May 2003, pp. 1-6.
65. Y. Tatusuta, S. Koso, H. Abe, M. Urata, H. Ohsaki, et al., "Development of SMES for Power System Control," IEEE Transactions on Applied Superconductivity, Vol. 14, No. 2, 2004, pp. 693-698. doi:10.1109/TASC.2004.830033
66. S. Nomura, T. Shintomi, S. Akita, T. Nitta, R. Shimada and S. Meguro, "Technical and Cost Evaluation on SMES for Electric Power Compensation," IEEE Transactions on Applied Superconductivity, Vol. 20, No. 3, 2010, pp. 1373- 1378. doi:10.1109/TASC.2009.2039745
67. Q. Li, "Status and Future Prospect of SMES for Grid Applications," Advanced Microgrid Concepts and Technologies Workshops, Brookhaven National Laboratory, Washington DC, 2012.
68. F. Crotogino, K. U. Mohmeyer and R. Scharf, "Huntorf CAES: More than 20 Years of Successful Operation," The Solution Mining Research Institute Spring Meeting, Orlando, 23-25 April 2001.
69. Electric Power Research Institute, "Compressed Air Energy Storage: 1994," EPRI Brochure BR-102936.
70. W. Liu, Y. Yang, W. Zhang, G. Xu and Y. Wu, "A Novel Hybrid-Fuel Compressed Air Energy Storage System for China's Situation," Proceedings of the International Conference of Efficiency, Cost, Optimization, Simulation Environmental Impact Energy Systems, Perugia, 26-29 June 2012, pp. 1-16.
71. C. Bullough, C. Gatzen, C. Jakiel, M. Koller, et al., "Advanced Adiabatic Compressed Air Energy Storage for the Integration of Wind Energy," Proceedings of European Wind Energy Conference, EWEC, London, 22-25 November 2004, pp. 1-8.
72. C. Linnemann and M. W. Coney, "The Isoengine: Realization of a High-Efficiency Power Cycle Based on Isothermal Compression," International Journal of Energy Technology and Policy, Vol. 3, No. 1-2, 2005, pp. 66-84. doi:10.1504/IJETP.2005.006740
73. Ridge Energy Storage & Grid Services L. P., "The Economic Impact of CAES on Wind in TX, OK and NM," Texas State Energy Conservation Office, Austin, 2005.

74. P. Breeze, "Power Generation Technologies," Elsevier, Boston, 2005.
75. P. Brown, J. Lopes and M. Matos, "Optimization of Pumped Storage Capacity in an Isolated Power System with Large Renewable Penetration," IEEE Transactions on Power Systems, Vol. 23, No. 2, 2008, pp. 523-531. doi:10.1109/TPWRS.2008.919419
76. J. G. Gonzalez, R. D. L. Muela, L. Santos and A. Gonzalez, "Stochastic Joint Optimization of Wind Generation and Pumped-Storage Units in an Electricity Market," IEEE Transactions on Power Systems, Vol. 23, No. 2, 2008, pp. 460-468. doi:10.1109/TPWRS.2008.919430
77. A. Tuohy and M. O'Malley, "Impact of Pumped Storage on Power Systems with Increasing Wind Penetration," IEEE Power & Energy Society General Meeting, Calgary, 26-30 July 2009, pp. 1-8.
78. W. Henson, "Optimal Battery/Ultracapacitor Storage Combination," Journal of Power Sources, Vol. 179, No. 1, 2008, pp. 417-423. doi:10.1016/j.jpowsour.2007.12.083
79. H. Yoo, S. K. Sul, Y. Park and J. Jeong, "System Integration and Power Flow Management for a Series Hybrid Electric Vehicle Using Super-Capacitors and Batteries," IEEE Transactions on Applied Superconductivity, Vol. 44, No. 1, 2008, pp. 108-114. doi:10.1109/TIA.2007.912749
80. L. Shuai, K. A. Corzine and M. Ferdowsi, "A New Battery/Ultracapacitor Energy Storage System Design and Its Motor Drive Integration for Hybrid Electric Vehicles," IEEE Transactions on Vehicular Technology, Vol. 56, No. 4, 2007, pp. 1516-1523. doi:10.1109/TVT.2007.896971
81. M. B. Camara, H. Gualous, F. Gustin and A. Berthon, "Design and New Control of DC/DC Converters to Share Energy between Supercapacitors and Batteries in Hybrid Vehicles," IEEE Transactions on Vehicular Technology, Vol. 57, No. 5, 2008, pp. 2721-2735. doi:10.1109/TVT.2008.915491
82. O. Briat, J. M. Vinassa, W. Lajnef, S. Azzopardi and E. Woirgard, "Principle, Design and Experimental Validation of a Flywheel-Battery Hybrid Source for Heavy-Duty Electric Vehicles," IET Electric Power Applications, Vol. 1, No. 5, 2007, pp. 665-674. doi:10.1049/iet-epa:20060458
83. S. Lemofouet and A. Rufer, "A Hybrid Energy Storage System Based on Compressed Air and Supercapacitors with Maximum Efficiency

Point Tracking (MEPT)," IEEE Transactions on Industrial Electronics, Vol. 53, No. 4, 2006, pp. 1105-1115. doi:10.1109/TIE.2006.878323

84. T. Ise, M. Kita and A. Taguchi, "A Hybrid Energy Storage with a SMES and Secondary Battery," IEEE Transactions on Applied Superconductivity, Vol. 15, No. 2, 2005, pp. 1915-1918. doi:10.1109/TASC.2005.849333

Chapter 18

A Perspective on the Future of Distribution: Smart Grids, State of the Art, Benefits and Research Plans

Rosario Miceli, Salvatore Favuzza, Fabio Genduso

Dipartimento di Ingegneria Elettrica, Elettronica delle Telecomunicazioni, di Tecnologie Chimiche, Automatica e Modelli Matematici, Università di Palermo, Palermo, Italy

ABSTRACT

Currently, the design and operation criteria for electrical distribution networks are fastly changing due to some factors; among these, the progressive penetration of Distributed Generation (DG) is destined to cause deep changes in the existing networks, no longer considered as passive terminations of the whole electrical system. Moreover, the increasing application of Information Communication Technologies (ICT) will allow the implementation of the so called "smart grids", determining new interesting scenarios. In the paper the problems and the potential benefits of DG, the possible new electrical distribution system models and the major research projects on smart grids are faced and reported.

Citation: R. Miceli, S. Favuzza and F. Genduso, "A Perspective on the Future of Distribution: Smart Grids, State of the Art, Benefits and Research Plans," *Energy and Power Engineering*, Vol. 5 No. 1, 2013, pp. 36-42. doi: 10.4236/epe.2013.51005.

Keywords: Distributed Generation; Smart Grids

INTRODUCTION

Actually, the electrical distribution systems, overall Medium Voltage (MV) ones, and their design and operation criteria are subjected to deep changes, due to different factors. Among these, the main ones concern: the energy market liberalization, the new and complex energy governance policies, the environmental pollution reduction and sustainable development, the renewable energies development, the increase of energy efficiency, the costs reduction, the growth of the penetration of the so called Distributed Generation (DG). In particular, the forthcoming presence of DG in electrical distribution systems has strongly modified the nature of such systems; these systems, in fact, having, today, a radial topology and managed in a passive way (i.e. supplying energy from electrical power plants to end-users), are destined to reach an active role by means of the implementation of the typical functions of load management, demand side management, demand response and generation curtailment [1,2].

However the DG penetration determines some technical problems in electrical systems that must be faced and solved rapidly to exploit the potential benefits of DG and to really start the revision process aiming at the implementation of the so called smart grids [3,4].

The smart electrical distribution grids represent the needed evolution of the actual networks by means of a deeper implementation of automation functions, and a high level of Information and Communications Technologies (ICT) applications in order to increase the power quality and ancillary services, guaranteeing the security and economic/energetic efficiency in electric energy supplying [5].

After the analysis of the current scenario, characterized by the management and the problems solution determined from the above mentioned DG presence, the paper deals with the description of different possible models of smart grids (micro grids, active grids, local areas) putting in evidence the benefits and the necessary innovations which will be achieved in various fields involved in this important and fundamental evolution. Finally, to underline the importance of this issue, a brief presentation of the most important international research projects in this field are reported.

DISTRIBUTED GENERATION POSSIBLE BENEFITS AND PROBLEMATICS

The distributed generation can introduce in electrical distribution systems some potential benefits, such as:

- Flexibility and electrical load management [6];
- Coverage of the local load peaks;
- Diversification of the energy resources supply;
- Larger possibility of exploitation of renewable energy resources in favorable locations;
- Time deferral of investments aimed at the maintainance of transmission and distribution systems, at the building of new power plants and at the reinforcement of existing power plants and electrical systems;
- Electrical energy losses reduction [7,8].

On the contrary, the DG presence causes some technical problems that must be rapidly faced and solved like:

- The increase of short circuit currents;
- The increased complexity of automation and protection systems;
- The increased complexity of voltage regulation due to a modification of power flows;
- The unwanted MV systems islanding [9,10].

Increase of Short Circuit Currents

DG connection to distribution systems by means of synchronous and asynchronous generators causes a significant increase of short circuit currents (respect to passive systems). This issue can determine the overcomeing of dimensioning and sizing limits of circuit breakers and lines. The consequence is the necessity to substitute some conductors to adapt them to the bigger thermic solicitations and some circuit breakers to adapt them to higher breaking capacity.

Increased Complexity of Automation and Protection Systems

DG can determine the power flow inversion in some networks branches. This depends from type of DG, power size, connection points, loads, with important consequences on protection selectivity. In fact, it is possible to have unwanted interventions over un-faulted lines, see Figure 1, due to faults in adjacent lines connected to the same bus bar (loss of selectivity). Other problem concerns possible non-functioning of opening and re-closing temporized systems to eliminate transient and semipermanent faults or to isolate faulted lines with the aim of limiting the time of out of service.

So it is necessary:

- The redefinition and re-setting of protection systems;
- The verification of protection selectivity taking in account of intensity, versus and during time of fault currents;
- The verification of time interventions of protection devices with the aim of generators stability.

Increased Complexity of Voltage Regulation Due to a Modification of Power Flows

As known, without DG, the voltage regulation is based on loads and passive systems features: so the voltage profile is easily determinable with regard to radial structures and stable loads (unidirectional power flows—see Figure 2).

In presence of DG, the voltage regulation is more complex, because it depends on the sites, the sizes, the dispatching and the features of DG, as shown in Figure 3. A possible solution is a voltage regulation system based on a coordinated control of under load tap changers of HV/MV transformers and reactive power flows in distribution feeders.

Unwanted MV Systems Islanding

Today, in many countries, the MV systems islanding is prohibited, although it should constitute a significant potential advantage of DG. But islanding determines problems in the areas of safety, control and management. The safety aspects are related to permit repairing and maintaining interventions in safety conditions (electrical system out of voltage).

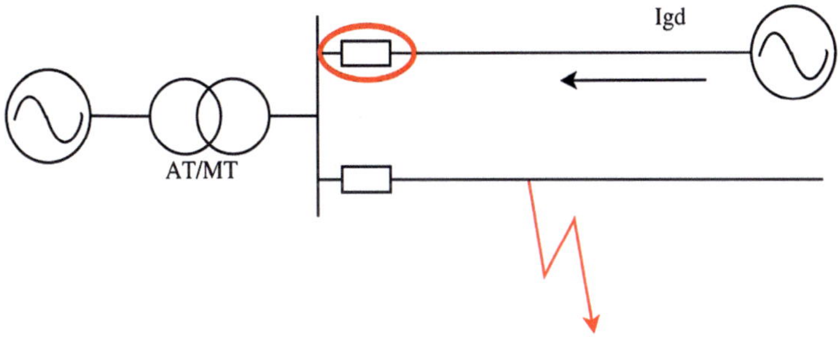

Figure 1: Loss of selectivity: unwanted intervention of a circuit breaker in an un-faulted line.

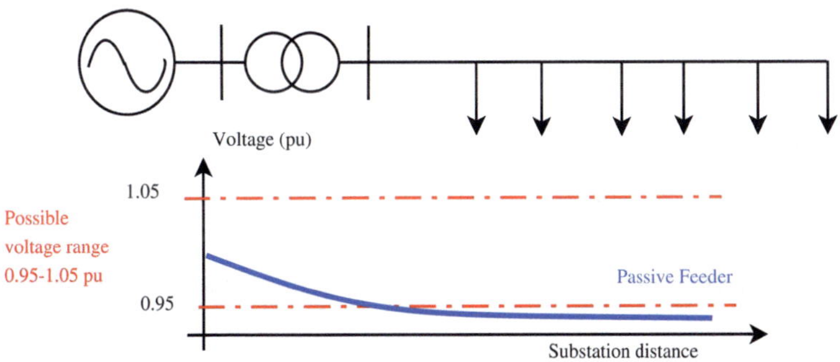

Figure 2: Voltage profile in a passive feeder.

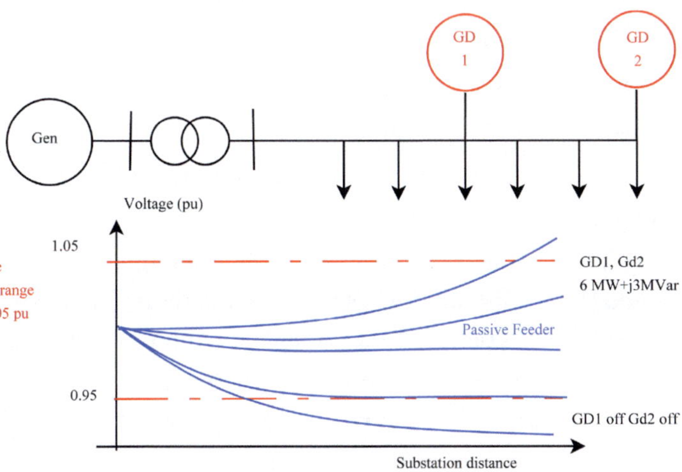

Figure 3: Different voltage profiles in an active feeder.

The control and management issues concern:

- The parallel connection of islanded system and electrical power system due to automatic opening and re-closing operation to eliminate transient and semipermanent faults;
- The ground arc extinction, with the consequence of unsuccessful of re-closing operation and the power quality (frequency and voltage variations [9,10].

CURRENT SCENARIO AND FUTURE EVOLUTION OF DISTRIBUTION SYSTEMS: THE SMART GRIDS

The classical scenario of electrical distribution systems, being passive termination of the whole electrical system, is characterized by radial topology, vertically integration with centralized generation, dispatch and control, unidirectional power flows, "connect and forget" logic for the loads, multi directional power flows in presence of DG.

The future scenario foresees an active system that guarantees connectivity through an increasing level of interaction with the consumers and meeting (at least in the future) the end users energy demands in terms of flexibility, economy and reliability, using, at the same time, the benefits of the energy market liberalization. So, it will be characterized by:

- Interconnected and meshed topology;
- Distributed logic;
- Full DG integration [11];
- Multi-directional power flows;
- Logic of integration of the loads taking large flexibility.

The major features concern:

- Larger reliability through the implementation of all the most advanced distribution automation functions (for example integrated Volt/VAR control, outage management, reconfiguration);
- Possibility to integrate the consumers and their behavior within the design and management of the network through the Demand Side Management (DSM) [12];
- Adoption of advanced communication technologies and automated controls, emergency and market demand response;

- Self-healing i.e. the possibility to detect, analyze and solve problems;
- Use of different technologies for energy generation and storage;
- Full usage of the opportunities offered by the electrical market.

To realize these goals, it is necessary to apply in electrical distribution systems an advanced distribution automation and specifically SCADA (Supervisory, Control and Data Acquisition) systems, developing and implementing:

- A monitoring of the electrical systems through sensors;
- A data transmission system (optical fibbers, PLC, GSM, WI-FI, etc.);
- A decision system and a network automation and remote control.

In this way it will be possible to realize the new electrical smart distribution grids. A smart grid is, in fact, an electrical system able to smartly integrate activities of all connected users—energy producers, consumers, prosumers—with the aim of distributing energy in an efficient, sustainable, reliable and economical favorable way [7].

The most important goal of the smart grids is to transform the functionality of the present electricity transmission and distribution grids so to provide a more useroriented service, enabling the achievement of the 20/20/20 targets and guaranteeing, in a competitive market environment, high security, power quality and economic efficiency of electricity supplying. But how and in which way the transition from actual scenario to the new one will proceed? It is hard question to answer.

At this moment it is possible to imagine for the new grids three different models, that would be, also, considered as integrated parts of smart grids, active grids, micro grids and virtual utilities [8,13].

Active Grids

An active grid is a network that does not only play the passive role of supplying the final consumers, but also in which the operator controls and/or rules the power required or generated by the loads or the generators, the bus voltages and the branch power flows. It is possible to assume an evolution in three different levels:

First-level: a simple local control of the generation at the connection point;

Second-level: a complete control system for all the distributed energy resources in the controlled area, realizing a coordinated dispatching and a voltage profile optimization (see Figure 4);

Third-level: creation of a strongly interconnected structure with a subdivision in cells ("local areas") responsible of their own management (protection, voltage regulation, etc.) that take part to the market, selling or buying energy to/from adjacent cells or from/to the transmission system (see Figure 5).

Micro Grids

A micro grid is a set of generators, loads and storage systems connected and able to operate independently from the electrical grid and that internally recreates the energy production and distribution system. In Figure 6 is reported an example of a micro grid presenting a micro grid separation device to performa the islanding operation, an energy manager connected to several power flow controllers and protection coordinators to control and manage the energy flowing in the micro grid branches and many different type sources to inlet energy not only from the main grid, but also to allow distributed generation.

It can be considered similar to the active network cell, since it is provided with a local control system that rules the exchanges of energy among the loads, generators and external network; moreover it can stay in intentional islanding configuration, disabling the loads that accept to be part of a "load curtailment" program [1].

Virtual Utilities

A virtual utility or virtual power plant realizes an optimized management and control of a set of distributed energy resources, in which all distributed generators, loads, storage systems are coordinated taking electric market signals into account. In conclusions, in all the possible imagined scenarios, a very important role will be played by the final users. The key for the development of smart grids concerns the active demand, i.e. the possibility for consumers to actively participate as actors in the electrical system management and control.

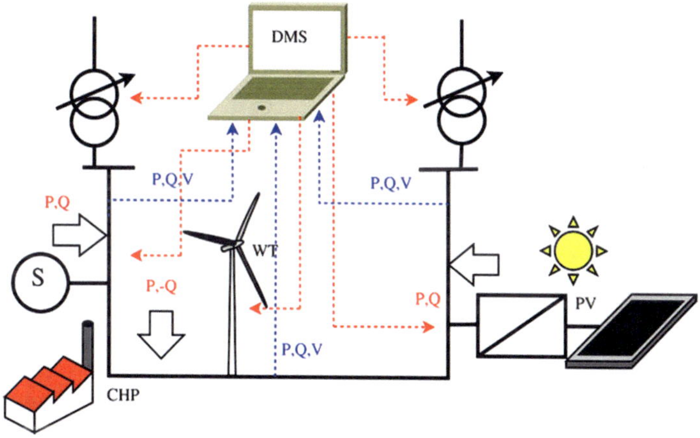

Figure 4: Decentralized control: coordinated dispatching and voltage profile optimization.

INTERNATIONAL RESEARCH PROJECTS ON SMART GRIDS

Many international projects and research activities are focused on smart grids, underlining the importance of this issue. The European Smart Grids Technology Platform [7] represent the strategic deployment document for Europe's electricity networks of the future. The mission is to theorize and promote a common vision about the future (2020 and over) of electric grids.

The platform presents a road map organized in six high priority areas:

1) Optimization of grids management: improvement of the cooperation between Transmission System Operators (TSO), Distribution System Operators (DSO), TSO and DSO, improvement of the grids monitoring and control;
2) Optimization of grids infrastructure: new infrastructures realization, mainly of transmission type, present infrastructures optimization, superconductor technologies development;
3) Large-scale integration of intermittent generation: development of grids to transport the energy generated by wind and photovoltaic plants, development of energy storage systems;
4) Information communication technology: development of easy, strong, safe and flexible communication infrastructures, Standardization of data types and transmission protocols;

5) Active distribution grids: spread the transmission grids services (power flow management, balance, contingencies analysis) to the distribution systems, grid topologies optimization, implementation of smart systems to the grid, communication infrastructures development;

6) New markets, users, energetic efficiency: development of new market models for the active demand integration, spread of bidirectional interface devices between users and market, knowledge boost towards the energy savings and the energetic efficiency improvement.

As known the EU Energy and climate package, that is the new European energetic strategy aiming at guaranteeing a sustainable development, electric market efficiency and quality and security of energy supply, previews some goals that must be gained in two steps: 2020 and 2050.

The first step concerns the 20-20-20 goals: reduction of 20% of greenhouse gases respect to 2005, adopted as reference year because of the start of European Emission trading (EU ETS);

20% of final consumptions of electrical energy produced by means renewables;

20% reduction of energy consumption. In 2008, to achieve these goals the European Council and Parliament have adopted the SET Plan—Strategic Energy Technology Plan, as planning and coordination document.

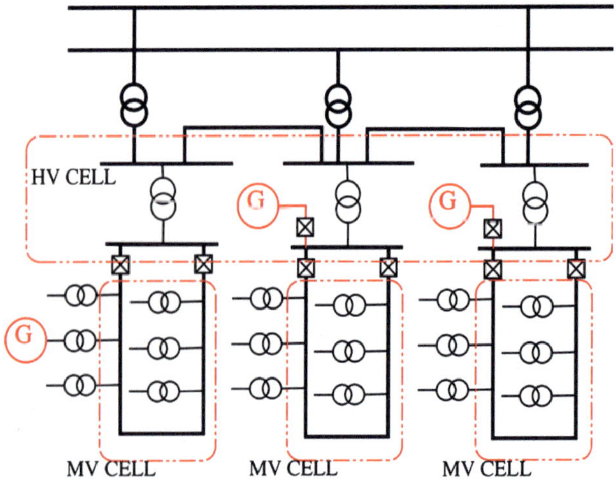

Figure 5: Distribution system organized by cells (local areas).

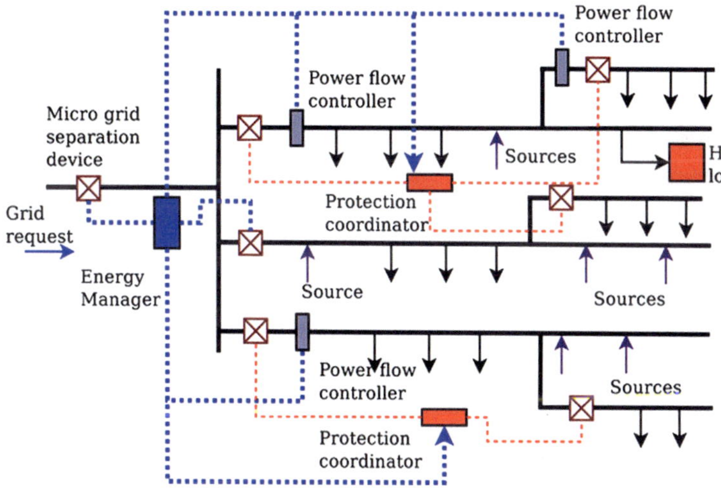

Figure 6: Example of a micro grid.

The SET Plan proposes six European Industrial Initiatives (EII), among which the smart grid. In fact, in June 2010, during the SET Plan Conference in Madrid, the European Electricity Grids Initiative (EEGI) has been presented, with a precise research and development program. The EEGI [14] is proposed by 7 Transmission System Operators (TSO) (Amprion, Elia, Red Electrica de Espana, RTE, Tennet, Transpower, 50 Hertz) and 7 Distribution System Operators (DSO) (CEZ, EON, Enel, Erdf, Iberdrola, RWE, Vattenfall). Key partners are the European Network of Transmission Systems Operators for Electricity (ENTSO-E) and the European Distribution Systems Operators Association for Smart Grids (EDSOSG). The initiative, characterized by a time duration of 8 years (2010-2018) and by a cost of 2 billions Euro, proposes a program of research, development and demonstration (RD&D) aiming at creating a new flexible, economic, efficient, sustainable, smart electrical system.

In particular, the major features are:

- Development and integration of innovative technologies for power systems and their validation in real conditions;
- Development of suitable solutions for other energetic initiatives (solar, wind) to increase renewables and distributed energy resources (DER);
- Creation of a strictly synergy among electrical systems operators, equipment and ICT manufacturers.

In particular the EEGI vision involves electrical energy producers, end-consumers and prosumers.

One of the most important item concerns the integration of new generation and consumption models with the particular goal of creating the smart consumer, by means of: the integration, in the control and management of the electrical systems, of the active demand; the development of the energetic efficiency of end-consumers and the development of new business opportunities for electric market actors, overall for end-users.

Concerning the smart consumer, the EEGI RD&D program proposes a specific cluster named "integration of smart customers", with two projects: Active Demand Response and Energy Efficiency from integration with Smart Homes. In 2008 to support and contribute in a coordinated way to the SET PLAN goals, 14 European research institutes have created the European Energy Research Alliance (EERA). Among all the joint research program proposed by EERA, the Smart Grids one, already started, foresees 4 sub-programs (SP), each subdivided in many research activities:

- SP 1—Network Operation: Adaptation of primary control cycle and automation, network monitoring system and ICT, ancillary services, fault and outage management, distributed generation management, load management;
- SP 2—Energy Management: Simulation and Analysis Model, System Operation with DER and ICT, Market Design with DER, System Design with DER;
- SP 3—Control System Interoperability: State of the art and terminology, Classification of Control Systems, Use Cases, Technical Communication Requirements, Standards and Protocols, Cyber Security;
- SP 4—Electric Energy Storage (EES) Technologies, Performance Testing of Technologies, Integration of Storage Resources to Smart Grids possible Services, Control Algorithms for Storage, Applications in Smart Grids, Economic and Technical Benefits of incorporating an EES onto Network.

In the area of smart distribution grids, a very important European research project, financially supported by the 7th Framework Program, is working: ADDRESS (Active Distribution network with full integration of Demand and distributed energy RESourceS). ADDRESS is a large scale R&D project, involving 25 partners (distributors, networks, ICT and trade operators, research institutes, industries) from 11 European countries, with a cost of 15.7 M (9 M funded). The vision concerns:

1) Flexibility—active demand and power flow optimization both at local and global level;
2) Reliability—technologies development for distributed control, safety and efficient management;
3) Accessibility—commercial and standard barriers removal, fully DG integration, RES, active demand;
4) Economy and Savings—energy consumption and sustainable development, bills amounts reduction.

Particularly interesting is the Italian project "Smart Distribution Network Operation" (SDNO) and developed by ENEL-Distribuzione, the most important Italian distribution operator. The finality of the SDNO project regards:

1) The evaluation of the generators and electrical system features to enable high penetration of DG, the definition and realization of new interface components and new network systems both hardware and software;
2) The testing of apparatus and systems in laboratories and in real electrical systems;
3) The promotion new national and international rules for connection criteria and electrical system control.

In more details, the project foresees the development or realization and testing of new functional requirements for automatic control protection systems, new Supervisory Control and Data Acquisition (SCADA) at HV/MV substation for MV system automatic control, DG dispatching and load control (DSM—demand side management). Another important aspect considered in the project is issued to new remote terminal units (RTU) in MV/LV substation for control and automation; new RTU for DG and load control (DSM); innovative rules for voltage regulation based on HV/MV transformers control and on DG dispatching.

CONCLUSIONS

All the hypothetical scenarios related to smart grids need evolution and development processes involving many aspects, which are today very interesting areas for studying and researching; in fact, the new challenges, that have to be faced, concern:

- Technical aspects: new criteria for electrical systems planning, design, control and management;
- Technological aspects: evolution of components, apparatus and systems (both hardware and software);
- Economical and policy-regulatory aspects: free markets, roles and responsibilities of all actors involved, connection rules, load shedding, generation curtailment, etc.;
- Social aspects: energy policies really sustainable, environmental impact reduction, energetic resources supplying diversification, to advantage renewable development, increase of power and services quality, lower costs for users (real competition).

A real cultural revolution has started, but, as already said by Albert Einstein"We can't solve problems by using the same kind of thinking we used when we created them".

ACKNOWLEDGEMENTS

Rosario Miceli, Salvatore Favuzza, Fabio Genduso (Italian Ministry of University and Research), by the "Sustainable Development and Energy Saving Laboratory" (SDESLAB) part of the UNINETLAB of the University of Palermo and financially supported by the project BeyWatch IST- 223888 funded by the European Community, web page: http://www.beywatch.eu/

REFERENCES

1. C. Cecati, F. Genduso, R. Miceli and G. R. Galluzzo, "A Suitable Control Technique for Fault-Tolerant Converters in Distributed Generation," IEEE International Symposium on Industrial Electronics (ISIE), L'Aquila 27-30 June 2011, pp. 107-112.
2. K. Knauss, C. Warren and D. Kearns, "An Innovative Approach to Smart Automation Testing at National Grid," Transmission and Distribution Conference and Exposition (T&D), 7-10 May 2012, pp. 1-8.
3. P. Chiradeja, "Benefit of Distributed Generation: A Line Loss Reduction Analysis," Transmission and Distribution Conference and Exhibition, Asia and Pacific, Bangkok, 2005, pp. 1-5.
4. D. L. Jia, X. L. Meng and X. H. Song, "Study on Technology System of Self-Healing Control in Smart Distribution Grid," 2011 International Conference on Advanced Power System Automation and Protection, Beijing, 16-20 October 2011, pp. 26-30. doi:10.1109/APAP.2011.6180379
5. A. O. Di Tommaso, S. Favuzza, F. Genduso and R. Miceli, "Development of Diagnostic Systems for the Fault Tolerant Operation of Micro-Grids," International Symposium on Power Electronics, Electrical Drives, Automation and Motion, Palermo, 14-16 June 2010, pp. 1645-1650.
6. F. Genduso, R. MIceli and G. R. Galluzzo, "Flexible Power Converters for the Fault Tolerant Operation of MicroGrids," XIX International Conference on Electrical Machines (ICEM), Palermo, 6-8 September 2010, pp. 1-6.
7. C. Eu, "European Smartgrids Technology Platform-Vision and Strategy for Europe Electricity Networks of the Future," European Commission, 6-8 September 2006.
8. M. Samotyj and B. Howe, "Creating Tomorrow's Intelligent Electric Power Delivery System," 18th International Conference and Exhibition on Electricity Distribution, Palo Alto, 6-9 June 2005, pp. 1-5.
9. K. Jennett, C. Booth and M. Lee, "Analysis of the Sympathetic Tripping Problem for Networks with High Penetrations of Distributed Generation," International Conference on Advanced Power System Automation and Protection (APAP), Glasgow, 16-20 October 2011, pp. 384- 389. doi:10.1109/APAP.2011.6180432

10. M. Hagh, N. Ghadimi, F. Hashemi and S. Zerbadst, “New Islanding Detection Algorithm for Wind Turbine,” 10th International Conference on Environment and Electrical Engineering (EEEIC), Ahar, 8-11 May 2011, pp. 1-5.
11. A. O. Di Tommaso, F. Genduso, G. R. Galluzzo and R. Miceli, “Computer Aided Optimization via Simulation Tools of Energy Generation Systems with Universal Small Wind Turbines,” 3rd IEEE International Symposium on Power Electronics for Distributed Generation Systems (PEDG), 25-28 June 2012, pp. 570-577. doi:10.1109/PEDG.2012.6254059
12. F. Aalamifar, H. Hassanein and G. Takahara, “Viability of Powerline Communication for the Smart Grid,” 26th Biennial Symposium on Communications (QBSC), Kingston, 28-29 May 2012, pp. 19-23. doi:10.1109/QBSC.2012.6221343
13. M. Loddo, “Pianificazione e Gestione Delle Reti Attive,” Ph.D. Thesis, University of Cagliari, Cagliari, 2008.
14. ENTSO-E, “Roadmap 2010-2018 and Detailed Implementation Plan 2010-12,” European Commission, The European Electricity Grid Initiative (EEGI), 2010.

INDEX

T

U

V

W

Z

Printed in the United States
by Baker & Taylor Publisher Services